4th International Conference on Environmentally Sustainable Animal Industry (ICESAI 2023)

BIO Web of Conferences Volume 81 (2023)

Malang, Indonesia
25 October 2023

Editors:

Veronica Margareta Ani Nurgiartiningsih
Joey Jai-Wei Lee
Abdul Razak Alimon
Wantamas Jantasin
Eko Widodo
Marjuki

ISBN: 978-1-7138-8654-9

Printed from e-media with permission by:

Curran Associates, Inc.
57 Morehouse Lane
Red Hook, NY 12571

Some format issues inherent in the e-media version may also appear in this print version.

No changes have been made to the content of these proceedings. There may be changes to pagination and minor adjustments for aesthetics.

Printed with permission by Curran Associates, Inc. (2024)

For additional information, please contact EDP Sciences – Web of Conferences at the address below.

EDP Sciences – Web of Conferences
17, Avenue du Hoggar
Parc d'Activité de Courtabœuf
BP 112
F-91944 Les Ulis Cedex A
France

Phone: +33 (0) 1 69 18 75 75
Fax: +33 (0) 1 69 28 84 91

contact-edps@webofconferences.org

Additional copies of this publication are available from:

Curran Associates, Inc.
57 Morehouse Lane
Red Hook, NY 12571 USA
Phone: 845-758-0400
Fax: 845-758-2633
Email: curran@proceedings.com
Web: www.proceedings.com

TABLE OF CONTENTS

FACULTY OF ANIMAL SCIENCE
Universitas Brawijaya

Scopus | Web of Conferences

Mode of Conference :
Hybrid

THE 4TH ICESAI

International Conference on Environmentally Sustainable Animal Industry

October 25, 2023
Faculty of Animal Science
Universitas Brawijaya

"Green Economy and Smart Farming System to Support Sustainable Livestock Industry"

Keynote Speakers

Ir. Syafri, MBA
Poultry Industry, Charoen Pokphand Foundation Indonesia

Prof. Joey Jai-wei Lee
Department of Tropical Agriculture and International Cooperation, NPUST, Taiwan

Prof. Hendrawan Soetanto
Faculty of Animal Science Universitas Brawijaya, Indonesia

Prof. Simon Oosting
Animal Production System, Wageningen University & Research, The Netherlands

Invited Speakers

Prof. Hasliza Abu Hassim
Faculty of Veterinary Medicine, Universiti Putra Malaysia, Malaysia

Prof. Kecha Kuha
Faculty of Animal Science and Agricultural Technology, RMUTL, Thailand

Dr. Wantamas Jantasin
Faculty of Animal Science and Technology Maejo University, Thailand

TOPIC
- Animal Products and Sustainable Post-Harvest Technology
- Feed Resources and Technology
- Local Animal Resources
- Socio-Economic Studies and Community Development
- Sustainable Animal Production Systems

IMPORTANT DATES
- Registration and Abstract Submission Extended : 23 May - 7 September 2023
- Abstract Acceptance Extended : 17 September 2023
- Full Paper and Poster Submission : 14 - 28 September 2023
- Presentation Material Submitted : 3 October 2023
- 4th ICESAI 2023 : 25 October 2023

PUBLICATION
Selected manuscript will be published in Scopus-indexed Proceedings (Bio Web of Conference)"

SECRETARIAT
International Relation Office
Faculty of Animal Science, Universitas Brawijaya
Jl. Veteran, Malang 65145, Jawa Timur, Indonesia

CONTACT INFO
We're happy to answer any question.
+62-898-8013-588 : Danung Nur Adli, S.Pt., M.Pt.M.Sc.
icesai@ub.ac.id

REGISTRATION FEE
- Online (Certificate)
  - Presenter : 300.000 IDR or 20 USD
  - Presenter student : 200.000 IDR or 15 USD
  - Participant : 150.000 IDR or 10 USD
  - Participant student : 100.000 IDR or 5 USD
- Offline (Seminar kit, coffee break, lunch, and certificate)
  - Presenter : 500.000 IDR or 35 USD
  - Presenter student : 400.000 IDR or 30 USD
  - Presenter Poster : 350.000 IDR or 25 USD
  - Presenter Poster student : 300.000 IDR or 20 USD
  - Participant : 250.000 IDR or 20 USD
  - Participant student : 200.000 IDR or 15 USD

PUBLICATION FEE
Publication in proceedings (paid upon acceptance) : 2.000.000 IDR or 150 USD

PAYMENT
Bank Name : Bank Mandiri
Bank Account Number : 144-001-614-9608
Bank Account Name/Beneficiary : Asri Nurul Huda
Swift/BIC : BMRIIDJA

icesai.ub.ac.id

Statement of Peer review

In submitting conference proceedings to *Web of Conferences*, the editors of the proceedings certify to the Publisher that

1. They adhere to its **Policy on Publishing Integrity** in order to safeguard good scientific practice in publishing.
2. All articles have been subjected to peer review administered by the proceedings editors.
3. Reviews have been conducted by expert referees, who have been requested to provide unbiased and constructive comments aimed, whenever possible, at improving the work.
4. Proceedings editors have taken all reasonable steps to ensure the quality of the materials they publish and their decision to accept or reject a paper for publication has been based only on the merits of the work and the relevance to the journal.

Title, date and place of the conference

The 4th International Conference on Environmentally Sustainable Animal Industry (ICESAI 2023)

October 25, 2023

Faculty of Animal Science, Universitas Brawijaya, Malang 65145, Indonesia

Proceedings editor(s):

Prof. Veronica Margareta Ani Nurgiartiningsih
Prof. Joey Jai-Wei Lee
Prof. Abdul Razak Alimon
Dr. Wantamas Jantasin
Dr. Eko Widodo
Dr. Marjuki

Date and editor's signature

November 30, 2023
Prof. Veronica Margareta Ani Nurgiartiningsih

Influence of feed supplement based on *Andrographis paniculata* and irradiated chitosan on *in vitro* fermentation and methane production of selected forage

Wahidin Teguh Sasongko[1,2], *Anuraga* Jayanegara[3], *Dewi Apri* Astuti[3], *Akhmad* Rasyid Syahputra[4], *Slamet* Widodo[2], and *Teguh* Wahyono[5*]

[1]Graduate School of Nutrition and Feed Science, IPB University, Bogor 16680, Indonesia
[2]Research Centre for Animal Husbandry, National Research and Innovation Agency (BRIN), Bogor 16915, Indonesia
[3]Departement of Nutrition and Feed Technology, Faculty of Animal Science, Bogor Agricultural University
[4]Research Centre for Radiation Process Technology, BRIN, South Jakarta 12440, Indonesia
[5]Research Centre for Food Technology and Processing, BRIN, Gunungkidul 55861, Indonesia

Abstract. The Wallacean region of Indonesia, specifically the Nusa Tenggara and South Sulawesi regions in Indonesia, carry a large population of cattle. Therefore, it is necessary to evaluate methane production from forage-based feed in this region. The objective of this study was to determine the effect of feed supplement containing *Andrographis paniculata* and irradiated chitosan on the *in vitro* fermentation and methane production, of forage-based rations in this region. This study evaluates five forages that are mainly used as feed ingredients by smallholder farmers nnamely, (Sorghum forage, *Pennisetum purpureum cv. Mott*, *Pennisetum purpupoides*, *Indigofera zollingeriana* and cocoa pod husk. That five forages were supplemented with *Andrographis paniculata* and irradiated chitosan. There were a total of 10 treatments with four replications. The samples were subjected to *in vitro* analysis using the gas production technique. There was no differences between the treatment with and without feed supplements on *in vitro* gas production. There was no interaction between feed supplement addition and forage type. Feed supplement addition influence pH ($P<0.05$) and single chain fatty acids (SCFAs) value ($P<0.01$). The addition of feed supplements with added *Andrographis paniculata* and irradiated chitosan in the *Pennisetum purpupoides* forage can reduce the production of enteric methane ($P<0.05$). The results suggest that, except for sorghum forages, the addition of feed supplements tended to reduce methane gas emissions from forage samples. Moreover, except for pH and SCFAs, *In vitro* gas production and digestibility of forage from eastern Indonesia were not affected by the addition of feed supplement.

* Corresponding author: tegu021@brin.go.id

1 Introduction

Minimising methane emissions is one of the most important ruminant nutrition objectives. Because ruminants generate greenhouse gas (GHG) emissions that are associated with climate change, the environmental sustainability of ruminant farming practices has been heavily criticised [1]. The data on livestock sector methane emissions in Indonesia are continually updated to ensure their accuracy. According to these data, methane gas emissions have increased each year [2]. Therefore, Indonesian livestock stakeholders must take a role in reducing methane emissions. The Wallacean part of Indonesia, specifically the Nusa Tenggara and South Sulawesi regions, is a major resource of large ruminants [3]. Therefore, it is necessary to evaluate methane production from forage-based feed in this region. Various farming methods, modifications to diets, feed additives, chemical methanogenesis modulators, probiotics, immunisation against the rumen microbiome, selective breeding, and genetic techniques are now being used to reduce methane emissions from ruminants [4]. The most recent advances in the understanding of methane production have led to a development of feed additives that may reduce methane emissions. Among these additives are tannin and saponin [1, 5, 6], 3-Nitroxypropanol [7], seaweed [8], chitosan [9], and fatty acids [10].

Indonesia is rich in fishery byproducts and herbal production, and has the potential to produce additives that reduce methane emissions. Chitosan can be used as methane-reducing compounds because it is antimicrobial and affects the fermentation process [11]. Chitosan may be utilised as a natural rumen modifier because it modifies rumen fermentation in desirable methods, such as increasing propionate and decreasing acetate, which is desirable for higher energy synthesis with possibly minimised methane emission [9]. As herbs, due to the availability of active compounds in *Andrographis paniculata*, it can be used as a supplement's ingredient [12, 13]. As ruminant feed supplements, it is necessary to evaluate the effectiveness of these two component combinations. Previous study reported that feed supplements based on herbs and chitosan could reduce methane emission without reducing forage digestibility [14]. However, to our knowledge, there are no studies that have evaluated the use of this feed supplement on some common forages from Wallacea region in Indonesia. Therefore, the aim of the experiment was to determine the effect of feed supplement with addition of *Andrographis paniculata* and irradiated chitosan on *in vitro* fermentation and methane production in forage-based rations.

2 Experimental Procedure

2.1 Sample preparation

Sorghum forage, *Pennisetum purpureum cv. Mott*, and *Pennisetum purpureum* were collected from the field laboratory, Siwabessy Science and Technology Area, National Research and Innovation Agency, Jakarta, Indonesia. *Indigofera zollingeriana* was collected from the field laboratory, Ciawi, Ministry of Agriculture Republic Indonesia. Cocoa pod husk was obtained from the field laboratory, Faculty of Animal Science, Haluoleo University, Kendari, Southeast Sulawesi. Samples of forage were dried at 60oC for 72 h and then ground into 1 mm particles using a hammer mill. In preparation for nutrient and *in vitro* analysis, dry samples were kept in plastic containers at 4°C. Feed supplement containing Andrographis paniculata and irradiated chitosan was obtained from the Research Center for Animal Husbandry, BRIN.

2.2 Nutrient and fibre analysis

The content of organic matter (OM), ash, crude protein (CP), and ether extract (EE) were analysed according to Association of Official Analytical Chemist method [15]. Neutral detergent fibre (NDF), and acid detergent fibre (ADF) were analysed according to Ankom procedure protocol methods 6 [16] and 5 [17]. Hemicellulose was estimated using equations, as follows:

$$\text{Hemicellulose (\%)} = \text{NDF (\%)} - \text{ADF (\%)}$$

2.3 Experimental design

This study investigates five forages that are primarily utilized as feed ingredients by smallholder in Wallacea area, Eastern Indonesia Region (Sorghum forage, *Pennisetum purpureum cv. Mott*, *Pennisetum purpureum*, *Indigofera zollingeriana*, and cocoa pod husk). That five forages were supplemented with feed supplement with added Andrographis paniculata and irradiated chitosan (10 g/kg DM). There were in total ten treatments with four replications.

2.4 *In vitro* analysis

In vitro gas production was determined according to the Hohenheim gas test procedure [18] with slight modifications. Rumen fluids were collected from two fistulised ongole crossbreeds' bulls (approximate average 350 kg live weight) cared at Animal Nutrition field laboratory, Faculty of Animal Science, Gadjah Mada University, Yogyakarta, Indonesia. A 300 mg (DM basis) of each sample was weighed into a 100 ml glass syringe (Fortuna, Germany). Media (rumen fluid + buffer) consisting of 33% rumen fluid was prepared and put into a syringe (40 ml of each sample). The glass syringe was incubated at 38.5-39.5oC for 48 h. *In vitro* gas production was observed after 3, 6, 9, 12, 24, 36 and 48 h of incubation. Total gas production values were adjusted by following Ørskov and McDonald's equation [19] as follows: P=a+b (1-e^(-ct)), where P is the gas production at time t (ml), a is the *in vitro* gas production from the soluble fraction (ml), b is the *in vitro* gas production from the insoluble fraction (ml), c is the gas production rate constant, (a+b) is the potential gas production (ml), and t is the incubation time (h). After 48 h incubation, samples-rumen-buffer fluids were collected for pH, NH3 and single chain fatty acids (SCFAs) observation. Metabolizable energy (ME), *in vitro* organic matter digestibility (IVOMD) and microbial protein (MP) were calculated as follows [18, 20, 21]:

$$\text{ME (MJ/kg DM)} = 2.2 + (0.136 \times \text{GP}) + (0.057 \times \text{CP}) + (0.0029 \times \text{EE})$$
$$\text{IVOMD (\%)} = 14.88 + (0.889 \times \text{GP}) + (0.45 \times \text{CP}) + (0.0651 \times \text{ash})$$
$$\text{MP (g/kg IVOMD)} = \text{IVOMD} \times 19.3 \times 6.25$$

Where: GP is gas production at 12 h (ml/300 mg DM); CP is crude protein (% DM); and EE is ether extract (% DM).

2.5 Methane production estimation

Methane (CH_4) emission estimation was calculated based on SCFAs proportions by following equation [22]:

$$\text{Methane emission (mM)} = (\text{acetate proportion} \times 0.5) + (\text{butyrate proportion} \times 0.5) - (\text{propionate proportion} \times 0.25)$$

Methane production was converted to mM/100 mg IVOMD.

2.6 Statistical analysis

The data were analysed based on a completely randomised design with two factors. The first factor was forage type, followed by feed supplement addition as the second factor. SPSS version 25.0 was used for analysis of variance on the data. The Duncan's Multiple Range Test was used to evaluate all mean comparisons [23].

3 Result and discussion

Nutrient and fibre composition from the five samples of forages are presented in Table 1. All nutrient and fibre compositions differed significantly ($P<0.01$) between forage types. Pennisetum purpureum cv. Mott had the greatest ($P<0.05$) ash concentration (17.43%DM) and the lowest OM concentration. Cocoa pod husk had the lowest ($p<0.05$) concentration of EE (0.81 %). *Indigofera zollingeriana* demonstrated the highest level of CP (25.22 %), followed by Pennisetum purpureum cv. Mott (12.81%), and *Pennisetum purpupoides* (11.01%). In contrast, sorghum forage and cocoa pod husk had the lowest CP ($P<0.05$). In terms of fibre composition, the concentration of NDF and ADF in *Indigofera zollingeriana* was lowest ($p<0.05$), by 50.55 and 31.64%, respectively. Even though Sorghum forage had the highest NDF and ADF, there are no significant differences with *Pennisetum purpupoides*. Cocoa pod husk and *Indigofera zollingeriana* had lower hemicellulose content than other forages ($p<0.05$).

Table 1. Nutrient and fibre composition forage samples

	Nutrient and fibre composition (% DM)						
Forage	**Ash**	**OM**	**EE**	**CP**	**NDF**	**ADF**	**Hemi**
Sorghum forage	10.23^{a}	89.77^{c}	4.21^{b}	7.25^{b}	73.01^{c}	47.74^{c}	25.27^{b}
Pennisetum purpureum cv. Mott	17.43^{c}	82.57^{a}	3.46^{b}	12.81^{d}	71.77^{c}	43.37^{b}	28.40^{c}
Pennisetum purpupoides	15.31^{b}	84.69^{b}	3.79^{b}	11.01^{c}	72.51^{c}	45.44^{bc}	27.07^{bc}
Indigofera zollingeriana	10.51^{a}	89.49^{c}	3.86^{b}	25.22^{e}	50.55^{a}	31.64^{a}	18.90^{a}
Cocoa pod husk	10.56^{a}	89.44^{c}	0.81^{a}	6.45^{a}	58.78^{b}	42.21^{b}	16.57^{a}
SEM	0.812	0.812	0.309	1.550	2.042	1.257	1.118

DM (dry matter), OM (organic matter), EE (ether extract), CP (crude protein), NDF (neutral detergent fiber), ADF (acid detergent fiber), Hemi (hemicellulose).
Different superscripts in the same row indicate significant differences ($P < 0.05$), Standard Error of the Means (SEM).

Table 2. *In vitro* gas production and gas characteristics of forage and feed supplement treatment (ml/300 mg DM)

Forage	FS	Incubation time (h)							gas characteristics	
		3	6	9	12	24	36	48	a+b	c
Sorghum forage	-	12.08^{a}	18.03^{ab}	24.34^{a}	31.87^{a}	53.40^{a}	64.26^{ab}	70.39^{ab}	85.21^{bc}	0.038^{a}
Sorghum forage	+	10.14^{a}	15.74^{a}	21.51^{a}	29.03^{a}	49.50^{a}	59.81^{a}	65.41^{a}	79.31^{ab}	0.038^{a}
Pennisetum purpureum cv. Mott	-	11.94^{a}	20.13^{bc}	30.11^{b}	43.22^{bc}	66.47^{b}	80.88^{d}	88.26^{fg}	102.91^{d}	0.042^{ab}
Pennisetum purpureum cv. Mott	+	13.75^{ab}	22.38^{c}	31.33^{b}	43.95^{bc}	65.63^{b}	81.90^{d}	89.36^{g}	103.69^{d}	0.042^{ab}
Pennisetum purpupoides	-	12.59^{a}	21.20^{bc}	29.82^{b}	41.09^{b}	62.30^{b}	73.57^{c}	79.53^{cde}	89.00^{c}	0.048^{b}
Pennisetum purpupoides	+	13.45^{a}	20.18^{bc}	31.12^{b}	41.72^{b}	63.09^{b}	74.19^{c}	80.08^{cdef}	89.06^{c}	0.049^{b}
Indigofera zollingeriana	-	17.44^{bc}	27.64^{de}	36.52^{c}	46.71^{cd}	64.31^{b}	72.70^{c}	77.47^{bcd}	80.15^{ab}	0.067^{c}
Indigofera zollingeriana	+	17.27^{bc}	26.83^{d}	36.56^{c}	45.62^{bcd}	61.72^{b}	70.11^{bc}	74.80^{bc}	76.86^{a}	0.069^{c}
Cocoa pod husk	-	21.60^{d}	31.40^{ef}	39.54^{c}	47.84^{cd}	67.43^{b}	82.21^{d}	84.16^{defg}	90.37^{c}	0.061^{c}
Cocoa pod husk	+	20.50^{cd}	32.14^{f}	36.56^{c}	49.07^{d}	67.36^{b}	80.87^{d}	87.37^{efg}	90.77^{c}	0.060^{c}
SEM		0.655	0.884	0.961	1.054	0.983	1.228	1.311	1.535	0.002
Forage		**	**	**	**	**	**	**	**	**
Feed Supplement		ns	ns	ns	ns	ns	ns	ns	ns	ns
Forage x Feed Supplement		ns	ns	ns	ns	ns	ns	ns	ns	ns

FS (Feed supplement addition), a+b (optimum gas production), c (gas production rate constant.
Different superscripts in the same row indicate significant differences ($P<0.05$), Standard Error of the Means (SEM), ** ($P<0.01$), * ($P<0.05$)

It is necessary to examine nutrient profiles to identify nutritional characteristics that will influence feed fermentation. Due to its origin in the legume class, *Indigofera zollingeriana* has the greatest crude protein content. According to Kumalasari [24] experiment, *Indigofera zollingeriana* contains a protein content of between 25.17 and 26.44%. Sorghum forage, *Pennisetum purpupoides* and *Pennisetum purpureum cv.* Mott had higher fibre contents than other forages. According to Wahyono et al. [25], fiber components in forages have a negative correlation with digestibility and nutritive value. This complex combination may make foodstuff less digestible. Methane emissions are correlated with high hemicellulose and NDF concentration [26]. Low-nutrient forages can increase methane emissions since ruminants' digestive systems process them more slowly [14, 27].

In vitro gas production represents degradability characteristics from forage samples. *In vitro* gas production and gas characteristics of forage and feed supplement treatment are shown in Table 2. Significant differences were observed between forages for total gas production and gas characteristics ($P<0.01$). In contrast, there was no difference between the treatment with and without feed supplements. There was no interaction between feed supplement addition and forage type with regard to gas characteristics and total gas production. Cocoa pod husks (with and without feed supplement) had the highest *in vitro* gas production (3 – 36 h; $P<0.05$). Furthermore, after 36 h incubation, Pennisetum purpureum cv. Mott produced the highest gas production ($P<0.05$). Pennisetum purpureum cv. Mott (with and without feed supplement) also had the greatest optimum gas production (a+b). With regard to gas production rate constant (c), *Indigofera zollingeriana* and cocoa pod husk had the highest rate constant (P<0.05).

Previous research indicated that feed supplements containing irradiated chitosan and Andrographis paniculata influenced the characteristics of forage gas production [14]. Haryati et al. [11] also stated that the addition of feed additives chitin and chitosan has a significant impact on total gas production. However, these findings are not consistent with the findings presented by Yusuf et al. [12], who reported that the addition of *Andrographis paniculata* had no effect on the total gas production of basal feed. This difference may be related to variations in the nutrient composition of the forages. The findings of this study indicate that the addition of feed supplements has no negative effect on the digestibility of forage.

Table 3. *In vitro* fermentation characteristics of forage and feed supplement treatment

Forage	FS	ME (MJ/kg DM)	IVOMD (%)	MP (g/kg IVOMD)	pH	NH_3 (mg/100 ml)	SCFAs (mMol) C_2	C_3	C_4
Sorghum forage	-	9.89^{a}	47.14^{a}	56.86^{a}	7.00^{c}	44.77^{b}	47.67^{a}	13.75^{a}	7.37^{a}
Sorghum forage	+	9.36^{a}	44.62^{a}	53.82^{a}	7.03^{c}	44.49^{b}	47.36^{a}	13.11^{a}	7.14^{a}
Pennisetum purpureum cv. Mott	-	11.98^{bc}	60.21^{bc}	72.63^{bc}	6.93^{ab}	46.71^{b}	64.88^{c}	17.93^{bc}	9.60^{cd}
Pennisetum purpureum cv. Mott	+	11.87^{bc}	60.86^{bc}	73.41^{bc}	6.91^{a}	45.69^{b}	61.46^{c}	17.20^{b}	8.80^{b}
Pennisetum purpupoides	-	11.31^{b}	57.36^{b}	69.19^{b}	7.07^{c}	48.57^{b}	62.51^{c}	16.65^{b}	8.94^{bc}
Pennisetum purpupoides	+	11.42^{b}	57.92^{bc}	69.86^{bc}	6.92^{a}	48.70^{b}	53.70^{b}	14.00^{a}	7.55^{a}
Indigofera zollingeriana	-	12.39^{c}	68.45^{d}	82.57^{d}	7.04^{c}	60.07^{c}	70.97^{d}	18.75^{cd}	10.17^{d}
Indigofera zollingeriana	+	12.04^{bc}	67.48^{d}	81.39^{d}	7.05^{c}	58.14^{c}	63.64^{c}	16.82^{b}	9.06^{bc}
Cocoa pod husk	-	11.74^{bc}	61.00^{bc}	73.58^{bc}	6.90^{a}	38.85^{a}	66.70^{cd}	20.46^{e}	8.95^{bc}
Cocoa pod husk	+	11.73^{bc}	62.09^{c}	74.89^{c}	6.88^{a}	38.28^{a}	64.56^{c}	19.60^{de}	8.67^{b}
SEM		0.186	1.398	1.686	0.014	1.514	1.358	0.413	0.175
Forage		**	**	**	**	**	**	**	**
Feed Supplement		ns	ns	ns	*	ns	**	**	**
Forage x Feed Supplement		ns	ns	ns	**	ns	ns	ns	ns

FS (feed supplement); ME (metabolizable energy); IVOMD (*in vitro* organic matter digestibility), MP (microbial protein); NH_3 (ammonia); SCFAs (short chain fatty acids); C_2 (acetate); C_3 (propionate); C_4 (butyrate)
Different superscripts in the same row indicate significant differences (P<0.05), Standard Error of the Means (SEM), ** (P<0.01), * (P<0.05)

In vitro fermentation characteristics from forage samples and feed supplement treatment are shown in Table 3. Except pH, there was no significant interaction between forage type and feed supplement treatment. Significant differences were found between forages (pP<0.01). Feed supplement addition influences pH (P<0.05) and SCFAs value (P<0.01). Sorghum forage (with and without feed supplement) had the lowest ME, IVOMD and MP (P<0.05). *Indigofera zollingeriana* (with and without feed supplement) produced the highest IVOMD, MP and NH3 (p<0.05). The pH value ranges of ten treatments were from 6.88-7.05. Feed supplement addition in *Indigofera zollingeriana* and *Pennisetum purpupoides* could reduce C_2, C_3, and C_4 production (P<0.05). However, for other forages, feed supplement addiction treatment obtained numerically lower SCFAs concentration than without supplement treatment. Enteric methane estimation from forage-based ruminant rations, with and without supplementation are presented in Figure 1. Differences in forage samples and the addition of feed supplements influence methane production (P<0.05). Regarding methane production, there was no interaction between forage type and feed supplement treatment. The addition of feed supplements containing *Andrographis paniculata* and irradiated chitosan in the *Pennisetum purpupoides* sample reduced the production of enteric methane (P<0.05).

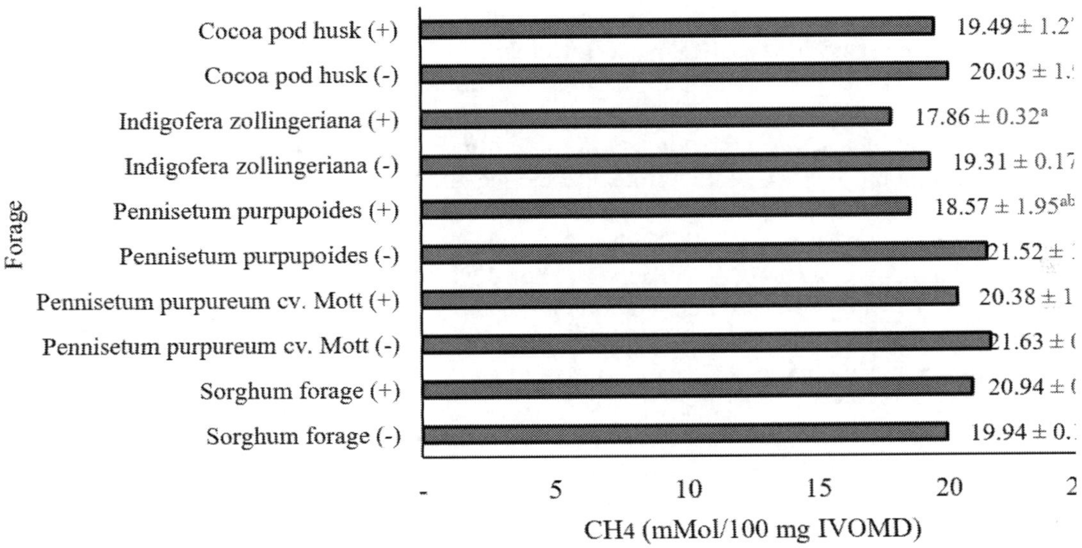

Fig. 1. Enteric methane estimation from forage-based ruminant rations, before (-) and after (+) supplementation. *In vitro* organic matter digestibility (IVOMD)

As a feed additive, chitosan may reduce methane production, enhance propionate production, decrease the acetate/propionate ratio, and contribute to improved animal performance [28]. Shah et al. [29] reported that chitosan has antibacterial properties that change the structural bacterial community, continuing the rumen fermentation pattern to propionate acid productions, which explains the reduced methane emission. This mechanism can be seen from changes in the SCFAs composition, particularly the propionate concentration shown in Table 3. However, chitosan's ability to reduce methane production is also influenced by the extraction method. Strongly hydrophobic chitosan had the best methane hydrate inhibition effectiveness and might extend the hydrate induction period [30]. It has been suggested that the bioactive compounds found in *Andrographis paniculata* play a role in methane reduction. Tannins present in this herb can be used as methane inhibitors [13].

4 Conclusion

In vitro gas production and digestibility of forages in this study were not affected by the addition of *Andrographis paniculata* and irradiated chitosan-based feed supplements. However, the addition of feed supplements changed the pH value and characteristics of SCFAs. Except for sorghum forages, the addition of feed supplements tended to reduce methane gas emissions from some tested forage samples.

Acknowledgments

The authors acknowledge the IAEA Coordinated Research Project D31031 - 25116 for funding sources.

References

1. I.A. Aboagye, K.A. Beauchemin, Animals **9** (2019).
2. A. Mahmud, A. Prima, Livest. Anim. Res. **19** (2021).
3. Statistics Indonesia. Beef Cattle Population ByProvince (Heads). (2023).
4. B. Króliczewska, E. Pecka-Kiełb, and J. Bujok, Agr. **13** (2023).
5. A. Jayanegara, F. Leiber, and M.J. Kreuzer, J. Anim. Physiol. Anim. Nutr. **96** (2012).
6. Y. Widiawati, W. Puastuti, *The Effect of Condensed Tannin and Saponin in Reducing Methane Produced during Rumen Digestion of Agricultural Byproducts,* in Proceedings of International Seminar on Livestock Production and Veterinary Tech. 139–146 (2016).
7. A. Jayanegara, K.A. Sarwono, M. Kondo, Ital. J. Anim. Sci. **17** (2018).
8. B.M. Roque, M. Venegas, R.D. Kinley, PLoS One **16** (2021).
9. R.P. Harahap, S. Suharti, M. Ridla, Anim. Sci. J **93** (2022).
10. X. Sun, Q. Wang, Z. Yang, Fermentation **8** (2022).
11. R.P. Haryati, R.P. Harahap, A, Jayanegara, E.B. Laconi, *Evaluation of chitin and chitosan from insect as feed additives to mitigate ruminal methane emission*, in AIP Conf. Proc. 2120: 1–5 (2019).
12. A.I. Yusuf, M. Ebrahimi, Y.M. Goh, J. Anim. Vet. Adv. **11** (2012).
13. D. Priyoatmojo, Y. Maharani, D.J. Ansori, Anim. Healt. Prod. **9** (2021).
14. T. Wahyono, S. Widodo, A. Kurniawati, Y.N. Anggraeny, Y. Widiawati, M.N. Rofiq, A. Herliatika, D. Priyoatmojo, A.R. Syahputra, W.T. Sasongko, *Green medicated supplement (Green MS) can reduce enteric methane emission from forage-based ruminant rations: In vitro study*, in IOP Conf. Ser. Earth Env. Sci. **1133**: 0–8 (2023).
15. AOAC, Official Method of Analysis 2005.
16. Ankom, Neutral Detergent Fiber in Feeds. Filter Bag Technique - NDF Method 6 (2019)
17. Ankom, Acid Detergent Fiber in Feeds Filter Bag Technique - Method (2019)
18. K.H. Menke, Steingass, H. Anim. Res. Dev. **28** (1988)
19. E.R. Ørskov, I.J. Mcdonald, Agr. Sci. Cambrid. **92** (1979)
20. J.W. Czerkawski. An Introduction to Rumen Studies. (1986)
21. A. El-Waziry, F. Alkoaik, A. Khalil, Adv. Anim. Vet. Sci. **4** (2016).
22. Y. Widiawati, A. Thalib, Ind. J. Agr. **2** (2009)
23. R.G.D. Steel, J.H. Torrie. Principles and procedures of statistics. (1960).
24. N.R. Kumalasari, A.T. Permana, R. Silvia, *Interaction of Fertilizer, Light Intensity and Media on Maize Growth in Semi-Hydroponic System for Feed Production*, in The7th International Seminar on Tropical Animal Production, Yogyakarta (2017).

25. T. Wahyono, W.T. Sasongko, Y. Maharani, Adv. Anim. Vet. Sci. **9** (2021).
26. T. Wahyono, I. Sugoro, A. Jayanegara, Adv. Anim. Vet. Sci. **7** (2019).
27. M.A. Lee, A.P. Davis, M.G.G. Chagunda, Bio. **14** (2017).
28. A.S. Anggraeni, A. Jayanegara, E.B. Laconi, Czech J. Anim. Sci. **67** (2022).
29. A.M. Shah, I.H. Qazi, M. Matra, Ferment. **8** (2022).
30. Y. Zhang, Z. Li, X. Yang, Energies. **15** (2022).

Native vs broiler meat after gamma irradiation treatment: Color and surface image study

Sugeng Hariyadi[1,2], *Teguh* Wahyono[2*], *Angga Maulana* Firmansyah[2], *Andi* Febrisiantosa[2], *Endy* Triyannanto[3], *Waringin Margi* Yusmaman[4], and *Aisyatur* Rosyidah[5]

[1]Gadjah Mada University, Master Student, Graduate Program, Faculty of Animal Science, 55281 Sleman, Indonesia
[2]National Research and Innovation Agency (BRIN), Research Center for Food Technology and Processing, 55861 Gunungkidul, Indonesia
[3]Gadjah Mada University, Faculty of Animal Science, 55281 Sleman, Indonesia
[4]Indonesian Polytechnic of Nuclear Technology, Radiation chemistry laboratory, 55281 Sleman, Indonesia
[5]University of Jember, Faculty of Agricultural Technology, 68121 Jember, Indonesia

Abstract. In recent decades, food irradiation has been at the center of many different meat preservation investigations. In future decades, Indonesia will increase its exports of poultry meat products in response to an increasing trend. To our knowledge, no study has investigated the differences between native and broiler chicken meat post gamma irradiation treatment. Therefore, this study aimed to compare the effect of gamma irradiation on color, pH and surface image of native and broiler chicken meat. The treatments observed in this study were two types of chicken meat (native vs broiler) which were gamma irradiated at doses of 0, 5 and 10 kGy. Three replicates were used in all parameters except in surface image study by SEM and mineral analysis. Lightness (L^*) at day 0 and 14 was decreased due to irradiation ($p<0.05$), both in broiler and native chicken meat. After storage at day 14, native chicken meat after gamma irradiation had the highest redness (a^*) value ($p<0.05$). However, no significant differences were found in yellowness (b^*). Based on the various types of meat, the results reveal that there was no statistical difference in pH values between irradiation dosage. However, native chicken irradiated with 10 kGy had the highest pH value in 0 and 14 days of storage ($p<0.05$). Regardless of the kind of meat, gamma irradiation may affect the surface image of the sample. After irradiation treatment, it has been suggested that variations in muscle structure are responsible for differences in surface appearance. We concluded that, meat color, both in broiler and native chicken meat was affected by gamma irradiation treatment. However, no significant differences were found in pH values. Gamma irradiation may influence the surface image of the sample. Larger particles and aggregates were observed after treated with 5 and 10 kGy irradiation.

* Corresponding author: tegu021@brin.go.id

1 Introduction

From January to May of 2021, Indonesia's exports of poultry products reached 258,85 tonnes, an increase of 12.98% compared to the same period the previous year [1]. In future decades, Indonesia will increase its exports of poultry meat products in response to an increasing trend. Therefore, animal product preservation technology is required take advantage on this opportunity. The current demand of consumers is for clean and secure food that does not compromise the nutritional and sensorial qualities. Food safety is an international issue requiring a global response. Food chain approval is a complex process involving poultry, harvesting, transportation, slaughtering, packaging, sale, and consuming products from farms [2]. This moved the focus of food professionals towards the creation of non-thermal technologies for preservation that are green, secure, and environmentally friendly [3].

Ultrasound treatments, high-pressure hydrostatic processing, and gamma and irradiation treatments are the most widely used industrial applications with the potential to inhibit pathogenic microorganisms prevalent in meat products [4]. In recent decades, food irradiation, an emerging environmentally friendly technology, has been at the center of many different food preservation investigations [5]. Irradiation is an effective method of food preservation due to eliminates or minimizes pathogens, parasites, and pests without significantly altering the food's chemical composition, thus ensuring its safety [6]. A maximum dose of 10 kGy has been suggested for food irradiation, without any requirement for toxicological or nutritional evaluation. These doses have no effect on the overall flavor of foods [6]. Depending on the irradiation dose and the type of meat, applying irradiation on food products might result in free radicals, color changes, lipid oxidation, and an increase of unpleasant odours [7]. Additionally, the procedure breaks down some of the food's molecules, which can have an impact on its nutritional value and sensory qualities. To our knowledge, recently, there is no research that has investigated the differences between native and broiler chicken meat post gamma irradiation treatment. Therefore, this study is a preliminary investigation through this issue. The purpose of this experiment was to compare the effect of gamma irradiation on color, pH and surface image of native and broiler chicken meat.

2 Experimental procedures

2.1 Sample preparation

Native and broiler chicken samples were purchased from a modern market in Gunungkidul, Yogyakarta. About 20–30 g of minced meat was anaerobically packed by polypropylene bags using vacuum sealer. Samples were stored at a temperature -18°C for 48 h before irradiated.

2.2 Experimental design

The treatments observed in this study were two types of chicken meat (native vs broiler) which were gamma irradiated at doses of 0, 5 and 10 kGy. The first factor was meat type and the second factor was irradiation dose. The observed parameters are color, pH, surface image, and mineral content. Color (L^*, a^*, and b^*) and pH value were evaluated after 0, 7, and 14 d of refrigerated storage (2°C). Three replicates were used in all parameters except in surface image study by SEM and mineral analysis.

2.3 Gamma irradiation

Gamma irradiation was performed at Irradiator Laboratory, which is under the Indonesian Polytechnic of Nuclear Technology, National Research and Innovation Agency. Under chilled conditions, the Gamma Cell 220 Upgraded (Izotop, Hungary) was applied to irradiate samples with doses of 0, 5, and 10 kGy. During radiation, the dose rate was 2.81 kGy/h.

2.4 Color

The surface color values of the untreated and irradiated meat samples were determined using Chromameter Konica Minolta (Konica Minolta, Tokyo, Japan). Color values expressed as lightness (L^*), redness (a^*), and yellowness (b^*) by using average from three samples replicate.

2.5 pH

pH value of untreated and irradiated meat was measured using pH meter (Eutech Instruments PC700) from meat homogenate. The homogenate meat was prepared by blending approximately 4 g of meat with 8 ml destilled water.

2.6 Scanning Electron Microscope (SEM)

Untreated and irradiated meat was dried in the oven for 120 h at 55°C. Using a Scanning Electron Microscope (SEM), surface area characteristics of dry meat were evaluated. As a conductor, the sample was placed in a gold- and palladium-coated plate before observation. Observations conducted at 15kV accelerated energy. At 3000 magnification, the obtained images were compared and observed in detail.

2.7 Mineral measurement

The mineral content was determined using X-Ray Fluorescence (XRF). Meat sample was placed in the pellet press to be completely vacuumed before being exposed to 50 kV X-rays. The instrument (Panalytical epilson 4) needs about 8 minutes to measure a sample.

2.8 Statistical Analysis

The data was evaluated using a completely randomized design with two-factor. The first factor was meat type and the second factor was irradiation dose. SPSS version 25.0 was applied for analysis of variance on the data. The Duncan's Multiple Range Test was used for evaluating all mean comparisons [8].

3 Results and discussion

Color values, expressed as lightness (L^*), redness (a^*), and yellowness (b^*) are shown in Figure 1. In present study, differences in meat type affected the overall value of the meat's color ($p<0.05$), except for parameters L^* day 0 and b^* day 14. There was no interaction between meat type and dose of irradiation. Gamma irradiation can decrease the value of parameters L^* (0 and 14 d) and a^* (0 and 7) ($p<0.05$). However, the results for parameter b lack a consistent pattern. Nevertheless, the yellowness (b^*) of native chicken meat is greater than broiler chicken ($p<0.05$).

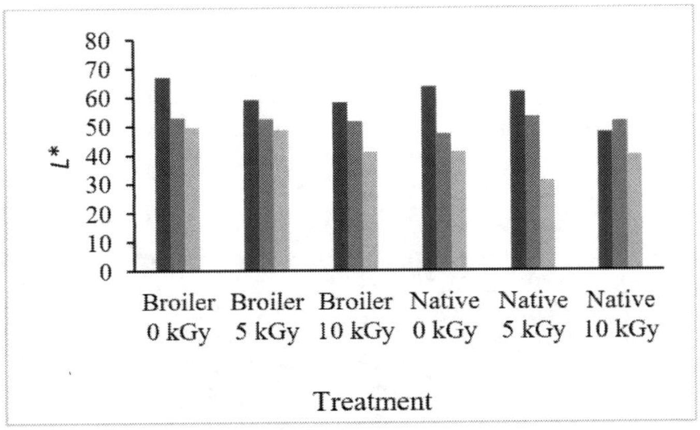

(a)

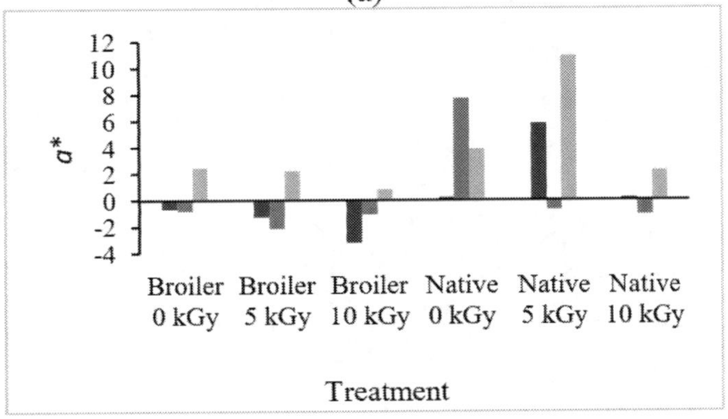

(b)

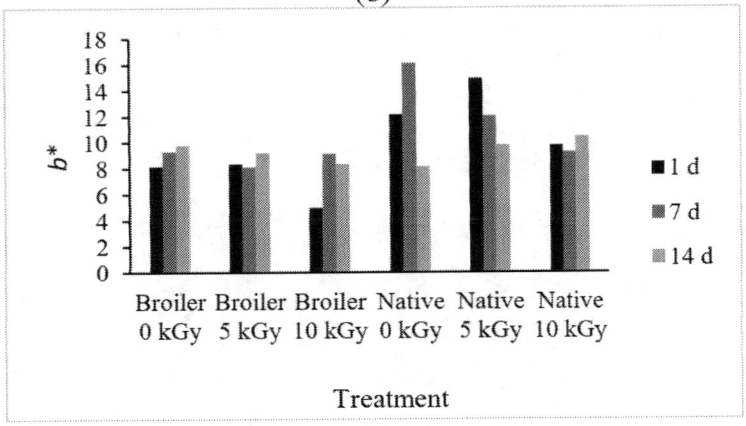

(c)

Fig. 1. Lightness (L^*; a), redness (a^*; b), and yellowness (b^*; c) values of broiler and native chicken meat after gamma irradiation treatment and storage at 0, 7, and 14 d.

In contrary, Arshad et al. [9] reported that L^* and a^* value of poultry breast was increased by irradiation. This difference may be due to differences in the part/composition of the irradiated meat. In present study, chicken meat with skin were utilized. High lipid concentrations in raw materials make them more vulnerable to oxidation and discoloration [10]. Ahn et al. [11] demonstrated that animal species, irradiation dose and muscle type influence the color variations in irradiated meat. Khalid et al. [12] reported that variable b^* (yellowness) was the main factor for the color variability, since chicken meat was basically yellow. Iron, as myoglobin molecules, are sensitive to chemical alterations as well as modifications in energy input, which altered the structure of the irradiated meat during refrigeration [10, 13]. Due to differences in pigmentation, the value of a^* is greater in native than broiler meat. Oxymyoglobin is associated with greater a^* values than myoglobin or metmyoglobin (14).

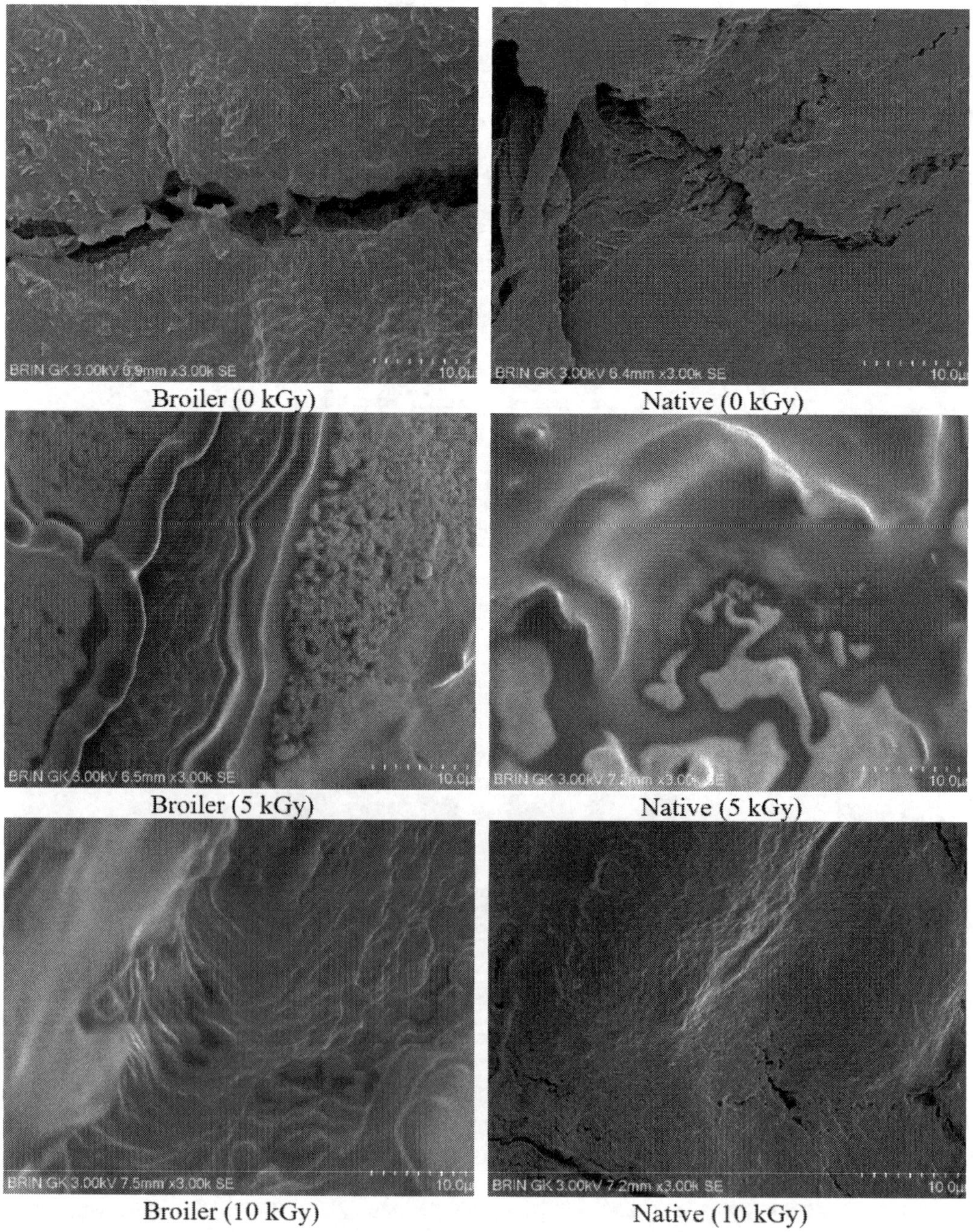

Fig. 2. Scanning electron microscopy (SEM) of untreated and gamma irradiated chicken meat at 3000x magnification.

Figure 2 present SEM image of the surfage image study of broiler and native meat after gamma irradiation treatment at 3000x magnification. Regardless of the kind of meat, gamma irradiation may affect the surface image of the sample. The irradiation process causes meat differences that are only detectable with the SEM test. After irradiation treatment, it has been suggested that variations in muscle structure are responsible for differences in surface appearance. Previous study reported that the low dose (range of 2.2 to 2.9 kGy) of gamma irradiation induced significant textural increasing of cooked chicken breast meat, restriction of sarcomere width, and physical breakdown of myofibril units in muscle fibers [15]. Changes in the protein structure of the meat may also contribute to irradiation-induced visual changes. In different samples, the scanning electron microscopy illustrations revealed that initially, whey protein had a fairly uniform structure, however after gamma radiation, aggregated particles formed [16]. Furthermore, Wang et al. [16] reported that Scanning electron micrographs of samples of whey protein treated with 15 and 25 kGy indicated larger particles and aggregates, particularly for samples treated with 25 kGy.

Table 1. pH value of broiler and native chicken meat after gamma irradiation treatment.

Meat type	**Irradiation dosage (kGy)**	**Storage period (d)**		
		0	**7**	**14**
Broiler	0	6.28d	6.15d	6.12a
	5	5.73a	6.37f	6.37b
	10	5.98b	5.94b	6.14a
Native	0	6.18c	6.23e	6.53c
	5	6.24d	5.80a	6.32b
	10	6.52e	6.07c	6.59d
SEM		0.060	0.045	0.044
Meat type		*	ns	**
Irradiation dosage		ns	ns	ns
Meat type vs Irradiation dose		ns	ns	ns

Different letters in the same row indicate significant differences ($p < 0.05$), Standard Error of the Means (SEM).

The pH value of broiler and native chicken meat samples irradiated at doses of 5 and 10 kGy during 14-day storage are shown in Table 1. Based on the various types of meat, the results reveal that there was no statistical difference in pH values between irradiation dosage. However, native chicken irradiated with 10 kGy had the highest pH value in 0 and 14 days of storage ($p<0.05$). Whereas at 7 days of storage, broiler meat irradiated at 5 kGy had the greatest pH value of 6.37 ($p<0.05$). The pH value at 0 and 14 days of storage varies based on the type of meat ($p<0.05$). There is no interaction between meat type and irradiation dosage.

In our study, gamma irradiation dosage had no significant effect on pH value. This may be the result of the inhibition of bacterial growth, which may delay the synthesis of alkaline chemicals, thereby sustaining a stable pH level during storage [17]. Previous research investigations also revealed similar results [18–20]. It should be noted that in this study, chicken meat contained no additional ingredients. In several studies, pH changes in post-irradiated meat were induced by interactions between radiation and additional treatments, such as the addition of chitosan [20] and herbs [21]. The pH value is also represented by the color of chicken meat; the lower the pH, the paler the color [22]. At 0 and 14 days of storage, it is interesting to note that, native chicken meat exposed to 10 kGy radiation produced a higher pH than broiler meat ($p<0.05$). This will affect the appearance of color and consumer acceptability. Normal breast pH is between 5.91 and 6.02, whereas pH 5.88 to 5.84 indicates a pale, soft, exudative breast, and pH 6.35 to 6.4 indicates a dark, firm, dry-looking breast [23]. However, the difference in pH value may also be influenced by differences in the part of meat and the slaughter age of the chicken [24].

Table 2. Mineral content of broiler and native chicken meat after gamma irradiation treatment.

	Broiler			Native		
	Irradiation dose (kGy)					
Macro-minerals (%)	0	5	10	0	5	10
P	13.18	11.88	11.61	12.53	11.34	11.86
S	12.47	13.49	13.94	15.33	15.54	15.64
Cl	11.64	11.04	11.69	11.92	10.62	11.17
K	49.99	51.49	50.10	46.98	45.89	44.34
Ca	10.55	9.99	10.49	10.79	14.41	14.66
Micro-minerals (ppm)						
Zn	0.1	0	0.1	0	0.1	0.1
Cu	0	0	0	0	0	0
Mn	0	0	0	0	0	0
Fe	0	0	0.1	0	0.1	0.1
Co	0	0	0	0	0	0
Mo	109.2	100.1	92.9	105.9	115.6	88.3
I	528.6	366.9	499.9	515	297.6	329.1
Se	0	0	0	0	0	0
Cr	0	0	0	0	0	0

As a mineral source food, changes in mineral content must be evaluated after gamma irradiation. The results of mineral assays for untreated and irradiated chicken meat are provided in Table 2. The effect of irradiation on the mineral content of chicken is typically different for each macro and micro mineral component. In both types of meat, irradiation treatment tends to increase the content of P and S. After irradiation, the mineral K content of broiler meat tends to increase, whereas it tends to decrease in native meat. In contrary, Ca mineral content in broiler meat tends to decrease following irradiation, whereas it tends to increase in native meat. Regardless of the type of meat, the content of micromineral I tends to decrease after radiation. The effects of irradiation on mineral composition might be explained more precisely, with some studies indicating a decrease in certain minerals and others indicating no substantial changes [25]. We predict that the change in mineral composition after gamma irradiation may be the result of a decrease or elimination in the microbial population. However, more research is necessary. Various studies have demonstrated that gamma irradiation can reduce the populations of aerobic bacteria [9], coliform [26], *Escherichia coli* [3], salmonella [27], yeast, and fungi [28].

4 Conclusion

The gamma irradiation treatment changed the color value of both broiler and native chicken meat. However, no significant differences in pH values were observed. Gamma irradiation may affect the sample's surface image. Following irradiation with 5 and 10 kGy, larger particles and aggregates were observed. This research is essential to determine the differences between native and broiler meat after irradiation treatment, in the prospect of exports.

References

1. Ministry of Agriculture. Ministry of Agriculture: This Year Export Volume of Indonesian Poultry Products Increases. Available from: https://ditjenpkh.pertanian.go.id/berita/1322-kementan-tahun-ini-volume-ekspor-produk-unggas-indonesia-naik [cited 2023 Jun 20]. (2021).
2. N.D. Wahyono, M.M.D. Utami. J. Phys. Conf. Ser. **953** (2018).
3. H.B. Jadhav, U.S. Annapure, R.R. Deshmukh. Front. Nutr. **8** (2021).
4. W. Khalid, A. Maggiolino, J. Kour, M.S. Arshad, N. Aslam, M.F. Afzal, et al. Front. Nutr. **9** (2023).
5. A.A Fallah, E. Sarmast, M. Ghasemi, T. Jafari, A. Mousavi Khaneghah, M. Lacroix. Food Chem. **405** (2023).
6. M.M.B.B.S. Xavier, R.M. Franco, M.C.L. Souza, Duque S da S, W.T.C. Esteves. J. Bioenergy Food Sci. **5**, 4 (2018).
7. R. Kaur, L. Kaur, T.B. Gupta, J. Singh, J. Bronlund. J. Food Sci. **87**, 10 (2022).
8. R.G.D. Steel, J.H. Torrie. Principles and procedures of statistics. (1960).
9. M.S. Arshad, Z. Amjad, M. Yasin, F. Saeed, A. Imran, M. Sohaib, et al. Int. J. Food Prop. **22**, 1 (2019).
10. P.P. Brito, H. Azevedo, K.M.V.A.B. Cipolli, H.T Fukuma, G.B. Mourao, C.V. Roque, et al. J. Food Sci. **76**, 2 (2011).
11. D.U. Ahn, I.S Kim, E.J. Lee. Poult. Sci. **92**, 2 (2013).
12. W. Khalid, M.S. Arshad, M. Yasin, A. Imran, M.H. Ahmad. Int. J. Food Prop. **24**, 1 (2021).
13. G.H. Zhou, X.L. Xu, Y. Liu. Meat Sci. **86**, 1 (2010).
14. S.J. Millar, B.W. Moss, M.H. Stevenson. Meat Sci. **55** (2000).
15. K.S. Yoon. Meat Sci. **63** (2003).
16. X.B. Wang, C.N. Wang, Y.C. Zhang, T.T. Liu, J.P. Lv, X. Shen, et al. J. Dairy Sci. **101**, 6 (2018).
17. R.F. Baptista, C.E. Teixeira, M. Lemos, M.L.G. Monteiro, H.C. Vital, E.T. Mársico, et al. Poult. Sci. **93** (2014).
18. M. Al-Bachir, Y. Othman. Innov. Rom Food Biotechnol. **13** (2013).
19. K. Hwang, H. Kim, D. Song, Y. Kim, Y. Ham, J. Lee, et al. Radiat. Phys. Chem. **106** (2015).
20. P. Hassanzadeh, H. Tajik, S. Mehdi, R. Rohani, M. Moradi, M. Hashemi, et al. Radiat. Phys. Chem. **141** (2017).
21. M. Al-Bachir, S. Farah, Y. Othman. Radiat Phys Chem. **79**, 8 (2010).
22. B.S. Hertanto, C.D.A. Nurmalasari, A.M.P. Nuhriawangsa, M. Cahyadi, L.R. Kartikasari. IOP Conf. **102**, 1 (2018).
23. M.J.V. Sumague, O.M. Del Rosario, W.T. Tan, D.M.O. Santiago, F. Flores, A.F.C. Algar, et al. Philipp J. Sci. **145**, 4 (2016).
24. M. Ali, S.Y. Lee, J.Y. Park, K.C. Nam. Food Sci. Anim. Resour. **41**, 5 (2021).
25. R. Indiarto, A.N. Irawan, E. Subroto. Foods. **12**, 9 (2023).
26. G. Li, P. Lin, Y. Li, Y. He, Z. Liu. Food Sci. Nutr. **11**, 4 (2023).
27. M. Acharjee, I. Islam, R. Noor. Int. Food Res. J. **26**, 4 (2019).
28. A.A. Fallah, S.S. Saei-dehkordi, M. Rahnama. Radiat. Phys. Chem **79**, 10 (2010)

Evaluation of the physical properties of slow release urea based on irradiated chitosan as a feed supplement

Akhmad Rasyid Syahputra[1*], *Ade Lestari* Yunus[1], *Asti* Nilatany[1], *Oktaviani* Oktaviani[1], *Nunung* Nuryanthi[1], *Wahidin Teguh* Sasongko[2], and *Taufiq* Bachtiar[3]

[1]Research Center for Radiation Process Technology, National Research and Innovation Agency, BRIN, Jakarta, Indonesia
[2]Graduate School of Nutrition and Feed Science, IPB University, Bogor, Indonesia
[3]Research Center for Environmental and Clean Technology, National Research and Innovation Agency, BRIN, Bandung, Indonesia

Abstract. A favorable candidate for innovative feed supplement ingredients that can optimize nutrient utilization and reduce environmental impacts could be a slow-release urea based on irradiated chitosan. Chitosan, a biopolymer derived from chitin, has shown immense potential in various applications due to its unique physicochemical properties and biocompatibility. The findings of this study shed light on the promising prospects of irradiated chitosan as a feed supplement ingredient for slow-release urea formulations. Slow-release gel is composed by starch, acrylamide, polyvinyl alcohol and irradiated chitosan. Copolymer gels are treated by Co-60 gamma-ray with 5 kGy and 10 kGy absorbed doses. The lowest gel viscosity escalation is around four thousand times and the highest is more than seventy thousand times. As time of immersion and acrylamide raise in per vary from dose absorbed, gel swelling capacity also increases, start from 33.13 g/g at 5 minutes rise to 164.71 g/g at 720 minutes. Gel fraction from 5 kGy nearly increases two times from 53.57 g/g to 125 g/g at 10 kGy.

1 Introduction

In the dynamic and ever-evolving domain of livestock nutrition, the quest for innovative and sustainable feed supplementation strategies remains at the forefront. Among the many challenges faced by animal nutritionists and agricultural scientists, optimizing the utilization of urea as a cost-effective source of non-protein nitrogen in animal diets stands out. The potential of incorporating irradiated chitosan as a novel matrix for slow-release urea, aiming

* Corresponding author: akhmadrasyids@gmail.com

to improve nutrient utilization in livestock and contribute to the sustainability of modern animal farming practices has been investigated.

The choice of irradiated chitosan as the central component in this investigation stems from its remarkable properties as a biopolymer. Chitosan, derived from chitin, is renowned for its biocompatibility, biodegradability, and non-toxic nature. By subjecting chitosan to controlled irradiation, its physical and chemical properties can be tailored to enhance its functionality as a carrier for urea. This innovative approach not only promises controlled urea release within the digestive tract of animals but also opens doors to environmentally friendly and economically viable solutions for the livestock industry.

Moreover, the need for slow-release urea formulations is underscored by the ecological concerns related to excessive nitrogen excretion from animal waste. Conventional urea supplementation can result in nitrogen losses to the environment, contributing to pollution and ecosystem disruption. In light of these challenges, this study explores the potential of slow-release urea based on irradiated chitosan to mitigate nitrogen waste and reduce the environmental footprint of animal farming, aligning with by the investigate slow-released urea substances from physical properties evaluation side.

2 Experimental Procedures

2.1. Chitosan irradiation

Chitosan powder was packed in polyethelene plastics 20x20 cm^2. Chitosan is irradiated using a Co-60 γ-radiation source with an activity 53.13 kCi and 75 kGy absorbed dose [1] in a natural rubber irradiator (IRKA) at the Research Center for Radiation Process Technology (PRTPR), National Research and Innovation Agency, BRIN, Jakarta, Indonesia.

2.2. Sample preparation

The experimental samples consisted of two types which were sequentially irradiated at absorbed doses of 5 and 10 kGy inside a γ Co-60 irradiator. Each sample was prepared in two variations, each comprising starch, poly vinyl alcohol (PVA), irradiated chitosan, and acrylamide. They were placed in a glass beakers and then 100 ml of distilled water was added, followed by stirring for 30 minutes at 90°C until a thick solution was formed. The solution was allowed to stand at room temperature for 60 minutes. The solution was packed in PE plastics 20 x 20 cm^2.

Table 1. Synthesis formula of gel copolymer gamma irradiation induced

Sampel	S	Aa	PVA	C	γ
KK-1	5 g	3 g	3 g	5 ml	5 kGy
KK-2	5 g	5 g	3 g	5 ml	5 kGy
KK-3	5 g	3 g	3 g	5 ml	10 kGy
KK-4	5 g	5 g	3 g	5 ml	10 kGy

S (starch), Aa (acrylamide), PVA (poly vinyl alcohol), C (chitosan), γ (absorbed dose)

2.3. Viscosity measurement

The gel copolymer that had been allowed to stand at room temperature and its viscosity was measured using a AMETEK Brookfield™ viscometer with a LV-2 needle type at 6 rpm ((Brookfield Engineering, Middleborough, MA, USA), Finder Factor (FF) 50 for the gel copolymer before irradiation, and LV-4 needle type at 0.6 rpm, FF 10^4 for irradiated copolymer. The viscosity values of the gel copolymer before and after irradiation were compared to determine whether a copolymerization process has occurred within the sample.

2.4. Swelling capacity test

A certain amount of gel copolymer was taken and placed in a thin cloth (tea bag) with a known constant weight. Subsequently, each gel was immersed in distilled water [2] at intervals of 5 to 7200 minutes. Weight measurements are taken at 5, 10, 20, 25, 30, and 7200 minutes by separating the unabsorbed distilled water by hanging the bag. Gel swelling capacity was calculated using the following equation.

$$ESV\ (Equilibrium\ Swelling\ Value) = \frac{Wt - Wo}{Wo} \times 100\% \quad (1)$$

Wt = *weight of the gel sample undergoing swelling at t minutes*
Wo = *weight of the dry gel*

2.5. Gel fraction test

Approximately ±3 grams of irradiated gel were packed in a wire mesh and immersed in hot water (±90°C) for 8 hours [3]. They were then dried in an oven at 60°C until a constant weight was achieved. The gel fraction was calculated using the following equation.

$$Gel\ Fraction = \frac{Wt}{Wo} \times 100\% \quad (2)$$

Wt = *weight of gel before immersion*
Wo = *weight of dry gel after immersion*

3 Results and Discussion

The use of chitosan in agriculture is widely recognized as a growth promotor and antibacterial agent, capable of controlling the release rate of easily lost nutrient elements. Chitosan has a high molecular weight and strong intermolecular and intramolecular hydrogen bonds. Chitosan is derived from the natural polymer chitin, which consists of 2-amino-2-deoxy-D-glucose and 2-acetamido-2-deoxy-D-glucose [4][5]. In this study, white to yellowish chitosan powder was used. A sample of 200 grams of chitosan was placed in a 10 x 10 cm^2 PET container, and then the sample was irradiated using gamma rays from a Co-60 source. The dose rate used for the irradiation of chitosan was 9.3 kGy/hour. The total dose obtained for the irradiation of chitosan was 75 kGy. The purpose of irradiating chitosan was to reduce its molecular weight significantly [6], resulting in what is called oligochitosan [7]. Chitosan degradation is commonly used to transform chitosan into oligochitosan (o-chitosan) with better solubility in water. Chitosan with lower molecular weight and higher water solubility has several improved biological, chemical, and physical properties compared to regular chitosan [8], such as antibacterial activity [9] antifungal activity [10], and antitumor activity [11]. The irradiation dose of 75 kGy is based on a study conducted by El-sawy et al. in 2010 [12], where chitosan was irradiated with doses ranging from 20 to 200 kGy. The results showed significant changes in the viscosity of chitosan powder as the irradiation dose increased [13]. A substantial decrease in molecular weight occurred at doses between 20 and 120 kGy. The radiation dose of 70-200 kGy resulted in very low degradation rates [14], which is a characteristic of ionizing radiation on biodegradable polysaccharide materials, leading to a decrease in the average molecular weight [15]. The irradiation dose of 20 kGy resulted in an average molecular weight of 7.106 g/mol, 40 kGy resulted in 6.106 g/mol, 80 kGy resulted in 3.106 g/mol, and doses of 75-200 kGy exhibited relatively stable molecular weights ranging around 2.106 g/mol. The irradiated chitosan molecules undergo cleavage of β-(1-4) glycosidic bonds, leading to a reduction in their average molecular weight [16].

3.1. Gel copolymer viscosity

Table 2, showed that all doses of gamma irradiation resulted in a significant increase in viscosity values. Gamma irradiation can influence the viscosity of a starch-acrylamide-PVA gel by promoting cross-linking, which generally increases viscosity [17][18]. The lowest increase was 5234.77 times for sample KK-1 with a radiation dose of 5 kGy, and the highest (measurable) increase is 71596.83 times for sample KK-4 with a radiation dose of 10 kGy.

Table 2. Average viscosity of gels copolymer

Before Co-60 gamma irradiation gel	
Sample	**Average viscosity, η (10³.50 cPs)**
KK-1	1.23
KK-2	1.31
KK-3	1.94
KK-4	1.84
After Co-60 gamma irradiation gel copolymer	
Sample	**Average viscosity, η (10³.10⁴ cPs)**
KK-1	32.2
KK-2	193.7
KK-3	558.3
KK-4	658.7

Samples KK-1 to KK-4 were copolymer gels formed through irradiation [19], which were not very physically dense and could be used for blending with urea fertilizer. However, despite of KK-1 to KK-4 gel copolymer, the viscosity values after the radiation process were extremely elevated [20], forming highly dense gels [21] that cannot be mixed with urea fertilizer. Radiation doses higher than 10 kGy are too high, resulting in gels with excessively large cross-linking densities, causing a narrowing of the space between networks for water absorption [22]. Based on the viscosity measurements of the irradiated copolymer gel samples at various doses, it can be concluded that doses of 5 and 10 kGy are the optimum dose for the formation of gel copolymer that can be blended with urea fertilizer. At the lower radiation dose of 2 kGy, the hydrogel was not formed, likely due to the lack of the necessary activation energy required to begin the reaction. Typically, when dealing with an aqueous solution, gamma rays serve to break apart water molecules, forming hydrogen and hydroxyl free radicals that serve as initiators for the reaction [23].

3.2. Swelling capacity of gel copolymer

The development capacity of the synthesized gel copolymer is determined by its ability to absorb water. The swelling ability of gel will decrease as the concentration of ions that dissociate in water increases [24]. When the charge concentration inside the gel is greater than the charge concentration in the external solution, it will cause the gel to expand [25–27]. The results of water absorption capacity for the 5 kGy irradiated gel can be seen in Figure 1a. Swelling capacity of the 5 and 10 kGy irradiated gel can be observed from minute 5 to minute 720. The swelling capacity of gel raised up generally with increasing immersion time and increasing acrylamide concentration [28][29][30]. During the immersion period of 15-720 minutes, sample KK-1 to KK-4 reaches near constant with a swelling capacity at 20-30 minutes. This is because gels were added with 3-5% w/v acrylamide, causing the $-NH_2$ water-attracting groups, which contain hydrogen bonds, to become saturated at minute 20 due to the low concentration [31][32] [33]. Sample KK-1, with the addition of 3% w/v acrylamide, experienced an increase in swelling capacity from minute 5 to minute 30, reaching 121.25 g/g. After being left to stand for 720 minutes, the swelling capacities of KK-1 and KK-2 decreased to 87.87 g/g and 51.07 g/g, respectively. This decrease is possible because during the 720-minute standing period, cross-links between acrylamide molecules and cross-links between acrylamide and polyvinyl alcohol (PVA) molecules broke, causing some acrylamide and PVA molecules to dissolve in water, thereby reducing the swelling capacity [34][35]. During the swelling process, water continuously move into the gel. This process is caused by the difference in osmotic pressure between the gel and water [36][37][38]. The greater the difference in osmotic pressure, the greater the swelling capacity [39].

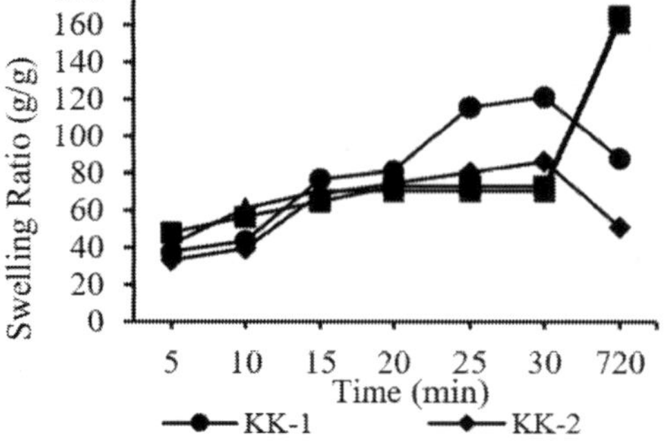

(a) Swelling capacity of irradiated gel

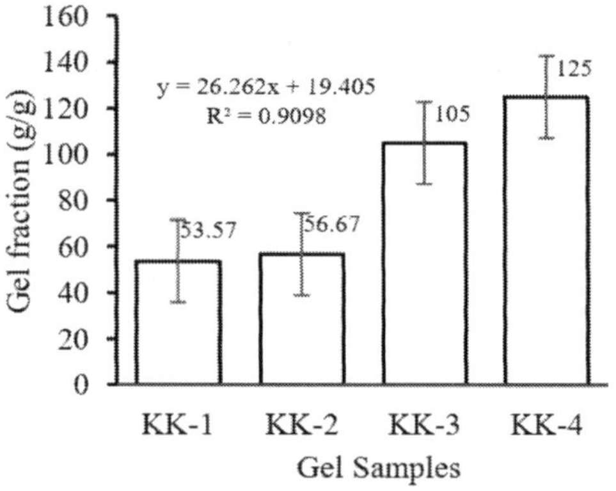

(b) Gel fraction percentages of irradiated gel

Fig 1. Physical properties evaluation of 5 and 10 kGy gel

3.3. Gel fraction percentages

The fraction of gel values was determined to understand the influence of gamma rays on cross-linked molecules in copolymer gel [40][41]. The gel fraction increases with higher radiation doses and monomer concentrations [42]. A higher monomer concentration increases the likelihood of cross-linking within the gel. In general, the effect of radiation on the solution system makes it difficult to reach 100% gel fraction value. This was because of the presence of oxygen in the water/environment molecules interfered with the polymerization process and the possibility of gel degradation at excessively high radiation doses [43]. The addition of PVA to the gel can increase polymer tacticity [44][45] which means chemical structure of PVA has reactive tertiary protons near the -OH group, making it easier to detach. The gel fraction values for samples KK-1 to KK-4 increase proportionally with the added acrylamide concentration. More acrylamide will increase likelihood of acrylamide and PVA cross-linking, which can improve the physical properties of the copolymer gel [46][47]. Sample KK-4, with a 5% acrylamide concentration, had the highest gel fraction value of 125 g/g, with none of the copolymer gel fraction lost at 90°C for 8 hours. The gel fraction values for samples KK-3 and KK-4 also increased proportionally with the added acrylamide concentration. Sample KK-3, with a 3% w/v acrylamide concentration, had a gel fraction value of 105 g/g, suggesting that there was no gel fraction lost from the gel until after 8 hours. Figure 1b. showed KK-3 and KK-4 had gel fractions exceeding 100 g/g, suggesting that acrylamide and polyvinyl alcohol compounds are strictly cross-linked [46], forming a strong polymer bond that is resistant to high temperatures (the polymer does not easily break down). A high gel fraction value indicates that copolymer gel KK-3 and KK-4 can still absorb water even at high temperatures [48] following 8 hours of immersing.

4 Conclusion

Synthesis of irradiated gel copolymer (slow release urea) based on chitosan as feed supplements ingredients is affected by dose of Co-60 gamma irradiation. Nevertheless, the addition of acrylamide and poly vinyl alcohol also modified water absorption capacity and cross-linking percentage of gels. The physical properties of this gel are qualified as chitosan-based supporting materials which used as a feed supplement ingredient.

The authors acknowledge the HITN (Nuclear Innovation Technology Product) 2023 and IAEA Coordinated Research Project D31031 – 25116 (Nuclear and related techniques to measure the impact of type of feeding and production system on greenhouse gas (GHG) emissions and livestock productivity) for funding source.

5 References

1. D. Darwis, D. S. Pangerteni, B. Abbas, T. Puspitasari, N. Nuryanthi, A. Nurrochmad, and C. M. Airin, Macromol. Symp. **391**, (2020)
2. T. Jayaramudu, H. U. Ko, H. C. Kim, J. W. Kim, and J. Kim, Materials (Basel). **12**, (2019)
3. C. Chang, B. Duan, J. Cai, and L. Zhang, Eur. Polym. J. **46**, (2010)
4. T. Thirugnanasambandan and S. C. B. Gopinath, Process Biochem. **130**, (2023)
5. F. A. Vicente, B. Bradić, U. Novak, and B. Likozar, Biopolymers **111**, (2020)

6. D. Xuan Du, B. Xuan Vuong, and H. D. Mai, Adv. Mater. Sci. Eng. **2019**, (2019)
7. V. Ngoc Boi, N. Thi My Trang, D. Xuan Cuong, V. Thi Hoan, and L. Hai, World J. Food Sci. Technol. **4**, (2020)
8. I. Aranaz, A. R. Alcántara, M. C. Civera, C. Arias, B. Elorza, A. H. Caballero, and N. Acosta, Polymers (Basel). **13**, (2021)
9. F. Zhang, G. Ramachandran, R. A. Mothana, O. M. Noman, W. A. Alobaid, G. Rajivgandhi, and N. Manoharan, Saudi J. Biol. Sci. **27**, (2020)
10. P. Poznanski, A. Hameed, and W. Orczyk, Molecules **28**, (2023)
11. P. Zou, X. Yang, J. Wang, Y. Li, H. Yu, Y. Zhang, and G. Liu, Food Chem. **190**, (2016)
12. N. M. El-Sawy, H. A. Abd El-Rehim, A. M. Elbarbary, and E. S. A. Hegazy, Carbohydr. Polym. **79**, (2010)
13. I. Zainol, H. M. Akil, and A. Mastor, Mater. Sci. Eng. C **29**, (2009)
14. M. Mahmud, M. I. Naziri, N. Yacob, N. Talip, and Z. Abdullah, in *AIP Conf. Proc.* (2014)
15. K. Wiszumirska, D. Czarnecka-Komorowska, W. Kozak, M. Biegańska, P. Wojciechowska, M. Jarzębski, and K. Pawlak-Lemańska, Materials (Basel). **16**, (2023)
16. M. A. García, N. de la Paz, C. Castro, J. L. Rodríguez, M. Rapado, R. Zuluaga, P. Gañán, and A. Casariego, J. Radiat. Res. Appl. Sci. **8**, (2015)
17. J. N. BeMiller, in *Starch Food Struct. Funct. Appl.* (2017)
18. Z. Sui and X. Kong, *Physical Modifications of Starch* (2018)
19. A. W. M. El-Naggar, M. M. Senna, and S. A. Khalil, J. Appl. Polym. Sci. **91**, (2004)
20. M. Navaf, K. V. Sunooj, B. Aaliya, C. Sudheesh, P. P. Akhila, S. Sabu, A. Sasidharan, and J. George, Radiat. Phys. Chem. **201**, 110459 (2022)
21. Erizal, Indones. J. Chem. **12**, (2012)
22. A. Ortega, S. Valencia, E. Rivera, T. Segura, and G. Burillo, Gels **9**, (2023)
23. M. M. Bhuyan, M. Islam, and J.-H. Jeong, Gels **9**, (2023)
24. L. R. Shivakumara and T. Demappa, Turkish J. Pharm. Sci. **16**, 252 (2019)
25. S. Bashir, M. Hina, J. Iqbal, A. H. Rajpar, M. A. Mujtaba, N. A. Alghamdi, S. Wageh, K. Ramesh, and S. Ramesh, Polymers (Basel). **12**, (2020)
26. E. Czarnecka and J. Nowaczyk, Int. J. Mol. Sci. **22**, (2021)
27. A. Mahmoud, A. Fahmy, A. Naser, and M. A. Saied, Sci. Rep. **12**, 22017 (2022)
28. W. Bai, B. Ji, L. Fan, Q. Peng, Q. Liu, and J. Song, Polymers (Basel). **15**, (2023)
29. S. L. Favaro, F. De Oliveira, A. V. Reis, M. R. Guilherme, E. C. Muniz, and E. B. Tambourgi, J. Appl. Polym. Sci. **107**, (2008)
30. S. Noppakundilograt, N. Pheatcharat, and S. Kiatkamjornwong, J. Appl. Polym. Sci. **132**, (2015)
31. F. S. El-banna, M. E. Mahfouz, S. Leporatti, M. El-Kemary, and N. A. N. Hanafy, Appl. Sci. **9**, (2019)
32. X. Y. Wu, S. W. Huang, J. T. Zhang, and R. X. Zhuo, Macromol. Biosci. **4**, (2004)
33. C. M. Yeng, H. Salmah, and S. Sung Ting, in *Adv. Mater. Res.* (2013)
34. M. Wang, J. Bai, K. Shao, W. Tang, X. Zhao, D. Lin, S. Huang, C. Chen, Z. Ding, and J. Ye, Int. J. Polym. Sci. **2021**, 2225426 (2021)
35. J. Yu, S. Gao, D. Hou, P. Wang, and G. Sun, J. Phys. Chem. B **124**, 6095 (2020)
36. M. J. Krafcik and K. A. Erk, Mater. Struct. Constr. **49**, (2016)
37. J. Wang, S. Gao, J. Tian, F. Cui, and W. Shi, Water **12**, (2020)
38. J. L. Holloway, A. M. Lowman, and G. R. Palmese, Acta Biomater. **9**, 5013 (2013)

39. S. Saber-Samandari, M. Gazi, and E. Yilmaz, Polym. Bull. **68**, (2012)
40. D. R. Barleany, Jayanudin, A. S. Utama, U. Riyupi, H. Alwan, R. S. D. Lestari, A. B. Pitaloka, M. Yulvianti, and Erizal, Mater. Today Proc. (2023)
41. E. Erizal and R. C. Rahayu C, Indones. J. Chem. **9**, (2010)
42. T. Fekete, J. Borsa, E. Takács, and L. Wojnárovits, Chem. Cent. J. **11**, (2017)
43. W. Qiao, G. Zhang, P. Jiang, and H. Pei, Gels **9**, (2023)
44. G. W. Oh, I. W. Choi, W. S. Park, C. H. Oh, S. J. Heo, D. H. Kang, and W. K. Jung, J. Appl. Polym. Sci. **139**, (2022)
45. G. A. Mahmoud, D. E. Hegazy, and H. Kamal, Arab J. Nucl. Sci. Appl. **47**, (2014)
46. A. S. Tarek Mansour Mohamed and G. A. Mahmoud, Polym. Technol. Mater. **62**, 866 (2023)
47. M. M. El-Toony, Egypt. J. Chem. **60**, (2017)
48. Y. Tang, Z. Yu, L. Tam, A. Zhou, and D. M. Li, Mater. Today Commun. **33**, 104834 (2022)

The use of tomato peel nanopowder as a natural antioxidant in low-fat mayonnaise

Herly Evanuarini[1*], *Agus* Susilo[1], and *Dedes* Amertaningtyas[1]

[1]Animal Product and Technology Department, Faculty of Animal Science, Universitas Brawijaya Malang 65145, East Java, Indonesia

Abstract. The tomato processing industry produces waste products that are not utilized and add to the problem of environmental pollution. Tomato peel still contains high bioactive components, tomatoes as the main source of beta-carotene and lycopene are expected to improve food quality. Owned antioxidant activity can reduce the oxidation process in food products. Low-fat mayonnaise is a type of low-fat mayonnaise that has a fat content of less than 40%. The purpose of this study was to investigate the effect of adding tomato peel nanopowder to low-fat mayonnaise based on emulsion stability, viscosity, moisture content, fat content, antioxidant activity, and fatty acid profile. Low-fat mayonnaise was prepared using sunflower oil, vinegar, and egg yolk by giving 1%, 2%, and 3% tomato peels nanopowder treatment, and without the addition of control which was repeated 5 times. A completely randomized design was used as the research design. Analysis of variance was used as statistical analysis and continued with Duncan's test if there were significant or very significant differences. The results of the study adding nanopowder tomato peel with different percentages had a very significant effect ($p<0.01$) on emulsion stability, viscosity, water content, fat content, antioxidant activity, and the presence of various types of fatty acids in low-fat mayonnaise. The addition of tomato peel nanopowder can increase the stability of the emulsion, viscosity, and antioxidants, as well as reduce the water content and fat content. This study concludes that 3% tomato peel nanopowder is capable of being a natural antioxidant that can improve the physicochemical quality of low-fat mayonnaise.

1 Introduction

Mayonnaise is the result of an emulsion of oil in water made by mixing egg yolks, vegetable oil, vinegar, and other optional ingredients. Mayonnaise is one of the sauces that is widely used in the food industry. Traditional mayonnaise contains as much as 70 – 80% fat, which causes concerns about excessive fat intake. The fact that occurs in society is that the amount of fat consumed is positively correlated with the number of chronic diseases such as obesity and cardiovascular disease [1]. Therefore, strategies are needed for developing food products to reduce the amount of fat but not change the taste and texture

* Corresponding author: herlyfptub@ub.ac.id

attributes. The development of low-fat mayonnaise is a challenging thing to be able to provide taste, color, and texture attributes that are almost the same as full-fat mayonnaise. If there is a reduction in fat droplets, it will affect the texture, viscosity, and stability of the emulsion in the emulsion system. Research that has previously developed includes the strategies applied, including the use of fruit and vegetable peel waste as a stabilizer in reduced-fat mayonnaise [2], [3] and the use of pumpkin as a stabilizer and source of antioxidants for reduced-fat mayonnaise [4]. This can be adopted for the development of alternative low-fat mayonnaise products.

Tomato (*Lycopersion esculentum*) is a type of fruit cultivated throughout the world. In general, tomatoes are used as a basic ingredient for making fresh sauces, tomato sauce, pasta, and processed into fresh juice as well as selling peeled tomatoes without peel. The by-products of processing fresh tomatoes are tomato peel, tomato seeds, and a mixture of peel and seeds called tomato pomace. This waste can disturb the environment due to the lack of use of by-products, considering that tomatoes have a high water content which can cause the growth of microorganisms to grow faster. Efforts to utilize tomato waste include renewable energy in biodiesel, product packaging, cereal industry, and can be used in the food industry [5–7]. Previous research stated that tomato peel contains bioactive compounds in the form of carotenoids, dietary fiber, phenols, unsaturated fatty acids, and many lycopene which can prevent non-communicable diseases in humans [8, 9]. Carotenoids are natural pigments in fruit and vegetables to produce red, yellow, and orange colors which are rich in antioxidants [10]. Carotenoids can improve the immune system and reduce the risk of degenerative diseases. Carotenoids as provitamin A bioactive compounds can increase product shelf life and sensory properties. This can streamline tomato processing and increase the added value of the product in food science. In addition, 50% of tomato peel is fiber which includes cellulose, hemicellulose, lignin, and pectin [11]. Lycopene in tomato peel as a coloring agent and antioxidant compound [6, 12, 13]. Tomato peel will be transformed into nanopowder to increase absorption and reduce particles when applied to food products. The functional properties of tomato peel are considered to be an antioxidant agent in low-fat mayonnaise.

This research investigates the effect of using tomato peel nanopowder as a natural antioxidant in low-fat mayonnaise. The parameters observed were emulsion stability, viscosity, water content, fat content, antioxidant activity, and fatty acid profile in the best treatment.

2 Materials and methods

2.1 Materials

The material used in this research was tomato peel which was separated from fresh tomatoes. Fresh tomatoes are found at the Malang City Traditional Market. The ingredients for making mayonnaise are vegetable oil, vinegar, sugar, salt, ground pepper, and mustard from local supermarkets. Meanwhile, chicken eggs are from smallholder farmers in Junrejo, Batu City.

2.2 Methods

Low-fat mayonnaise was prepared using sunflower oil, vinegar, and egg yolk by giving 1%, 2%, and 3% tomato peels nanopowder treatment, and without the addition of control which was repeated 5 times. A completely randomized design was used as the research design and are presented in Table 1.

Table 1. Experimental design and research treatment

Treatments	**Descriptions**
Control	Full-fat mayonnaise control with 70% oil
TP1	Low-fat mayonnaise with 30% oil and 1% tomato peels nanopowder
TP2	Low-fat mayonnaise with 30% oil and 2% tomato peels nanopowder
TP3	Low-fat mayonnaise with 30% oil and 3% tomato peels nanopowder

2.3 Preparation of Tomato Peels Nanopowder

Making tomato peel nanopowder [24] starts by separating tomato peel from fresh tomatoes. Place the tomato peels on a baking sheet lined with parchment paper. Put the tomato peels in a hot oven at 60°C ± 2 for 6 hours. The heating process also involves turning the tomato peel every 3 hours. The dried tomato peel is refrigerated for 1 × 24 hours. The cooled dried tomato peel is put into a dry mill to be ground and filtered to 300 mesh. Furthermore, nanopowder tomato peel can be applied to low-fat mayonnaise products.

2.4 Low-fat Mayonnaise Preparation

Mayonnaise is prepared using a mixer at 2000 rpm for 10 seconds to mix sugar (2%), salt (1%), ground pepper (0.5%), and mustard (1%). After that, it is homogenized with egg yolk (20%), vinegar (5%), and vegetable oil for full fat (70%) and low fat (30%) treatment. Stirring using a mixer at 2000 rpm ± 3 minutes. The final step is to enter the tomato peel treatment with nanopowder TP1 (1%), TP2 (2%), and TP3 (3%) stirring for 30 seconds. Mayonnaise samples were put into glass jars and stored at room temperature 26°C ± 2°C [4].

2.5 Data analysis

All tests were carried out with five replications and produced data in the form of mean ± standard deviation. Calculation of statistical data using two-way ANOVA IBM SPSS Statistics 25.0 (IBM SPSS, Inc). Duncan's follow-up test with a level of 0.05 was carried out whether or not there was a significant difference in the data.

3 Results and Discussion

3.1 Emulsion stability

The results showed that tomato peel nanopowder had a very significant effect ($p<0.01$) on low-fat mayonnaise. The higher the percentage of added tomato peel nanopowder, the stability of the low-fat mayonnaise emulsion increases. The average of emulsion stability of low-fat mayonnaise can be seen in Table 2.

Table 2. The average emulsion stability of low-fat mayonnaise

Treatment	Emulsion Stability (%)
C	$95.00^{b} \pm 0.82$
TP1	$92.25^{a} \pm 0.96$
TP2	$93.00^{ab} \pm 0.82$
TP3	$96.00^{b} \pm 0.82$

Note: $^{a\ and\ b}$ in the same column show a very significant effect ($p < 0.01$)

Emulsion stability is one of the important parameters in emulsion products. Emulsion instability will occur if the oil and water ratio is incorrect or the emulsifier concentration is insufficient. The emulsion stability of low-fat mayonnaise with nanopowder tomato peel is almost the same as full-fat mayonnaise. Nanopowder tomato peel can act as a stabilizer between oil and water in the mayonnaise emulsion system. Previous research stated that tomato peels contain sufficient amounts of carbohydrates to be used in food product applications [15]. Apart from that, tomato peel is a source of dietary fiber and has bioactive components that are attractive for application in the food industry as a functional food [16]. Pectin is a heterosaccharide polymer part of food fiber that has hydrocolloid properties, it can absorb water [13]. Nanopowder tomato peel can bind water and play a role in the continuous phase of the low-fat mayonnaise emulsion, thereby producing a stable emulsion. Previous research has emulsion stability of mayonnaise with watermelon rind is 87.05 – 97.79% [2].

3.2 Viscosity

The results showed that the addition of tomato peel nanopowder had a very significant effect ($p<0.01$) on low-fat mayonnaise. The higher the percentage of added tomato peel nanopowder, the viscosity of low-fat mayonnaise increases. The average of viscosity of low-fat mayonnaise can be seen in Table 3.

Table 3. The average viscosity of low-fat mayonnaise

Treatment	Viscosity (cP)
C	$3645^{c} \pm 60$
TP1	$3235^{a} \pm 93$
TP2	$3420^{b} \pm 42$
TP3	$3800^{d} \pm 62$

Notes: $^{a,\ b,\ c,\ and\ d}$ in the same column show a very significant effect ($p < 0.01$)

The increase in viscosity in low-fat tomato peel nanopowder mayonnaise is due to the pectin content in tomato peel. The viscosity in TP3 and C is almost the same, where treatment C is control with 70% oil, while low-fat mayonnaise only uses 30% oil. Pectin can replace oil and become a stabilizer so that it can increase the viscosity of low-fat mayonnaise. Previous research stated that tomato peel contains pectin which can be used as a stabilizer in food [11]. Hydrocolloids in pectin can act as a thickening agent which can reduce the fluidity of oil so that the network structure of oil in water in mayonnaise increases the viscosity of mayonnaise [17]. Other research also states that if more oil is used, it can increase the viscosity of mayonnaise [18]. Previous research about reduced-fat mayonnaise using watermelon rind flour have a viscosity around 3177.5 – 3805 cP [2].

3.3 Moisture content

The results showed that tomato peel nanopowder had a very significant effect ($p<0.01$) on low-fat mayonnaise. The higher the percentage of added tomato peel nanopowder, the lower the moisture content of the low-fat mayonnaise. The average of moisture content of low-fat mayonnaise can be seen in Table 4.

Table 4. The average moisture content of low-fat mayonnaise

Treatment	Moisture content (%)
C	$25.31^{a} \pm 0.28$
TP1	$24.52^{b} \pm 0.23$
TP2	$23.31^{c} \pm 0.47$
TP3	$22.29^{d} \pm 0.18$

Notes: [a, b, c, and d] in the same column show a very significant effect ($p < 0.01$)

The decrease in moisture content is caused by nanopowder tomato peel which plays a role in binding and absorbing water. This is caused by the complex carbohydrates in the nanopowder tomato peel interacting with water. Tomato peel has a high starch content so it can increase the adhesiveness of food products [15]. Starch has large hydroxyl groups, so it has a high water absorption capacity [19]. The absorption of water by starch granules will increase viscosity and water is trapped in the starch granules [20]. Research on reduced-fat mayonnaise with the addition watermelon rind flour have a moisture content around 15.34 – 19.28% [2].

3.4 Fat content

The results showed that the addition of tomato peel nanopowder had a very significant effect ($p<0.01$) on low-fat mayonnaise. The higher the percentage of added tomato peel nanopowder, the lower the fat content of the low-fat mayonnaise. The average of fat content of low-fat mayonnaise can be seen in Table 5.

Table 5. The average fat content of low-fat mayonnaise

Treatment	Fat content (%)
C	$70.68^{c} \pm 0.25$
TP1	$36.35^{b} \pm 0.42$
TP2	$34.90^{ab} \pm 0.62$
TP3	$34.05^{a} \pm 0.18$

Notes: $^{a, b, c}$ in the same column show a very significant effect ($p < 0.01$)

The fat content in low-fat mayonnaise with the addition of nanopowder tomato peel decreased along with the increase in treatment percentage. This is because tomato peel has a low-fat content. Reducing the fat content in low-fat mayonnaise by replacing the oil with modified additional ingredients in the form of flour [21]. Previous research also reported that the addition of ingredients rich in carbohydrates can reduce the fat content in mayonnaise [2]. Another similar research is the addition of arrowroot starches as a fat substitute in low-fat mayonnaise [18]. Low-fat mayonnaise with different stabilizer has fat contant around 49.2% [25].

3.5 Antioxidant activity

The results showed that the addition of tomato peel nanopowder had a very significant effect ($p<0.01$) on low-fat mayonnaise. The higher the percentage of added tomato peel nanopowder, the antioxidant activity of low-fat mayonnaise increases. The average of antioxidant activity of low-fat mayonnaise can be seen in Table 6.

Table 6. The average antioxidant activity of low-fat mayonnaise

Treatment	Antioxidant Activity (%)
C	$2.88^{a} \pm 0.39$
TP1	$11.09^{b} \pm 0.95$
TP2	$13.48^{c} \pm 0.33$
TP3	$16.49^{d} \pm 0.40$

Notes: $^{a, b, c, \text{and } d}$ in the same column show a very significant effect ($p < 0.01$)

The higher the addition of nanopowder tomato peel, the more antioxidant activity increases. This is caused by the presence of antioxidants and bioactive compounds in tomato peel. Tomato peel contains carotenoids as a source of coloring and also functions as an antioxidant [10]. Apart from that, tomato peel contains phenol which can be used as an antioxidant [7]. Carotenoids and phenols in tomato peel increase the antioxidant activity of low-fat mayonnaise. The peel of fruits such as tomatoes, grapes, and blueberries has strong antioxidants [22].

3.6 Fatty acid profile

The research results showed that in low-fat mayonnaise with the addition of 3% nanopowder tomato peel, there were various types of fatty acids. Types of fatty acids low-fat mayonnaise can showed in Table 7.

Table 7. Fatty acid profile of low-fat mayonnaise

Types of fatty acids	**(%)**
Myristic Acid (C 14:0)*	0.1057
Myricroleic Acid (C 14:1)*	0.0183
Pentadecanoic Acid (C 15:1)*	0.0099
Palmitic Acid (C 16:0)*	3.9962
Palmitoleic Acid (C 16:1)*	0.2500
Heptadecanoic Acid (C 17:0)*	0.0194
Heptadecanoic Acid (C 17:1)*	0.0146
Stearic Acid (C 18:0)*	1.6957
C-Linoleic Acid (C 18:2 W6C)***	17.9180
Linolenic Acid (C 18:3 W3)***	0.0088
Arachidic Acid (C 20:0)***	0.0626
Eicosenoic Acid (C 20:1)***	0.0528
Eicosadienoic Acid (C 20:2)***	0.0176
Arachidonic Acid (C 20:4 W6)***	0.2597
Oleic Acid**	10.8249
Linolenic Acid***	0.0431
Linoleic Acid***	17.9180
Omega 3 Fatty Acids	0.2915
Omega 6 Fatty Acids	18.2022
Omega 9 Fatty Acids	10.8249

Description: *Saturated Fatty Acids, **Singularly Unsaturated Fatty Acids, ***Polyunsaturated Fatty Acids

The fatty acids identified in low-fat mayonnaise with the addition of nanopowder tomato peel are mostly produced by the vegetable oil used and egg yolks. The fatty acids detected were classified into 3 types is saturated fatty acids, monounsaturated fatty acids, and polyunsaturated fatty acids. The types of fatty acids are palmitic acid, oleic acid, and linoleic acid, respectively. Palmitic acid is one of the main components that is always present in fatty acids. Essential fatty acids can function for the body and promote growth. In addition, unsaturated fatty acids can speed up the oxidation process resulting in a short product shelf life. Lipid oxidation is caused by the amount of unsaturated fatty acids which will react directly with oxygen [23]. However, the presence of tomato peel nanopowder in low-fat mayonnaise can act as an antidote to free radicals or as a natural antioxidant agent so that it can inhibit oxidative reactions.

4 Conclusion

3% nanopowder tomato peel can be a natural antioxidant in low-fat mayonnaise. The quality of low-fat mayonnaise is increased by the addition of nanopowder tomato peel.

This research was funded by the Faculty of Animal Science Universitas Brawijaya under the "Doktor Lektor Kepala" scheme with contract number 2045.2/UN10.F05/PN/2023.

References

[1] J. J. DiNicolantonio, S. C. Lucan, J. H. O'Keefe, *The Evidence for Saturated Fat and for Sugar Related to Coronary Heart Disease*, Progress in Cardiovascular Diseases, **58**, 464–472 (2016)

[2] H. Evanuarini, D. Amertaningtyas, D. T. Utama, A. R. Safitri, *The Use of Watermelon Rind Flour as Stabilizer for Reduced Fat Mayonnaise*, Jurnal Ilmu dan Teknologi Hasil Ternak, **15**, 172–182 (2020)

[3] H. Evanuarini, A. Susilo, *Optimization of Reduced Fat Mayonnaise by Using Apple Peel Flour as a Stabilizer*, Jurnal Ilmu dan Teknologi Hasil Ternak, **16**, 174–180 (2021)

[4] H. A. Nidhal, H. Evanuarini, I. Thohari, *Characteristics of reduced fat mayonnaise using pumpkin flour (Cucurbita moschata) as fat replacer*, in *E3S Web of Conferences*, EDP Sciences, (2022)

[5] S. A. Nemes, K. Szabo, D. C. Vodnar, *Applicability of agro-industrial by-products in intelligent food packaging*, Coatings, **10** (2020)

[6] G. A. Martau, M. Mihai, D. C. Vodnar, *The use of chitosan, alginate, and pectin in the biomedical and food sector-biocompatibility, bioadhesiveness, and biodegradability*, Polymers, **11** (2019)

[7] I. F. Strati, V. Oreopoulou, *Recovery of carotenoids from tomato processing by-products - A review*, Food Research International, **65**, 311–321 (2014)

[8] M. Grabowska, *Let food be your medicine: Nutraceutical properties of lycopene*, Food and Function, **10**, 3090–3102 (2019)

[9] A. C. da Silva, N. Jorge, *Bioactive compounds of the lipid fractions of agro-industrial waste*, Food Research International, **66**, 493–500 (2014)

[10] K. Szabo, *Evaluation of the Bioactive Compounds Found in Tomato Seed Oil and Tomato Peels Influenced by Industrial Heat Treatments*, (2021)

[11] F. Jiang, Y. Lo Hsieh, *Cellulose nanocrystal isolation from tomato peels and assembled nanofibers*, Carbohydr Polym, **122**, 60–68 (2015)

[12] C. Chung, G. Smith, B. Degner, D. J. McClements, *Reduced Fat Food Emulsions: Physicochemical, Sensory, and Biological Aspects*, Crit Rev Food Sci Nutr, **56**, 650–685 (2016)

[13] E. D. Ngouémazong, S. Christiaens, A. Shpigelman, A. Van Loey, M. Hendrickx, *The Emulsifying and Emulsion-Stabilizing Properties of Pectin: A Review*, Compr Rev Food Sci Food Saf, **14**, 705–718 (2015)

[14] AOAC, *Official method of Analysis*, 18th Edition. Washington DC: Method 935.14 (2005)

[15] K. Kaboré, *Effects of growing zones on nutritional and bioactive compounds of by-products of two tomato cultivars*, J Agric Food Res, **10** (2022)

[16] P. I. akubor, A. U. Owuse, *Chemical Composition, functional and biscuit making properties of tomato peel flour*, South Asian Journal of Food Technology and Environment, **06**, 874–884 (2020)

[17] W. Wang, *Physicochemical Properties, Stability and Texture of Soybean-Oil-Body-Substituted Low-Fat Mayonnaise: Effects of Thickeners and Storage Temperatures*, Foods, **11** (2022)

[18] J. J. Park, I. F. Olawuyi, W. Y. Lee, *Characteristics of low-fat mayonnaise using different modified arrowroot starches as fat replacer*, Int J Biol Macromol, **153**, 215–223 (2020)

[19] D. Donmez, L. Pinho, B. Patel, P. Desam, O. H. Campanella, *Characterization of starch–water interactions and their effects on two key functional properties: starch*

gelatinization and retrogradation, Current Opinion in Food Science, **39**, 103–109 (2021)

[20] X. Lu, R. Ma, J. Zhan, F. Wang, Y. Tian, *The role of protein and its hydrolysates in regulating the digestive properties of starch: A review*, Trends in Food Science and Technology, **125**, 54–65 (2022)

[21] A. Carcelli, G. Crisafulli, E. Carini, E. Vittadini, *Can a physically modified corn flour be used as fat replacer in a mayonnaise?*, European Food Research and Technology, **246**, 2493–2503 (2020)

[22] B. E. Ştefanescu, K. Szabo, A. Mocan, G. Crisan, *Phenolic compounds from five ericaceae species leaves and their related bioavailability and health benefits*, Molecules, **24** (2019)

[23] R. Domínguez, M. Pateiro, M. Gagaoua, F. J. Barba, W. Zhang, J. M. Lorenzo, *A comprehensive review on lipid oxidation in meat and meat products*, Antioxidants, **8** (2019)

[24] Yun, D., C. Li, Z. Wang, F. Xu, D. Chen, J. Liu, *Preparation of cost-effective and hydrophobic freshness indicating labels based on passion fruit peel powder and stearic acid*, Food Bioscience, **53** (2023)

[25] Than, W. T., S. Win, *The Effects of Various Food Stabilizers on Quality Evaluation of Low-Fat Mayonnaise*, Research Journal, **11**, 190-197 (2020)

Association of Gonadotropin-Releasing Hormone Receptor (GnRHR) gene (c.18085 T>G) and litter size in Ettawa crossbreed goat

Nur Hidayah[1], *Ari* Ardiantoro[1], *Wike Andre* Septian[1], *Guruh* Prasetyo[1], *Heri* Damayanti[1], *Ahmad Zarkasi* Efendy[1], and *Suyadi* Suyadi[1*]

[1]Faculty of Animal Science, Universitas Brawijaya, Malang 65145, Indonesia

Abstract. In the present study, the association of Gonadotropin-Releasing Hormone Receptor gene (GnRHR) and litter size were investigated using the PCR-RFLP method as a genetic marker candidate for litter size in 35 individuals from Ettawa crossbreed goats. This research has two steps conducted in the laboratory and in the field. Data on litter size and blood samples were collected from the field. In the laboratory was carried out DNA isolation, amplification, and restriction. Molecular analysis of DNA isolation were precise, the enzyme Msp1 (C*CGG) with a product length of 706 bp was used for amplification and restriction analysis, releaved TT genotype in GnRHR gene fragment. In the exon 3 of chromosome 6, the single nucleotide polymorphism (SNP) c.18085 T>G was found. Ettawa crossbreed goat in calculations using the hardy-weinberg law, insignificant results are obtained because the value is more than > 0.05 (db; 1) = 0.50. The genotypes and allele frequencies found in the Ettawa Crossbreed Goat GnRHR genes are monomorphic and the GnRHR gene did not associate with litter size. The conclusion is that GnRHR exon 3 gene (SNP) c.18085 T>G did not associate with litter size and can not be used as a genetic marker on Ettawa crossbred goats.
Keyword: Ettawa crossbreed goat, Litter size, GnRHR gene, Msp1 enzyme, PCR-RFLP

1 Introduction

Goats were reared for several reasons, including their abilities to adapt in diverse ecosystems (frigid, harsh, and arid zones), adaptive under low input environments, tolerance to heat stresses, diseases, and parasites, and their abilities to provide socioeconomics, cultures, and religious benefits, as well as cash readily for immediate needs and meaningful income among poor households. Indigenous goats have been part of rustic subsistence in many developing countries [1].

Etawah-crossbreed goats are one of the most important types of dual-purpose goats in tropical areas, especially Indonesia [2]. It is a crossbreed between an Etawah goat and an Indonesian local goat [3]. Ettawa Crossbreed goat is known as dairy production. These goats

* Corresponding author: suyadi@ub.ac.id

are often crossbreed with other local goats of Indonesia through reproductive technology, such as artificial insemination (AI) or natural mating. goats have been maintained for a long time, so that they are considered to be local goats with better genetic quality than other local goats. Milk production of ECDG can reach 1.5-3 liters per day [4]. Goat's milk fat and protein is quite high at 6.08% and 4.48% [5]. The average litter size of Ettawa crossbreed goats and birth weight (kg) were 1.89±0.66 and 3.84±0.73 respectively.

Litter size has a significant impact on farmers' economic gains, which can determine the availability of the number of goats in a livestock business. Goats can have several kids at same time, which is known as the litter size. Each goat's litter size is impacted differently by genetic changes that accelerate ovulation and increase the number of offspring produced per pregnancy [6]. The genetics of litter size emphasizes the crucial of major fecundity genes in goats. There needs to be an effort to increase the number of litter sizes in ettawa crossbreed goats to produce optimal litter size. Development of Ettawa Crossbreed goats which have the goal to enhance the potential goat that exists in the country. One of the most important factors to increase the number of litter sizes is the selection of goats.

The last decade has seen an incredible increase in the use of application of molecular genetics for identification of gene loci and chromosomal areas containing single nucleotide polymorphisms (SNPs) that affect commercially significant traits in the selection of livestock [7]. The use of marker assisted selection technology enables precise selection of advantageous sequence variants that have been demonstrated to improve performance, hence removing the majority of the uncertainty involved in traditional phenotypic selection [8]. Gonadotropin-Releasing Hormone Receptor (GnRHR) is one of the candidate genes for goat litter size under marker-assisted selection.

Gonadotropin Releasing Hormone Receptor (GnRHR) is a crucial regulator of the reproductive system, which causes the pituitary gland to produce and release the luteinizing hormones (LH) and follicle-stimulating hormones (FSH) [9]. GnRHR is responsible for stimulating the synthesis and to release of pituitary gonadotropins (leutenizing and follicle stimulating hormones). These gonadotropins in turn control ovulation and the development of gonadal steroids because of its critical role in regulating the activities of hypothalamo-pituitary-gonadal axis.

Etawa crossbreed dairy goats have been maintained for a long time, so that they are considered to be local goats with better genetic quality than other local goats. Etawa crossbred Dairy Goat (ECDG) is a dual-purpose goat that can produce milk such as Etawa goat and produce meat such as Kacang goats [5]. Milk production of ECDG can reach 1.5-3 liters per day [6]. Goat's milk fat and protein is quite high at 6.08% and 4.48% [7]

There has not been any information about the association of GnRHR genetic variations with litter size on ettawa crossbred goats. Given that the GnRHR gene is linked to luteinizing hormone and follicle-stimulating hormone release, it may be a candidate gene for litter size on goats. The objectives of this study were to identify polymorphisms GnRHR gene and evaluate the associations between GnRHR gene and litter size on ettawa crossbreed goats.

2 Materials and Methods

2.1 Materials

Materials in this study used 35 Ettawa Crossbreed goat and the polymorphism in the gonadotropin-releasing hormone receptor (GnRHR) gene was detected by PCR-RFLP. The characteristics of goats in this study were actively reproductive, given birth and healthy. The research was conducted from January 2023 to Mei 2023. This observation was carried out in different locations. For collecting blood samples and gathering information on litter size, the location is a field at UPT PT and HMT Singosari in Malang. The other location is the laboratory at the Livestock Biotechnology Laboratory, Faculty of Animal Science, Brawijaya University, Malang for DNA extraction and analysis of blood samples of Ettawa crossbreed goats.

The first step on this study was collecting phenotypic data (litter size) and taking blood samples of each Ettawa Crossbreed goat in UPT PT HMT Singosari Malang. Ettawa Crossbreed goat was taken a blood sample in the neck, precisely in the jugular vein as much as 5 ml. The second step in the laboratory includes DNA analysis (DNA isolation, Amplification, and Restriction). DNA isolation to separate DNA from white blood cells. In the beginning, research is carried out by taking blood samples of each goat as much as 5 mL with venoject in the area of the jugular vein in the neck region. Laboratory materials alkohol 70%, Genomic DNA Mini Kit (Geneaid), nucleus Free Water (NFW) (Promega), Buffer tango, gel agarose, ethanol absolute, 1X Green Master Mix (Promega), Restriction enzyme MspI (BioLabs), DNA marker, TBE 0,5X, Loading blue dye, Nucleic Acid Dye Diamond.

The equipments were used in this study including hand glove, centrifuge, tube rack, hot plate (BIAB), beaker glass (IWAKI), tip (Axygen scientific), micropipette (Select BioProducts), tube (Axygen scientific), Eppendorfvortex tubes (Wiggens vortex3000), microcentrifuge (Force mini SBC-140-3), centrifuge (HETTICH Mikro 185), machine for PCR (Bio Rad T100 Thermal cycler), analytical balance ohaus (OHAUS Pioneer), spatula, microwave (Panasonic NN-SM32HM), blue light gel doc (Gite 965 GW), sprayer, spoon, electrophoresis tool (Bio-Rad mupit-ex), incubator (Memmert IN), nanodrop spectrophotometer, electrophoretic gel system, gel printer, well comb, electric heater.

This genetic characterization consists of several stages, including DNA isolation, DNA qualification, and PCR-RFLP. This study used genomic DNA Mini Kit (Geneaid) for the DNA isolation. Moreover, DNA qualification was carried out by electrophoresis using 1.5% agarose gel with diamond nucleic acid (Promega), with a voltage of 100 volts for 35 minutes. Gel doc visualized the electrophoresis results and showed the thickness of the DNA band subjectively. Gene candidate amplification using primer pairs was shown in Table 1. below:

Table 1. Length, Location and Sequence of primary pairs

Gene	Length	Location	Sequence	GenBank
GnRHR	706	Exon 3	F: $^{5'}$CACCAGGTTGAATTACGAT$^{3'}$	CM004567.1
			R: $^{5'}$AGGTCTGTGTTTCTCAGTG$^{3'}$	

The amplification procedure was carried out in accordance with the PCR steps in Table 2. In the PCR reaction, the total volume of each sample was 15 µl, consisting of forward and reverse primers of 0.25 µl each, Nucleous Free Water (NFW) of 6.5 µl, Gotaqgreen of 7 µl, and DNA samples of 1 µl.

The results of the amplification were obtained by electrophoresis on 1.5% agarose gel stained with diamond nucleic acid (Promega), at 100 volts for 35 minutes, and then used UV light from gel doc to visualize the result.

The amplified band was cut with the Msp1 restriction enzyme from Thermoscientific brand according to the gene locus with the following composition of enzyme 0,4 µl, 10X buffer BSA (Buffer tango) volume 0,7 µl, 0,9 µl Free Nucleus Water, with 5 µl of PCR product DNA.

Agarose gel stained (2%) with diamond nucleic acid (Promega) was used for the electrophoresis, which took place at 100 volts for 45 minutes before being seen with a gel doc and UV light. DNA Ladder from Thermoscientific brand determined the genotype identification of each sample based on the size and pattern of 100 bp.

Table 2. Optimal temperature, time and cycle of PCR stages

Stages	Gonadotropin-Releasing Hormone Receptor gene (GnRHR)		
	Temperature (^{0}C)	Time (Hour:minute:second)	Cycle
Pre-denaturation	94	00:05:00	1x
Denaturation	94	00:00:30	
Annealing	59	00:00:45	35x
Extention	72	00:01:00	
Final Extention	72	00:05:00	1x
Enzym restriction Mpsl	37	16:00:00	

2.2 Data Analysis

2.2.1 *Genotype and allele frequency*

A total of 200 eggs (50 eggs per storage duration) were used in this experiment. Eggs were stored in egg trays with the large end positioned upward. The temperature and relative humidity during storage were 24-26 °C and 60-70%, respectively. At the end of each storage period, eggs were subsequently transferred to a forced-draft incubator. The temperature and relative humidity for incubation were set at 37.7-37.9 °C and 50-60%, respectively, throughout the whole period. Regular hourly turning at a 90° angle occurred during the setter phase. At the 18th day of embryonic development, candling was performed to remove clear eggs, while those with viable embryos were transitioned from turning trays to hatching baskets and subsequently moved to the hatcher.

$$x_{ii} = \frac{nii}{N} \quad (1)$$

$$x_i = \frac{2n_{ii} + \Sigma_{i \neq j} n_{ij}}{2N} \quad (2)$$

Description :
x_{ii}= Genotype frequency
nii = Observed genotypes
N = Total population |
xi = The allele frequency of the samples,
nij = Sample genotype ij,
N = Total sample.

3 Results and discussion

GnRHR genes identified using Msp1 cutting enzymes with CC↓GG cutting site are in Figure 1. The description three cut sites of 233 bp, 473 bp, and 706 bp were produced using the Gonadotropin-Releasing Hormone Receptor (GnRHR) gene and the Msp1 cutting enzyme in the Ettawa Crossbreed goat. To distinguish the differences between the individuals used, a DNA restriction pattern (cutting) is anticipated to be produced by the use of restriction enzymes.

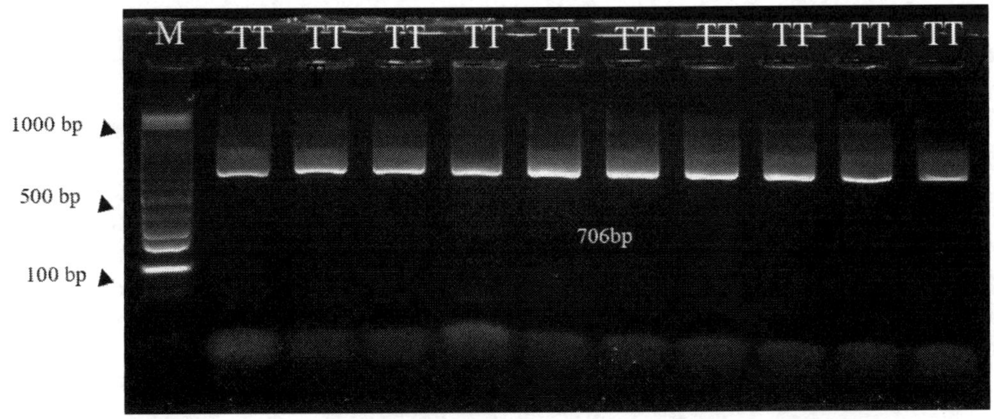

Fig. 1. Visualization of GnRHR|Msp1 RFLP results of Ettawa Crossbreed Goat

Table 3. Genotype and allele frequency of Ettawa Crossbreed Goat

Breed		n	Genotype Frequency			Allele Frequency	
			TT	TG	GG	T	G
Ettawa Goat	Crossbreed	35	1	0	0	1	0

The restriction enzyme will cut the DNA at a known location if there are a cutting site causing the DNA sequence to separate into DNA bands [16]. DNA strands will be cut in this investigation utilizing the restriction enzyme Msp1, which is 706 bp. Three fragments, 233, 473, and 706 will be added to the sequence if there is a mutation or alteration in the nucleotide sequence. The C*CGG nucleotide sequence will not be restricted if there is no mutation. It will produce one fragment, measuring 706 bp with the TT genotype.

According to Figure 1, the results of cutting the GnRHR|Msp1 fragment indicated one strip of bands (706 bp). This shows that there is no truncation and no mutation or no change in nucleotide sequence resulting in the TT genotype and the T allele. The GnRHR gene (exon 3) was studied using PCR-RFLP, but the results were monomorphic and could not be related to litter size. The results of this study are the same as Ahlawat, Harma, Maitra, Tantia, Roy and Mandakmale [17] found a single genotype (TT) in Malabari and Black Bengal goats. It was also added to the study of [18] that the band results on the visible 140 bp fragments showed that the enzyme did not recognize the restriction site. Therefore, the genotype of all 100 samples are ++. PCR-RFLP results obtained by amplicon sequence analysis, confirmed no polymorphism at the FecB locus in Markhoz goats. Research on the GnRHR gene in exon 3 with PCR-RFLP produced monomorphisms and could not be associated with litter size.

On the other hand, on Exon 1 which was not observed in this study found polymorphisms, based on [18] the polymorphism in the gonadotropin-releasing hormone receptor (GnRHR) gene was detected by PCR single-strand conformational (PCR-SSCP) and DNA sequencing methods in 224 Boer goats. In the exon I of the goat GnRHR gene, two single nucleotide polymorphisms (SNPs), G891T and G757A, were discovered. Their relationships with litter size in Boer goats were assessed. A study of associations found that the mutations G891T and G757A significantly impacted litter size. Furthermore, polymorphism was detected in exon 1 in West African Dwarf goats [19]. Moreover, according to [20] the gonadotropin-releasing hormone receptor (GnRHR) gene polymorphisms were examined in the current study as a potential genetic marker for litter size in 40 Iraqi goats. Additionally, analyses of the CC and CA genotype nucleotide sequences revealed one mutation (A>T) in exon 2. The findings indicated that in the breeds of Iraqi goats, the CC genotype was related to larger litter sizes.

Table 3 demonstrates that in Ettawa Crossbreed goat in calculations using the hardy-weinberg law, insignificant results are obtained because the value is more than > 0.05 (db; 1) = 0.50. This suggests that in the sample that was used, the GnRHR gene cannot be used as a candidate gene marker.The goat Ettawa Crossbreed's genotype frequency for the GnRHR|Msp1 gene only one genotype (TT) and one allele (T) are present. The frequency of the T allele is one. A monomorphic allele, according to [21], has an allele frequency of 0.99 (99%) or less. The results of this study do not indicate diversity because there were very few samples and only came from one population. Limited genetic diversity in a livestock group may be caused by the selection process and the absence of new male introductions in a population [22].

4 Conclusion

Allele frequencies and genotypes of Ettawa crossbreed goats were monomorphic and could not be associated with litter size.

We would like to thank Laboratorium Biotechnology of Animal Science Brawijaya University, Livestock service office of UPT PT-HMT Singosari, Malang. Which fully supported this research and its collaboration.

References

1. A.B Febriana. S. Sutopoa, E. Kurniantoa and W. Widiyanto. Tropical Animal Science Journal, **45**(3), 255-269 (2022).
2. S. Suyadi, T.E. Susilorini, W.A. Septian, A. Furqon., C.D. Nugroho, R.F. Putri In IOP Conference Series: Earth and Environmental Science. IOP Publishing. 478, 012081 (2020).
3. N. W. Hafizuddin, K. Karja, L. Praharani and M. A. Setiadi. Trop. Anim. Sci. J. **43** 110–116 (2020).
4. N. Jamarun, R. Pazla, M. Zain and A. Arief. International Conference on Innovation In Research. IOP Conf. Series: Journal of Physics: Conf. Series. 1469 (2020).
5. Arief, N. Jamarun, Pazla and B. Satria. Int.J. Dairy Sci. (2018).
6. M. A. Abuzahra. I. Mustofa, M. Helmi Effendi, and M. Lamid. Asystematic review. F1000 Research. **12**(6), 1-12 (2023).
7. M. N. Bemji., A. M. Isa., E. M. Ibeagha Awemu and M. Wheto. Molecular Biology. **45**, 63–69. (2018).
8. I. Awemu, P. Kgwatalala, X. Zhao. Mamm Genome. **19**, 591–617. (2008).
9. W. Yang, K. Tang, C. Zhang, D. Xu, Q. Wen, and L. Yang. South Afr J Anim Sci. **41**(4), 398–402 (2011).
10. M. Nei and S. Kumar. New York (USA): Oxford University Press. (2000)
11. W. Mendenhall. R. J. Beaver, and B. M Beaver. Cengage Learning. (2012)
12. A.B Febriana. S. Sutopoa, E. Kurniantoa and W. Widiyanto.Tropical Animal Science Journal, **45**(3), 255-269 (2022).
13. Y. Ai, F. Zhang, C. Wang, R. Xie and Q. Liang. Chem. 117 (2019)

14. H. Wijaya, S. Darmawati,, A.I Kartika. In Prosiding Seminar Nasional Mahasiswa Unimus **1** (2018)
15. M. H. S. Budi, Depison and Gushairiyanto. Jurnal Ilmu-Ilmu Peternakan. **33**(1), 99-108 (2023).
16. J. T Spetiawan, A. Nuryanto, H. Pramono, K. Kusbiyanto and P. H. T. Soedibja. A Scientific Journal, **33**(2), 92-101. (2017).
17. S. Ahlawat, R. Sharma, A. Maitra, M. S. Tantia, M. Roy and S. Mandakmale. Indian Journal of Animal Sciences **84**(1), 39–44. (2014).
18. S.P Dogaheh , S. Z. Mirhoseini, V. Tufarelli , N. G. H. Zadeh, S. Badbarin, M. A. Colonna, A. Seidavi and M. Selvaggi. Animals, **10**,1582. (2020).
19. W. Yang, K. Tang, C. Zhang, D. Xu, Q. Wen, and L. Yang. South Afr J Anim Sci. **41**(4), 398–402 (2011).
20. M. N. Bemji., A. M. Isa., E. M. Ibeagha Awemu and M. Wheto. Molecular Biology, **45**, 63–69. (2018).
21. A.K.H.Taklan, A.F. Alwan, S. M. Nada. Medical Science, **4**,185-187 (2015).
22. M. Nei. Columbia University Press. (1987).
23. P. P Agung, S. Anwar, W. P. B Putra and S. Said. Pros Sem Nas Masy Biodiv Indon, **3**(3), 304-308. (2017).

Revealing the correlation between hatching weight and egg characteristics in Tukong chickens

Yuli Arif Tribudi[1,2], *Maria* Ulfah[3], *Dwi* Gusmalawati[4], *Aprilianna Putri Zahara Nafsina Luvita* Sari[5], *Dwi Nur Happy* Hariyono[6], *Alek* Ibrahim[7], and *Veronica Margareta Ani* Nurgiartiningsih[2*]

[1]Department of Animal Science, Universitas Tanjungpura, Pontianak 78124, Indonesia
[2]Department of Animal Science, Universitas Brawijaya, Malang 65145, Indonesia
[3]Department of Animal Science, IPB University, Bogor 16680, Indonesia
[4]Department of Biology, Universitas Tanjungpura, Pontianak 78124, Indonesia
[5]Department of Animal Breeding and Reproduction, Universitas Gadjah Mada, Yogyakarta 55281, Indonesia
[6]Department of Animal Science, Faculty of Agriculture, Universitas Khairun, Ternate 97719, Indonesia
[7]Research Center for Animal Husbandry, National Research and Innovation Agency (BRIN), Bogor 16911, Indonesia

Abstract. The Tukong chicken, renowned for its unique rumpless trait, is indigenous to West Kalimantan, Indonesia. This study endeavors to explore the intricate associations among hatching weight, egg weight, and egg index within the Tukong chicken breed. A total of 104 Tukong chicken eggs served as the primary research material, sourced from 10 female hens paired with 5 male counterparts. To ensure consistent data, the chickens were reared under uniform conditions and management practices. The parameters examined encompassed hatching weight, egg weight, egg length, egg width, and egg index. Comprehensive data analysis, including descriptive and correlation analyses employing SPSS version 25 software, unveiled specific metrics for the Tukong chickens in these dimensions: hatching weight (31.09±2.47 g), egg weight (42.65±4.59 g), egg length (49.84±2.33 mm), egg width (38.84±1.96 mm), and egg index (78.04±4.55%). Noteworthy findings include a robust and statistically significant positive correlation ($p<0.01$) between hatching weight and egg weight ($r=0.758$), egg length ($r=0.495$), and egg width ($r=0.522$). However, no significant correlation emerged between egg index and either egg weight or hatching weight. In conclusion, this research contributes valuable insights into the complex relationships governing hatching weight and various egg attributes in Tukong chickens. These findings promise to enhance egg quality and optimize hatching outcomes within Tukong chicken breeding programs.

* Corresponding author: vm_ani@ub.ac.id

1 Introduction

Indonesia boasts a rich repository of local chicken germplasm, which plays a pivotal role in bolstering economic resilience and national stability [1]. These indigenous chicken populations exhibit a blend of specific and non-specific characteristics coupled with unique morphological traits [2]. Specific local chicken breeds are characterized by distinct attributes, such as feather color or body shape, setting them apart from other breeds. One such remarkable example is the Tukong chicken, which hails from West Kalimantan and is renowned for its distinctive feature, the absence of a tailbone, rendering it rumpless [3]. Despite its uniqueness, the Tukong chicken remains a rare gem within Indonesia's local poultry landscape, meriting further investigation due to its limited population and dearth of production data.

Local chickens hold profound significance for rural communities as vital sources of animal protein, supplying both meat and eggs to satisfy dietary requirements [4,5]. Nonetheless, the relatively low production performance of these indigenous breeds poses a substantial challenge to their widespread adoption. Various strategies, including selective breeding, have been employed to enhance their productivity [6]. The criteria for selection encompass pivotal factors such as egg weight, egg size, and day-old chick (DOC) weight. Notably, the selection of high-quality eggs at the point of hatching assumes paramount importance in ensuring the production of healthy chicks.

Evidently, a robust correlation exists between egg weight and hatching weight in chickens. Egg weight exerts a direct influence on hatchability, embryo viability, hatching weight, and, by extension, the subsequent production of chicks [7–9]. In light of these considerations, this research endeavors to elucidate the relationship between egg weight, the egg index, and hatching weight in Tukong chickens. The outcomes of this analysis are poised to furnish valuable insights and guidelines for elevating the productivity of these unique local chickens, contributing to the broader goal of sustainable poultry farming in Indonesia.

2 Materials and methods

In this research, we utilized a total of 104 Tukong chicken eggs as our primary research material. These eggs were sourced from 10 female Tukong chickens, carefully matched with 5 male Tukong counterparts. The selected local breeds ranged in age from 8 to 12 months. To facilitate their well-being and maximize reproductive success, we accommodated these chickens in five flock cages. Each cage measured 1.5 x 2 meters, maintaining a mating ratio of 1:2. Adequate provisions were made for their nourishment and nesting requirements, including feeders, drinkers, and nesting spaces for egg-laying. Our chickens were fed a formulated diet twice daily (morning and evening) with the following composition: dry matter content at 89.21%, crude protein at 14.22%, crude fiber at 8.08%, and crude fat at 7.05%. Access to clean drinking water was provided ad libitum throughout the entire research period.

The tools employed in this study comprised calipers and electric scales. The research process consisted of three distinct stages: preparation, implementation, and data analysis. The preparation stage encompassed pre-research activities, which included the procurement of essential tools and materials. The implementation stage involved the collection of eggs and the natural incubation process for Tukong chicken eggs. The parameters under scrutiny in this study included egg weight, egg length, egg width, egg index, and hatching weight. Egg length and width were measured using calipers, while egg weight and hatching weight were determined using an electric scale. Hatching weight was ascertained by weighing the chick after its feathers had fully dried. The egg index was calculated by dividing the egg width by the egg length, multiplied by 100% [10]. The data collected were subjected to descriptive analysis and correlation analysis [11], performed using SPSS version 25 software.

3 Results and discussion

The descriptive analysis of egg measurements and hatching weight for Tukong chickens is presented in Table 1. In our study, the average egg weight was recorded at 42.65±4.59 grams, accompanied by an egg index of 78.04±4.55%. The observed egg weight for Tukong chickens in this study falls within the "normal" range, comparable to that of Kedu chickens in the small category, which exhibit an average egg weight of 42 grams [12]. In contrast, Kedu Red Combed chickens typically demonstrate a wider egg weight range, spanning from 45.80 to 47.14 grams [13]. Meanwhile, for Kedu Black chickens, the average egg weight varies throughout their development, with an initial period weight of 28.64 grams, reaching 35.69 grams at peak production, and stabilizing at 43.44 grams in parent hens aged 52 weeks [14]. It's important to note that the egg weight in our study also diverges from that of various other chicken breeds, including Kampung chickens (38.1 grams), Nunukan chickens (47.3 grams), Pelung chickens (47.6 grams), Gaok chickens (46.7 grams), Hubbard Classic chickens (ranging from 60.05 to 70.03 grams), Broiler chickens (40-60 grams), Nigerian local strain (41 grams), and Tswana chickens (with a range of 41.32 to 49.79 grams) [14–18]. These variations in egg weight can be attributed to a multitude of factors, such as breed, body weight, and age [19].

Table 1. Descriptive analysis of hatching weight and egg measurements

Traits	Minimum	Maximum	Mean	Standard Deviation
Hatching weight (g)	25.00	36.00	31.09	2.47
Egg weight (g)	32.00	52.00	42.65	4.59
Egg length (mm)	42.42	54.49	49.84	2.33
Egg width (mm)	27.95	47.73	38.84	1.96
Egg index (%)	52.19	95.31	78.04	4.55

In this study, the Tukong chicken egg index was calculated to be 78.04±4.55% (Table 1). The egg index serves as a vital metric for characterizing egg shape, with a higher value indicating a rounder egg, while a lower value signifies a more oval shape. Accurate determination of egg geometric parameters holds significant importance across various applications, encompassing egg processing, handling, transportation, and production forecasting. The egg index, as a general geometric descriptor, aids in categorizing eggs into sharp, normal (standard), or round shapes, with corresponding egg index values below 72%, between 72-76%, and above 76%, respectively [20]. Comparing our findings with those of other studies reveals interesting insights. The egg index of Tukong chickens in our study surpassed the reported index for eggs of Arab chickens, Local chickens, and Ketarra chickens, which stood at 76.53±3.03%, 74.34±3.63%, and 77.63±6.37%, respectively [21]. Conversely, it closely resembled the egg index of Superbrown layer chickens (78.08%) [20].

Moving on to the hatching weight of Tukong chickens in our study, we observed an average weight of 31.09±2.47 grams. When compared to other poultry breeds, the hatching weight of Tukong chickens falls within a distinct range. It exceeds the hatching weight of Kedu Red Combed chickens, which typically ranges from 30.33 to 30.92 grams [13], but remains lower than that of broiler chickens, which can range from 38.58 to 45.50 grams [9]. It's important to note that the relatively lower hatching weight in Tukong chickens can be attributed to their comparatively lower adult body weight. This, in turn, has an impact on both egg weight and hatching weight [3,22]. Furthermore, variations in egg weight and size, as well as chicken hatching weight, can be influenced by a multitude of factors including breed, rearing system, parent age, and diet [10]. These findings underscore the complexity of poultry biology and production, highlighting the importance of considering multiple factors when interpreting results and making informed decisions in the poultry industry.

Table 2. Pearson correlations (below diagonal) and its significance (above diagonal) between hatching weight and egg measurements

Traits	Hatching weight	Egg weight	Egg length	Egg width	Egg index
Hatching weight		**	**	**	Ns
Egg weight	0.758		**	**	Ns
Egg length	0.495	0.691		**	**
Egg width	0.522	0.687	0.279		**
Egg index	0.072	0.046	-0.577	0.620	

Note: Ns: Non-significant, **: significant at $p<0.01$ level

The correlation analysis conducted in this study, as summarized in Table 2, unveils several noteworthy associations between hatching weight and various egg characteristics in Tukong chickens. Notably, a positive correlation with a strong statistical significance ($r=0.758$, $p<0.01$) was observed between hatching weight and Tukong chicken egg weight. Similarly, positive and significant correlations ($p<0.01$) were identified between hatching weight and egg weight, as well as between hatching weight and the egg's linear dimensions, including egg length and egg width. These findings highlight the integral relationship between hatching weight and these critical egg attributes.

On the other hand, the egg index, representing the egg's shape, exhibited no significant correlation ($p>0.05$) with either egg weight ($r=0.046$) or hatching weight ($r=0.072$), with both correlation coefficients indicating a relatively weak association. It is worth noting that the egg shape index, defined as the ratio of egg width to length, holds significance in determining egg orientation during incubation and the embryo's mobility for efficient nutrient utilization. This observed weak correlation aligns with previous research conducted on Nigerian Indigenous chickens [23] and Tswana chickens [18], which also reported limited associations between egg weight and egg shape index.

The findings of our study reveal a positive, strong, and statistically significant correlation between egg characteristics and hatching weight in Tukong chickens. This outcome is consistent with research conducted on various other poultry breeds, including chickens [9,24], ducks [10], turkeys [25], and geese [26]. These collective results suggest that an increase in egg weight directly corresponds to an increase in hatching weight, indicating a proportional relationship. This correlation underscores the significance of hatching egg weight in the context of egg incubation, as it significantly influences the economic aspects of poultry farming [10,27].

However, it's important to note that while egg weight and hatching weight are closely correlated, their impact on post-hatching growth and eventual adult chick weight may vary. Egg size, particularly in terms of larger eggs, plays a pivotal role in chick performance, potentially leading to larger chicks compared to those hatched from smaller eggs [9,28]. These findings emphasize the multifaceted interplay between egg characteristics and hatching outcomes, which should be considered when optimizing poultry production strategies.

4 Conclusion

In conclusion, our study sheds light on the relationships between Tukong chicken egg attributes and hatching weight, offering valuable insights for poultry producers aiming to enhance egg quality and optimize hatching outcomes in Tukong chicken breeding programs. Further research can build upon these findings to explore additional factors influencing egg characteristics and their impact on overall poultry production.

References

1. M.H. Tamzil, B. Indarsih, Bull. Anim. Vet. Sci. **32**, 15–28 (2022)
2. I.W. Mahardhika, B. Daryono, A.A. Dewi, S. Hidayat, G. Firmansyah, P. Setyowati, U. Riswanta, M. Pratama M, J. Peternak. **17**, 6–16 (2020)
3. Y.A. Tribudi, M.H. Natsir, M. Ulfah, D. Gusmalawati, A.P.Z.N.L. Sari, D.N.H. Hariyono, P.W. Prihandini, V.M.A. Nurgiartiningsih, V M A, Worlds. Poult. Sci. J. **79**, 403–417 (2023)
4. F.O. Ajayi, B.O. Agaviezor, Nigeria Int. J. Livest. Prod. **7**, 141–144 (2016)
5. M.M. Fathi, I. Al-Homidan, O.K. Abou-Emera, A. Al-Moshawah, Worlds. Poult. Sci. J. **73**, 916–927 (2017)
6. T.A. Adedeji, S.R. Amao, A.D. Popoola, R.I. Ogundipe, Nigeria J. Biol. Agric. Healthc. **5**, 36–43 (2015)
7. S. Alkan, K. Karabag, A. Galic, M.S. Balcioglu, S. Afr. J. Anim. Sci. **38**, 231–237 (2008)
8. T. Çağlayan, M. Garip, K. Kirikçi, A. Günlü, Ital. J. Anim. Sci. **8**, 567–574 (2009)
9. M. Duman, A. Şekeroğlu, Rev. Bras. Cienc. Avic. **19**, 255–262 (2017)
10. T. Suselowati, E. Kurnianto, S. Kismiati, Sains Peternak. **17**, 24-30 (2019)
11. Z.N. Jannah, B.A. Atmoko, A. Ibrahim, M.A. Harahap, Panjono, Biodiversitas **24**, 3657-3664 (2023)
12. C. Ustadha, Sutopo, I.S. Kuswahyuni, Agromedia **34**, 65–71 (2016)
13. R. Sumbayak, E. Kurnianto, S. Kismiati, J. Sain Peternak. Indones. **15**, 138–141 (2020)
14. A.G. Nataamijaya, Bul. Plasma Nutfah **14**, 85-89 (2016)
15. S.S. Abiola, O.O. Meshioye, B.O. Oyerinde, M.A. Bamgbose, Arch. Zootec **57**, 83–86 (2008)
16. T.O. Allanah, J.C. Okonkwo, S.I. Omeje, Biol. Sci. **3**, 59–68 (2014)
17. J. Iqbal, S.H. Khan, N. Mukhtar, T. Ahmed, R.A. Pasha, J. Appl. Anim. Res. **44**, 54–64 (2016)

18. P. Kgwatalala, M. Molapisi, K. Thutwa, B. Sekgopi, T. Selemoge, S. Nsoso, Int. J. Environ. Agric. Res. **2**, 96–105 (2016)
19. M. Hammershøj, G.H. Kristiansen, S. Steenfeldt, Foods **10**, 1–17 (2021)
20. M. Sarica, H. Onder, U.S. Yamak, Int. J. Agric. Biol. **14**, 235–240 (2012)
21. O. Marlya, K. Kususiyah, D. Kaharuddin, Bul. Peternak. Trop. **2**, 103–111 (2021)
22. H. Nasri H, H. van den Brand, T. Najjar, M. Bouzouaia, J. Anim. Physiol. Anim. Nutr. (Berl). **104**, 257–268 (2020)
23. A. Yakubu, D.M. Ogah, R.E. Barde, Int. J. Poult. Sci. **7**, 579–585 (2008)
24. J. Iqbal, N. Mukhtar, Z.U. Rehman, S.H. Khan, T. Ahmad, M.S. Anjum, R.H. Pasha, S. Umar, J. Appl. Poult. Res. **26**, 183–191 (2017)
25. P. Hristakieva, M. Oblakova, N. Mincheva, M. Lalev, Slovak J. Anim. Sci **50**, 90–94 (2017)
26. M. Sreten, M. Milena, D.S. Mirjana, Indian J. Anim. Res. **52**, 497–501 (2018)
27. C.A. Mbajiorgu, N.O. Ramaphala, Indian J. Anim. Res. **48**, 209–213 (2014)
28. S. Vieira, J. Almeida, A. Lima, O. Conde, A. Olmos, Rev. Bras. Ciência Avícola **7**, 73–78 (2005)

GDF9 gene polymorphism and its relation to litter size in East Java Pote goat germplasm

Nida'ul Husna Imaniah[1*], *Tri Eko* Susilorini[2], *Kuswati* Kuswati[2], and *Rini Dwi* Wahyuni[2]

[1]Postgraduate Student Program Faculty of Animal Science, Universitas Brawijaya, Malang 65145, Indonesia
[2]Lecturer of Postgraduate Program Faculty of Animal Science, Universitas Brawijaya, Malang 65145, Indonesia

Abstract. Pote goat is one of the local Indonesian goats from Bangkalan, Madura, which has not been clumped. Procurement of Pote goat breeds as germplasm in enriching genetic diversity can be done by genetic selection. This study aims to identify the GDF9 gene associated with litter size traits of Pote goats. The materials used in this study were 48 female goats that had given birth with a range of permanent incicivi (0,1,2,3,4), and with healthy conditions. This research method uses survey methods and observations in the laboratory. Sampling was done by purposive sampling. Pote goat blood samples were taken from a smallholder farm in Soket Laok Village, Bangkalan Regency, Madura, East Java. GDF9 gene has a length of 490 bp. DNA fragments were amplified and genotyped using PCR-RFLP method using Msp1 restriction enzyme. This study resulted in two genotype types (AG, and GG), and two alleles (A and G). Genotype frequencies of AG, and GG were 0.40, and 0.60. The frequencies of alleles A and G are 0.40 and 0.80. The a.2912 A>G mutation was significant with the litter size. The results showed an association between GDF9 gene diversity found with litter size traits in Pote goats ($p<0.05$). Polymorphisms in the GDF9 gene are associated with litter size of Pote goats, and can be used as genetic markers for selection on the litter size trait of Pote goats.
Keywords: Polymorphism, Association, GDF9 gene, and Pote Goat

1 Introduction

The origins of the Pote goat breed are uncertain, but it is flourishing in the Bangkalan area of Madura, East Java. The Pote goat is held as a dual-purpose breed for meat and milk production due to its ability to fulfill both roles. The Pote goat has a similar body structure to the Saanen goat, but it is suspected that the Pote goat is the result of crossing and white color selection between the Peranakan Ettawa goat and the Kacang goat, and may also have the genetic properties of Etawah and Senduro dairy goats [3]. The Pote goats in Bangkalan, originating from areas between 2-100 meters above sea level, are typically fed a combination of eight plant varieties, including Moringa oleifera, Bambusa sp., and Artocarpus heterophyllus, as well as natural grass, which is a commonly used plant with

* Corresponding author: nidaulhusna23@gmail.com

relatively high density [17]. The adaptability of Pote goats to these food sources provides an advantage in which they can sustain and boost their population, leading to greater economic benefits for the people.

Procuring breeding stock is crucial for germplasm preservation in order to maintain the genetic resources of the region. These resources are unique and have adapted to their environment over time. Pote goats are also included in goats that have not been clumped and have not been well identified. Pote goats are traditionally raised in rural areas, but due to inadequate education on proper husbandry management within the local community, their maintenance remains difficult. According to [5] there are two ways to enhance livestock production: environmental improvements and genetic improvements. Morphological appearance remains a prevalent factor in livestock selection and characterization today. Environmental factors like nutrition and climate can influence livestock morphology.

Twin births may be affected by hereditary factors transmitted from the parents to their offspring. Productivity of the parent is directly associated with litter size; thus, the larger the litter size, the greater the parent productivity [16]. [18] assert that three factors determine the litter size of goats, namely, the number of eggs produced during lambing and ovulation, fertilization, and conditions during pregnancy, as well as embryo death. Reproduction is a significant characteristic that impacts the productivity and profitability of the goat industry [2]. Traditional selection results in slow genetic gains in reproductive performance due to the trait's low heritability and its expression over the animal's lifespan. Marker-assisted selection methods provide a more effective alternative to livestock breeding problems [14]. Litter size in Pote goats can be manipulated by specific genes known as fecundity genes [19]. Measuring productivity is possible by observing phenotypic characteristics, specifically litter size, and genotypic characteristics, such as gene markers. Certain genes have been discovered that regulate litter size, and they could be utilized as DNA molecular markers for litter size selection. Among these genes is GDF9, which plays a role in regulating goat litter size.

The GDF9 gene, which belongs to the TGF-β superfamily of fecundity genes, has been extensively researched in goats. This gene plays a crucial role during the early stages of folliculogenesis by facilitating growth and differentiation and is secreted by oocytes in mammals [6]. GDF9 plays a crucial role in mammalian folliculogenesis development due to its necessity for typical oocyte maturation and embryo development. Without GDF9, embryos will cease development until they reach the blastocyst stage [8]. This study aims to analyze the relationship between the GDF9 gene's polymorphism and litter size in Pote goats found in Bangkalan, Madura. The genetic information discovered in this study will serve as a foundation for developing selection strategies to enhance the genetic quality of Pote goats.

2 Materials and Methods

2.1 Materials

The research material used was 48 female Pote goats with a range Permanent incicvi (0,1,2,3,4) from smallholder farms. The Pote goats were healthy, disease-free, and previously parturient. A survey method was used to collect data on the litter size of Pote goats and the PCR-RFLP method was used for laboratory observation to detect GDF9 gene polymorphism. The first stage of the research was carried out by collecting litter size data and taking blood samples of Pote goats taken in the jugular vein as much as 5 ml. Then, the blood samples were analyzed at the Biotechnology Laboratory, Faculty of Animal Science, Universitas Brawijaya.

The second stage of research is the analysis conducted in the laboratory in the form of DNA Extraction, DNA Amplification (PCR), and Determination of GDF9 gene genotype by RFLP method. Research tools used during laboratory analysis were GD Column, 1.5 ml Tube, Beaker glass, Micropipet (Select Bio Products), Tip, Tube Rack, Centrifuge (HETTICH Micro 185), BioRad T100 Thermal Cycler, Mini Centrifuge Spindown, Microwave, Electrophoresis (BioRad Mupid-Exu), and Gel Documentation. Materials used during laboratory analysis were Genomic DNA Mini Kit (Promega), Agarose (1st Base Lot 1A1019FH9934), Ethanol absolute, Diamond Nucleic Acid Dye (Promega), Gel Loading Dye Blue, GoTaq Green Master Mix (Promega), Nuclease Free Water (Promega), TBE Buffer, and GDF9 gene primers (10 PM/μl).

DNA extraction in this study used the Genomic DNA Mini Kit (Promega), where DNA extraction is carried out as a process of separating DNA from other cell components such as proteins, carbohydrates, fats, and others (Hutami, et al., 2018). Then, followed by the process of DNA amplification or PCR, using primers. Primer sequences of the GDF9 gene in this study were designed based on the GenBank database with the access number: (ENSCHIG00000008167), consisting of forward primer (5'-GAATGAATGTGAGCTCCATGAC-3') and reverse primer (5'-CCCTTACATTGATAGATGCCAC-3). The primers were 491 bp long and located in exon 2.

DNA amplification has 3 stages, namely denaturation, annealing, and extension in accordance with PCR conditions on each target gene fragment. The results of DNA amplification (PCR), followed by the determination of GDF9 gan genotype by RFLP method. PCR products were cut using a restriction enzyme, Msp1. The results of all processes can be seen after the electrophoresis process using MUpid Exu electrophoresis machine..

2.2 Data Analysis

2.2.1 Genotype and Allele Frequency

Allele frequency (1) is the ratio or percentage of a particular allele at a locus. Allele frequency is calculated based on the formula of Nei and Kumar [10]. Genotype frequency (2) is the ratio or percentage of certain genotypes in a population. Genotype frequency is calculated based on the formula of Nei and Kumar [10].

$$x_i = \frac{2n_{ii} + \sum_{i \neq j} n_{ij}}{2N} \quad (1)$$

$$x_{ii} = n_{ii}/N \quad (2)$$

Description:
χi = i-th allele frequency
nii = number of individuals with genotype ii
nij = number of individuals with genotype ij
N = total sample
χii = frequency of the i-th genotype
Ni = number of individuals with genotype ii

3 Results and discussion

3.1. Amplification of Growth Differentiation Factor 9 (GDF9) Gene by PCR Method in Pote Goat.

Identification of the GDF9 gene can be achieved through the PCR (Polymerase Chain Reaction) method, an in vitro DNA synthesis and amplification technique. According to [9] PCR techniques can be employed to identify mutations in gene fragments, and during the amplification process of gene fragments, components called primers are required. GDF9 gene amplification results in Pote goats were electrophoresed using a 1.5% agarose gel and observed with UV light on the gel documentation **Figure 1.**

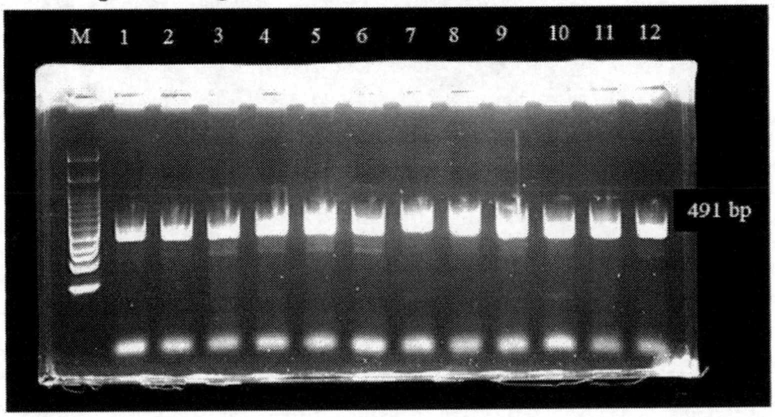

Fig 1. PCR result of GDF9 gene of Pote goat. Description: M: marker DNA leader Gene Aid 100 kb, 12 samples.

Based on the PCR results shown in Figure 2, it is evident that all GDF9 gene samples of Pote goats exhibit clearly visible DNA bands, indicative of proper DNA amplification or separation. According to [16] the purity of the DNA produced during the extraction, the selection of appropriate primers and the PCR conditions are important for good PCR results. The PCR results of the GDF9 gene in Pote goats exhibited targeted amplification of the desired fragment (491 bp) without any amplification of contaminant or unintended fragments.

The annealing temperature choice significantly impacts PCR success, as [11] reported that temperature affects the process of attaching the polymer to the DNA template, because if the annealing temperature is too high, the primer will not be able to attach to the template, and if the temperature is too low, the primer will not be able to attach to the specific attachment site, causing thermaplication of unwanted fragments.

Therefore, testing the temperature gradient is necessary because the selection of the correct annealing temperature depends on the primers utilized and the DNA target's characteristics. The optimal annealing temperature falls within the range of 50oC to 65oC. The optimal annealing temperature is in the range of 50oC to 65oC, in this study the temperature gradient used ranged from 58oC to 62oC, and at 62oC showed positive results and can then be used as a new annealing benchmark for samples that have not been PCR.

3.2. Restriction Result of GDF9 Gene by RFLP Method.

The restriction enzyme used in this study is the Msp1 enzyme, which recognizes and cuts the specific DNA sequence C^CGG. Cutting with MspI restriction enzyme produced allele A and allele G, with allele A size of 491 bp and allele G size of 140 bp. Pote goats obtained

2 genotype variations AG with 3 fragments (491 bp, 351 bp, and 140 bp), and GG genotype with 2 fragments (351 bp and 140 bp) Figure 2.

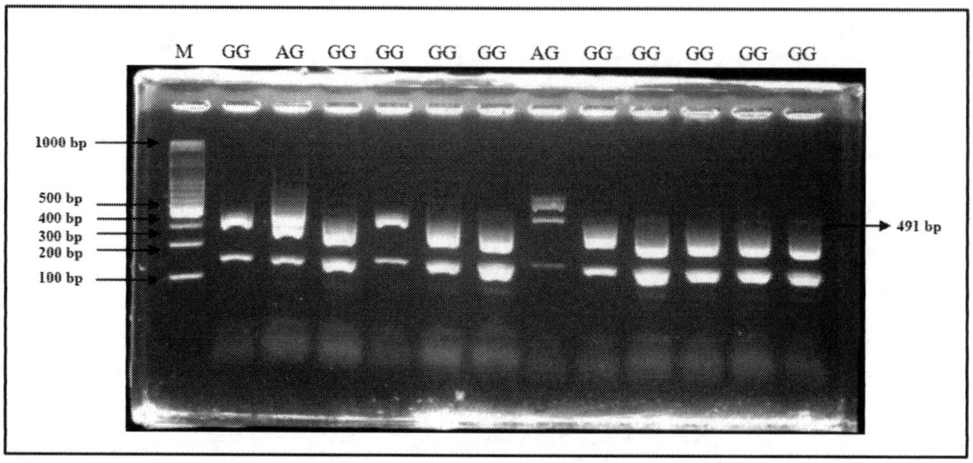

Fig. 2. RFLP visualization results of GDF9 Gene with MspI restriction enzyme on 2% agarose gel.

This study resulted in the location of MspI enzyme restriction results, namely there are 3 cut points located at 491, 351, and 140, with one mutation point at 491 bp. Indicates that there is a SNP point a.2912 A>G, which is a nucleotide change from A to G (GAT> GGT), namely Asparagine> Glycine. The type of mutation that occurs is called a missense mutation, which is a sequence variant, which changes one or more bases, resulting in a different amino acid sequence but the length is maintained. According to [13] missense mutation is a change in a genetic code that causes the amino acids bound to the polypeptide chain to change. [4] Identified a missense mutation in Brazilian Santa Ines sheep that caused substitution of phenylalanine with cysteine (F345C) in GDF9 protein named FecGSI.

Table 1. Allele Frequency and Genotype Frequency of GDF9 gene in Pote Goat

Genotype Frequency		**Allele Frequency**	
AG (19)	**GG (29)**	**A**	**G**
0.40	0.60	0.40	0.80

Based on the results of the study (Table 1) shows that the frequency of genotype GG is higher (0.60) than the genotype AG (0.40) while the frequency of allele G (0.80) is higher than allele A (0.40). The differences in the results of genotype frequency and allele frequency in the Pote goat population in this study indicate the existence of polymorphism. [2] Found that populations with allele and genotype frequencies less than or equal to 0.99 exhibit genetic diversity. Polymorphism is an indicator of this diversity, which provides populations with the capability to adjust to environmental changes.

3.3. GDF9 Gene Genotype Association to Litter Size of Pote Goats.

This study shows that GDF9 gene polymorphism is associated with litter size of Pote goats. Based on the results of the study in Table 2, the GDF9 gene genotype polymorphism showed a significant effect on the average litter size of Pote goats ($P<0.05$).

Table 2. Average litter size of Pote goats based on genotype

Genotype	n	Average *Litter Size*
AG	17	1.6765 ± 0.43088
GG	27	1.1852 ± 0,28244

Based on these results, it means that the GDF9 gene mutation is significantly associated with the litter size trait in Pote goats, and the genetic variant of the GDF9 gene is considered as a genetic marker for increased proliferation in Pote goats. In this study, the GDF9 gene heterozygous genotype AG had a higher average litter size than the homozygous genotypes GG. The results of research from [7] also show that homozygous segments are mostly associated with lower litter size groups, and there is an increase in the level of heterozygosity in higher litter size groups. This is in accordance with the results of [12] who discussed 5 different mutants from several sheep breeds that showed superiority in heterozygous genotypes of the BMP15 gene and GDF9 gene. These mutants increased ovulation rate and fecundity in the heterozygous genotype, but the homozygous genotype mutants had impaired oocyte development and maturation, resulting in underdeveloped ovaries and female infertility.

4 Conclusion

Polymorphisms in the GDF9 gene are associated with litter size of Pote goats in Bangkalan, Madura, East Java, and can be used as genetic markers for selection on the litter size trait of Pote goats.

We thank to Faculty of Animal Science Universitas Brawijaya for the financial support, and the farmers of Pote goat in Soket Laok Village, Bangkalan. This study is part of Hibah Doktor 2022 (Doctor Research Grant 2022).

References

1. A. Fleming, C. F. Baes, A. A. A Martin, T. C. S. Chud, F. Malchiodi, L. F. Brito, & F. Miglior. Journal of Dairy Science, 102(4), 3722–3734. (2019)
2. A. Gunawan, C. Sumantri, & R. Juniarti. Gen dan Keragaman Genetik Ternak. IPB Press. Bogor. (2017)
3. A. Rohman, K. Kuswati, and T. E. Susilorini. International Journal of Current Science Research and Review, 06(10). (2023)
4. B.D.M. Silva, E.A. Castro, C.J.H. Souza, S.R. Paiva, R. Sartori, M.M. Franco, H.C. Azevedo, T.A.S.N. Silva, A.M.C. Vieira, J.P. Neves & E.O. Melo. Animal Genetics. 42(1): 89–92. (2011)
5. E. Wiyanto, & A.Y. Putra. Jurnal Ilmiah Ilmu-Ilmu Peternakan. 23(1), 55-60. (2020).
6. J. A. Elvin, A. T. Clark, P. Wang, N. M. Wolfman, & Matzuk, M. M. Molecular Endocrinology, 13(6), 1035–1048. (1999).
7. J. J. Wang, T. Zhang, Q.M. Chen, R.Q. Zhang, L. Li, S.F. Cheng, W. Shen and C.Z. Lei. Frontiers in Genetics, 11, 510062. (2020).
8. J. Sudiman, M. L. Sutton-McDowall, L. J. Ritter, M. A. White, D. G. Mottershead, J. G. Thompson, & R. B Gilchrist. PLOS ONE, 9(7). (2014).

9. L. B. Ketut, Maitriani, I. Nengah Wirajana, D. Sagung, & C. Yowani. In Cakra Kimia (Indonesian E-Journal of Applied Chemistry). 3(2). (2015).
10. M. Nei, & S. Kumar. (2000). Molecular evolution and phylogenetics. Oxford University Press, USA.
11. N. P. D. Pertiwi, I. G. N. K. Mahardika, dan N. L. Watiniasih. Journal of Biology.19(2), 1-5. (2015).
12. P. W. Hedrick. Heterozygote Advantage. Journal of Heredity, 106(2), 141–154. (2015).
13. R. G. Almaida, Y. Oktanella, & G. Ciptadi. Journal of Tropical Animal Production, 21(2), 102-110. (2020)
14. R. Wakchaure, S. Ganguly, P.K Praveen, A. Kumar, S. Sharma, & T. Mahajan. J. Drug. Metab. Toxicol, 6(5), e127. (2015).
15. S. Rahayu, S. B. Sumitro, T. Susilawati, dan S. Soemarno. 2006. Biological Research Periodical. 12(1), 7-11.
16. Sudrajad, P., S. D. Volkandari, M. Cahyadi, A. Prasetyo, K. Komalawati, S. Wibowo, & S. Subiharta. Livestock and Animal Research, 19(1), 1-12. (2021).
17. T. E. Susilorini, K. Kuswati, R. D. Wahyuni, P. Surjowardojo, & S. Suyadi. AGRIVITA, Journal of Agricultural Science, 44(2). (2022).
18. T. Kostaman, & I. K. Sutama. JITV, 10(2), 106-112. (2005).
19. X. P. An, J. X. Hou, H. B. Zhao, G. Li, L. Bai, J. Y. Peng, Q. M, Yan, Y. X. Song, J. G. Wang, & B. Y. Cao. Animal Genetics, 44(2), 234–238. (2013).

Evaluation of *Gynura procumbens* cutting cultivation using miko-seedcookies

Welas Sri Mulyati[1*], *Panca Dewi Manu Hara* Karti[1], *Rima Shidqiyya Hidayati* Martin[1], and *Nur Rochmah* Kumalasari[1]

[1]Nutrition and Feed Technology, Faculty of Animal Science, IPB University, Jl. Agatis Kampus IPB Dramaga, West Java, Indonesia

Abstract. This research was conducted to evaluate the growth and production of *Gynura procumbens* with the miko-seedcookies addition usage. This research was use a randomized block design (RBD) with four treatments of miko-seedcookies diameter which included P0 (control), P1 (4 cm), P2 (5 cm), and P3 (6 cm) and five blocks which included K1, K2, K3, K4, K5. The research used 120 *Gynura procumbens* stem cuttings with an average diameter of 1.30 ± 0.13 cm and a planting length of 19.07 cm. The parameters were the percentage of germinating cuttings, plant height, number of leaves, leaf area, fresh weight, and mycorrhizal colonization. The results showed that the miko-seedcookies utilization increased plant height; number of leaves; fresh weight of shoot, mature leaves, edible stems; dry weight of mature leaves, edible stems; and mycorrhizae colonization. The use 6 diameter of miko-seedcookies could enhance the growth, production, and mycorrhizae colonization continuously.

1 Introduction

Longevity spinach or *Gynura procumbens* (*G. procumbens*) is known as a medicinal plant that grows wild in ditches, shrubs, and grasslands in Southeast Asian regions such as Malaysia, Thailand, and Indonesia, especially in Java Island at an altitude of 1-1200 m ASL and optimally at an altitude of 500 m ASL [1]. *G. procumbens* has great potential to be developed as a feed supplement and antibiotic in animal husbandry because of the active compound content in its leaves. *G. procumbens* leaf extract can inhibit the growth of *E. coli* bacteria [2].

Cultivation of *G. procumbens* with stem cuttings directly in the field is one of the vegetative plant cultivation techniques that is easy to do and less cost [3]. However, the use of this technique often fails because it is influenced by external factors such as soil fertility and weather as well as internal factors such as age and plant genetics [4]. The soil of the previous research site was a latosol soil with a sandy clay loam texture with an acidic soil pH value of 4.62, moderate soil C-organic 3.18%, low N-total 0.27%, moderate P_2O_5 32.09

* Corresponding author: welasmulyati@gmail.com

ppm, and low K 0.11 cmol/kg. The soil was dominated by sand 43.47%, dust 23.35%, and clay 33.18% [5]. This showed that the soil was less fertile due to nutrient deficiency.

Miko-seedcookies is an innovative planting media containing arbuscular mycorrhizal fungi (FMA). Arbuscular mycorrhizal fungus is a biological fertilizer as an inoculant made from living organisms that function to tether certain nutrients or facilitate the availability of nutrients in the soil for plants [6]. FMA can increase shoot dry weight production (34.04%), crude protein content (10.21%), phosphorus uptake (40%), N content (10.53%), N uptake (38.10%), and protein production (40.15%) in *Pennisetum purpureum* plants [7]. This planting medium is a place to grow plants in marginal areas such as soils with low productivity, potential and topography that are less favourable for plant growth. Therefore, the potential of *G. procumbens* is important. This study aimed to evaluate the growth and production of *G. procumbens* cultivated with the addition of miko-seedcookies as planting media.

2 METHODOLOGY

2.1 Research Procedure

The land used in this study was 12 m x 7.5 m with a plot size of 1 m x 0.5 m and a spacing of 0.5 m x 0.5 m. Cuttings were taken by cutting them with scissors with a length of 25 cm with an average diameter of 1.3 cm and an average mass of 37.36 g. The bottom of the cuttings was cut at an angle of 45°. The bottom of the cuttings was cut at an angle of 45°.

The miko-seedcookies growing medium was made by putting four liters of water into a bucket and adding 10 g of miko-seedcookies adhesive in the form of stiqfier and then formed flat rounds like cookies according to the diameter size. *G. procumbens* cuttings were stuck into the miko-seedcookies and penetrated the soil up to one-third of the cuttings.

Maintenance included watering, weeding, and fertilizing. Fertilizers were given in the form of NPK Mutiara 16-16-16 at a dose of 200 kg/ha in week 2 and urea at a dose of 100 kg/ha after four weeks of planting. Harvesting was done at the age of 12 weeks after planting (WAP). *G. procumbens* plants were trimmed to 30 cm and the roots, shoots, mature leaves, and edible stems were taken. The variables observed and measured in this study were the percentage of sprouted cuttings, number of leaves, leaf area, plant height, fresh weight of shoots, mature leaves, and edible stems, also percentage of root colonization.

2. 2 Research Design and Data Analysis

This research was carried out in the field with the experimental design used was a randomized blocked design (RBD) with 5 groups of research land positions based on sunlight reception and 4 treatments of miko-seedcookies planting media diameter enriched with arbuscular mycorrhiza, P0 (*G. procumbens* cuttings planted directly without using miko-seedcookies), P1 (using miko-seedcookies with a diameter of 4 cm (53 g)), P2 (using miko-seedcookies with a diameter of 5 cm (84 g)), and P3 (using miko-seedcookies with a diameter of 6 cm (130 g)). Data obtained from the research were analyzed and processed using analysis of variance (ANOVA) with SPSS Statistic 23 software and Microsoft Excel. Analysis of variance that significantly influenced, then further test with Duncan test at 5% level.

3 RESULTS AND DISCUSSIONS

3.1 Growability of *G. procumbens* stem cutting

The percentage of sprouted *G. procumbens* cuttings was not significantly different (p>0.05) after ANOVA test. Data on the percentage of sprouted cuttings was presented in Table 1.

Table 1. Sprouting percentage of *G. procumbens* cuttings

Age (WAP)	Treatments			
	P0	P1	P2	P3
1	86.67±6.24	96.67±3.33	90.00±6.67	100.00±0.00
2	93.33±4.08	100.00±0.00	100.00±0.00	100.00±0.00
3	93.33±4.08	100.00±0.00	100.00±0.00	100.00±0.00

P0: control (without miko-seedcookies), P1: use of miko-seedcookies diameter 4 cm, P2: use of miko-seedcookies diameter 5 cm, P3: use of miko-seedcookies diameter 6 cm.

The use of miko-seedcookies did not significantly affect the sprouting percentage of *G. procumbens* cuttings. Cuttings planted without using miko-seedcookies had a high sprouting rate of 93.33% in the third week, while cuttings using miko-seedcookies gave a sprouting percentage of 100% for all sizes. The presence of millet (*Setaria italica*) became a temporary host plant for arbuscular mycorrhizal fungi (FMA) on the miko-seedcookies. The millet would grow on the miko-seedcookies and its roots would first symbiotize with FMA before the roots of *G. procumbens* cuttings were fully formed at the beginning of planting.

3.2 Growth of *G. procumbens* Plant

G. procumbens plant growth could be observed through plant height and the number of leaves. The growth of *G. procumbens* could be seen in the graph below (Figure 1).

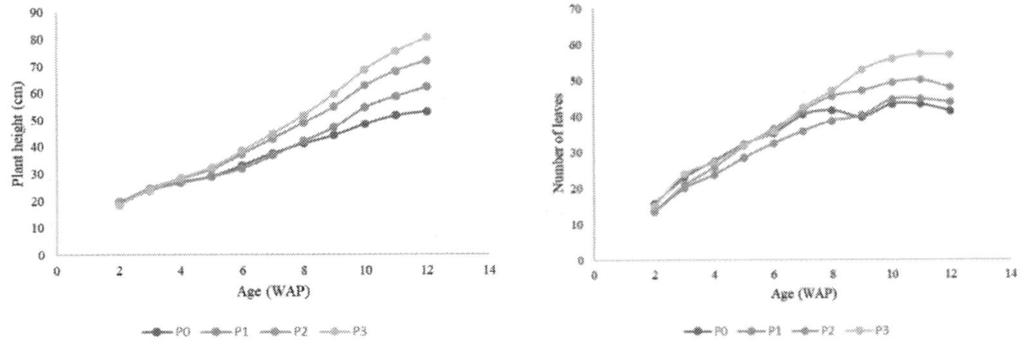

(a) (b)

Figure 1. Graph of plant height (a) and number of leaves (b)

The presence of FMA in mico-seedcookies could also increase the absorption of water and macro and micro nutrients. The greater the use of mico-seedcookies, the greater the FMA content which had a positive effect on plant height when compared to without the use of mico-seedcookies. FMA provides nutrient absorption efficiency due to the presence of external hyphae that expanded the nutrient absorption area and accelerated the movement of P to the roots [8].

The use of 6 cm diameter miko-seedcookies (P3) at 12 weeks after planting gave the best number of leaves. The greater amount of FMA causes the absorption area of nutrients, especially nitrogen (N) for the development of plant vegetative organs to be higher. In addition to helping the absorption of macro nutrients, FMA also helped the absorption of micro nutrients such as Zn and Cu which play a role in increasing the growth and development of new shoots [9]. Plant growth and the number of leaves increased if the process of protein synthesis that required nitrogen elements was not inhibited and FMA inoculation increased the absorption of N nutrients by plant roots [10]. After 12 weeks, the number of leaves decreased due to the high intensity of the sun in that week so that many leaves turned yellow and fell off. This is also due to the age of the plant, the higher the age of the plant, the more leaves that fall.

3.3 Production of *G. procumbens* Plant

G. procumbens production was evaluated through leaf area and fresh weight at the age of 12 weeks after planting. Leaf area of *G. procumbens* was not significantly different ($p>0.05$) after ANOVA test. However, the use of miko-seedcookies gave a very significant effect ($p<0.01$) on the fresh weight of shoots, mature leaves, and edible stems of *G. procumbens* after ANOVA test (Table 2).

Table 2. Fresh yield of *G. procumbens* plants at the age of 12 WAP

Fresh yield	Treatment			
	P0	P1	P2	P3
Leaf area (mm^2)	99.94±16.37	137.15±23.69	119.08±24.88	138.80±7.59
Shoot fresh weight (g $plant^{-1}$)	$5.49{\pm}2.60^{b}$	$6.60{\pm}1.92^{ab}$	$7.27{\pm}2.23^{a}$	$7.57{\pm}2.65^{a}$
Mature leaves fresh weight (g $plant^{-1}$)	$24.33{\pm}15.96^{b}$	$31.07{\pm}13.41^{b}$	$44.10{\pm}15.62^{a}$	$51.27{\pm}16.41^{a}$
Edible stem fresh weight (g $plant^{-1}$)	$3.96{\pm}4.89^{c}$	$9.14{\pm}8.47^{bc}$	$14.57{\pm}10.89^{b}$	$22.32{\pm}15.59^{a}$
Total fresh weight (g $plant^{-1}$)	$33.79{\pm}21.68^{d}$	$46.81{\pm}19.97^{c}$	$65.93{\pm}26.18^{b}$	$81.15{\pm}29.92^{a}$

Different superscript letters in the same row indicate highly significant ($p<0.01$). P0: control (without miko-seedcookies), P1: use of miko-seedcookies diameter 4 cm, P2: use of miko-seedcookies diameter 5 cm, P3: use of miko-seedcookies diameter 6 cm.

The treatment of miko-seedcookies diameter 6 cm (P3) gave a highly significant effect ($p<0.01$) and the best results for edible stem fresh weight and total fresh weight compared to other treatments. Photosynthesis required sufficient water and nutrients to produce abundant photosynthates. Adequate sunlight also played a role in photosynthesis. FMA could increase photosynthate because the hyphae on FMA could expand the absorption of water and nutrients. The presence of FMA increased the absorption of phosphate (P) which increased photosynthate which was positively correlated with fresh weight [11]. FMA in miko-seedcookies was effective for the host plant, *G. procumbens*, because it could increase plant dry weight, uptake of P and N nutrients and water used in the photosynthesis process. In a previous study, FMA inoculation increased shoot dry weight production (34.04%), crude protein content (10.21%), phosphorus uptake (40%), N content (10.53%), N uptake (38.10%), and protein production (40.15%) [7].

3.4 Root Colonization by Arbuscular Mycorrhizal Fungi

The observation results of FMA colonization on *G. procumbens* roots using 100x magnification microscope were presented in Table 3. The percentage of FMA colonization on *G. procumbens* roots was highly significant different ($p<0.01$) after ANOVA.

Table 3. Percentage of root colonization by arbuscular mycorrhizal fungi (%) at 12 WAP

Variables	Treatments			
	P0	P1	P2	P3
Colonization (%)	6.10 ± 6.14^{c}	44.11 ± 11.91^{b}	69.85 ± 8.47^{a}	77.17 ± 17.46^{a}
Colonization class [12]	Low	Medium	High	Very high

Different superscript letters in the same row indicate highly significant ($p<0.01$). P0: control (without miko-seedcookies), P1: use of miko-seedcookies diameter 4 cm, P2: use of miko-seedcookies diameter 5 cm, P3: use of miko-seedcookies diameter 6 cm.

The results of Duncan's further test showed that the treatment of miko-seedcookies diameter 5 cm (P2) and 6 cm (P3) gave a very significant effect ($p<0.01$) on the percentage of root colonization compared to the control (P0) and the use of miko-seedcookies diameter 4 cm (P1). The larger the size of miko-seedcookies, the percentage of colonization increased because the soil of the research site was less fertile [5] so that FMA in miko-seedcookies encourages more root infection [13]. The association of FMA with *G. procumbens* could be known by the presence of some typical structures in FMA colonization on the roots of its host plant observed through a microscope such as the structure of spores, vesicles, hyphae, and arbuscules (Figure 2).

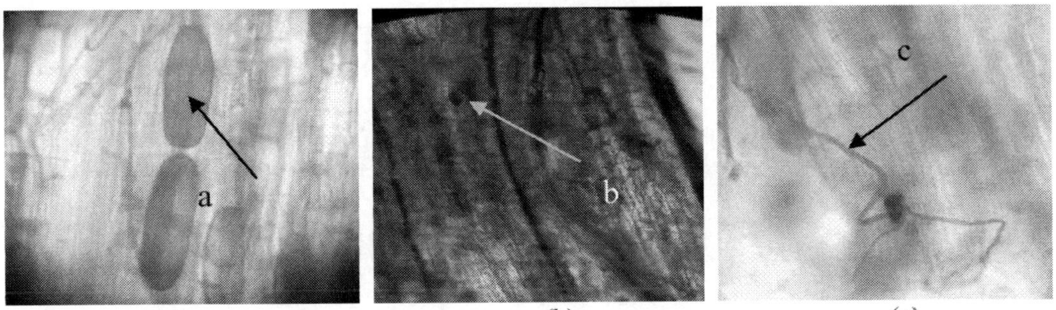

(a) (b) (c)

Figure 2. Structure of arbuscular mycorrhizal fungi on *G. procumbens* roots (100x magnification). (a): vesicles, (b): spores, (c): hyphae

4 Conclusion

Growth and production of *G. procumbens* increased with the addition of miko-seedcookies in terms of number of leaves, plant height, fresh weight (shoots, mature leaves, and edible stems), and mycorrhizal colonization. The best miko-seedcookies recommended for *G. procumbens* cultivation is 6 cm in diameter.

This research was supported by the Ministry of Education, Culture, Research, and Technology thorough the Program of Regular Fundamental Research 2023 with the grant number of 18813/IT3.D10/PT01.03/P/B/2023

Reference

1. R. P. J. Gultom, H. S. Siagian, *Potensial Farmakologi Tanaman "Gynura" Analisis Fitokimia dan Bioaktivitasnya* (Deepublish, Yogyakarta, Indonesia, 2019)
2. D. R. Siregar, P. M. Silitonga, IJCST-UNIMED **4**, 29-33 (2021)
3. K. P. Putri, Danu, JPHT **11**, 141-147 (2014)
4. N. F. N. Nurhasybi, D. J. Sudrajat. JPTH **8**, 107-120 (2020)
5. D. S. Banurea, Produksi Biomassa Hijauan dan Kualitas Nutrien Pucuk *Indigofera zollingeriana* yang Ditanam pada Kerapatan Tanaman yang Berbeda, Master Thesis, Institut Pertanian Bogor, Bogor, Indonesia, (2017)
6. P. R. O. Purba, N. Rahmawati, E. H. Kardhinata, A. Sahar, Agroekoteknologi **2**, 919-932 (2014)
7. P. M. D. H. Karti, I. Prihantoro, M. A. Setiana, TASJ **41**, 114-120 (2018)
8. I. Rosita, S. Wilarso, A. S. Wulandari, JST **8**, 96-102 (2017)
9. F. Fauziah, R. Wulansari, E. Rezamela, Agrikultura **29**, 26-34 (2018)
10. R. Prayudyaningsih, R. Sari, JPKW **5**, 37-46 (2016)
11. M. Abror, M. Mauludin, Nabatia **12**, 51-62 (2015)
12. A. P. Nusantara, Y. H. Bertham, I. Mansur, *Bekerja dengan Mikoriza Arbuskula* (IPB Press, Bogor, Indonesia, 2012)
13. R. P. Sianturi, D. Delvian, D. Elfiati, PFSJ **4**, 128-138 (2015)

Comparation between maize and sorghum silage quality: A meta-analysis study

Ferry Alifia Ratnaningtyas[1*], *Luki* Abdullah[1], and *Nur* Rochmah Kumalasari[1]

[1]Department of Nutrition and Feed Technology, IPB University, Bogor 16680, Indonesia

Abstract. Forage maize silage and sorghum silage might be comparable in terms of nutrient content, especially in fermentation quality. Many studies have been conducted on this matter, but no comprehensive analysis exists. Therefore, this study aimed to evaluate the nutrient content of both maize and sorghum silage through a meta-analysis method. A total of 190 articles relevant to the keyword were further selected until 17 scientific articles according to the determined criteria for this study. The articles were entered into the database. Hedges'g effect was used to measure the estimate with a 95% confidence interval (CI) using random-effects models. The results showed that the pH of forage sorghum silage was significantly ($p<0.05$, Estimate = 1.029, 95% CI = 0.404 to 1.653) higher than that of forage maize silage. And then, lactic acid (Estimate = -1.174, 95% CI = -1.797 to -0.551) and ammonia (Estimate = -3.180, 95% CI = -4.885 to -1.484) were significantly ($p<0.05$) lower than maize silage and no effect ($p>0.05$) in ammonia value. Based on this study, sorghum silage potentially a substitute for maize silage.

1 Introduction

Silage is one way of preserving forage with a lower risk of losing feed nutrient value due to unfavourable conditions. Almost all plants can be processed into silage, such as forage maize and sorghum, are currently being developed. Maize (*Zea mays* L.) is an ideal plant for making silage because it contains complete nutrients [1], high water-soluble carbohydrates, low buffering ability [2], and fresh forage production reaches 44.54 tons ha^{-1} [3]. The utilization of maize plants in making silage has been done widely. Previous research [1] found the pH of maize silage to be 3.64 with a nutrient content of 8.70% PK, 39.50% NDF, 23.50% ADF, 5.46% lactic acid, and 1.72% acetic acid. The use of maize forage is considered less economical in Indonesia because it requires more expensive costs for forage production and can only be harvested once in one production cycle. Therefore, alternative forages are needed to replace maize forage.

Sorghum (*Sorghum bicolor* (L.) Moench) is an alternative animal feed being developed as a substitute for maize. The cost of production of sorghum forage is IDR 268 kg^{-1}, which is 50% cheaper than the cost of maize production. Sorghum plants can be harvested 3-4 times in one cycle so that forage production is abundant. According to [4] sorghum forage

* Corresponding author: ferralfia14@gmail.com

production ranges from 38-45 tons ha^{-1}. The use of sorghum as raw material for silage is now being to be developed. Fermentative quality is one of the parameters that describe the success of the fermentation process.

The inconsistent result from numerous studies have been reported on the fermentative value of maize and sorghum silage. Therefore, scientific synthesis using a meta-analysis approach is needed to get a more comprehensive [5]. This study aims to evaluate the fermentative quality of maize and sorghum forage silage by integrating data from various related studies in a meta-analysis approach.

2 Method

2.1 Procedure

The material used in this research are Microsoft Excel 2019, Microsoft Word 2019, Publish or Perish 8, and OpenMEE. Data collected from various Scopus-indexed journal with several appropriate criteria. Literature searches were conducted on several scientific sources: Google Scholar, Science Direct, Research Gate, and Springer. The keywords used were "maize silage", "sorghum silage", and "comparation maize with sorghum silage", with the criteria that the articles used in the form of experimental research, published from 2008-2022, in English, and contained quantitative data in the form of mean, standard deviation, and a number of samples. A total of 190 potential articles were found, and then selected so that 17 articles were used in database (Table 1).

Table 1. Studies used in database.

Studi	Type of silage	Control silage	Stage		Ensilage
			Maize	Sorghum	
[1]	Sorghum, sorghum+ seed	Maize	*Maturity*	*Maturity*	72
[6]	Sorghum	Maize	*NA*	*NA*	42
[7]	Sorghum	Maize	*Maturity*	*Early bloom*	60
[8]	Sorghum	Maize	*Maturity*	*Early bloom*	15
[9]	Sorghum, sorghum+ xylanase	Maize	*Maturity*	*Bloom*	10
[10]	Sorghum	Maize	*NA*	*NA*	10
[11]	Sorghum	Maize	*Maturity*	*Maturity*	NA
[12]	Sorghum	Maize	*Dough*	*Dough*	NA
[13]	Sorghum	Maize	*Soft dough*	*Soft dough*	40
[14]	Sorghum, sorghum high sugar	Maize	*Milk stage*	*Milk stage*	120
[15]	Sorghum	Maize	*Maturity*	*Maturity*	60
[16]	Sorghum pegah 1, pegah 2, speedfeed 1, speedfeed 2	Maize	*Maturity*	*Soft dough*	136, 60
[17]	Sorghum	Maize	*Milk stage*	*Flowering*	30
[18]	Sorghum	Maize	*Milk stage*	*Soft dough*	30
[19]	Sorghum	Maize	*Milk stage*	*Milk stage*	40
[20]	Sorghum	Maize	*Dough*	*Milk stage*	60
[21]	Sorghum, DK67, Niquel, RBH, RBN	Maize	*Dough*	*Dough*	55

NA : Non Available

2.2 Data Analysis

The database was analysis using OpenMEE to see the effect size and heterogeneity values. The analysis was done with random effect models at a 95% confidence interval. The effect size value used Hedge's g to measure the distance comparison of fermentation quality compared to maize silage as a control. Data analysis results are presented in tables. Analysis of bias look at the value of Egger's test using the JASP application.

3 Result and Discussion

Descriptive statistical results of sorghum forage silage were higher in pH and acetic acid parameters and lower in lactic acid and ammonia parameters than maize forage silage. The results of descriptive statistics for fermentation quality are presented in Table 2.

Table 2. Descriptive statistic fermentation quality.

Silage	Parameters	N	Mean	SD	Min	Max
Maize	pH	27	3.73	0.21	3.51	4.31
	Lactic acid	25	6.29	2.58	3.00	13.60
	Acetic acid	23	2.06	1.51	0.92	8.50
	Ammonia	18	3.71	2.70	0.53	7.00
Sorghum	pH	27	3.86	0.28	3.47	4.37
	Lactic acid	25	5.63	3.02	2.28	14.50
	Acetic acid	23	2.10	1.58	0.79	7.80
	Ammonia	18	3.33	2.43	0.42	7.52

n: a number of observations, SD: standard deviation, Min: minimum, Max: maximum

The results of the analysis using the random effect model showed that the pH, lactic acid, and ammonia of sorghum forage silage were significantly different ($p<0.01$) from the control group of maize forage silage. However, the results were not significant in the acetic acid. Effect size and heterogeneity values using the meta-analysis are presented in Table 3.

Table 3. Meta-Analysis of fermentation quality

Parameters	N	Effect size			SE	p value	Heterogeneity		Egger's test
		Estimate	LB	UB			I^2	p value	
pH	27	1.029	0.404	1.653	0.319	0.001	79.889	<0.001	0.064
Lactic acid	25	-1.174	-1.797	-0.551	0.318	<0.001	72.837	<0.001	0.728
Acetic acid	23	-0.013	-0.678	0.653	0.340	0.970	74.938	<0.001	0.762
Ammonia	18	-3.180	-4.885	-1.484	0.870	<0.001	91.088	<0.001	0.002

The model is highly significant at $p<0.01$, significant at $p<0.05$, n: a number of observations, SE: *Standard Error,* LB: *Lower Bound,* UB: *Upper Bound*

Meta-analysis showed the results were significant, the pH value of forage sorghum silage was higher with the estimated (estimate=1.029, p=0.001, I^2 =79.889). The heterogeneity test resulted in a value of I^2 >50%, which means that this study used a very diverse study and there was no publication bias as seen from the Egger's test value greater than 0.05. The effect size is positive, indicating that high pH values of forage sorghum silage. The result of pH ranged from 3.47-4.37. This value is still classified as good according to [21], which states that good silage has a pH from 3.2-4.5. The difference in pH because the protein value in sorghum silage is higher than maize, it's causing buffering capacity. High buffering capacity inhibit decrease pH [2]. Silage with high crude protein will produce a high pH. In addition, the difference in pH is related to the content of water-soluble carbohydrates and starch. The water-soluble carbohydrates (WSC) content in forage sorghum is relative low at 6.7% [24] and in maize 10.16%. High WSC content in feed ingredients provides optimal conditions for lactic acid bacteria to increase carbohydrate fermentation into organic acids [2]. WSC is needed by bacteria in order to reduce pH to 3.5.

The meta-analysis resulted in an estimate value (Estimate = -1.174, $p<0.01$) indicating that sorghum silage contains significant lower lactic acid. This result was shown in the descriptive statistical analysis (Table 3), which was 6.29% vs 5.23%. The increase in lactic acid is in line with the decrease in pH [7]. However, the lactic acid content formed is still within the normal range according to [22] which is 3-18%.

Meta-analysis on ammonia showed an estimate value (-3.180, $p<0.001$), meaning that the experiment group was significantly lower than the control group. Based on Table 3, the ammonia content of sorghum forage silage is 3.33%. According to [15], the ammonia value of sorghum silage is low due to the content of tannin which can protect proteins from microbial degradation. The ammonia content in both types of silage treatment is classified as good because it is in the range of less than 7% total N [15], [22].

4 Conclusions

In summary, forage sorghum silage produces ideal fermentation quality, with lower ammonia, lactic acid, and acetate content compared to maize silage, but higher pH content. The fermentation quality of sorghum forage silage is still in the normal range, so sorghum forage silage can potentially substitute maize forage silage.

Acknowledgement to the Ministry of Education, Culture, Research, and Technology thorough the Program of Regular Fundamental Research 2023 with the grant number of 18813/IT3.D10/PT01.03/P/B/2023.

References

1. S. Colombini, G. Galassi, G. M. Crovetto, and L. Rapetti. *J Dairy Sci.* **95**, 8: 19-24 (2012)
2. Despal, I. G. Permana, S. N. Safarina, and A. J. Tatra. *Med Pet.* **34**, 1, 69–76 (2011)
3. E. P. Gamasari, I. Prihantoro, and M. Ridla. *JINTP.* **20**, 1, 1–6 (2022)
4. R. Sriagtula. (2016), [Daring]. Tersedia pada: https://repository.ipb.ac.id/handle/123456789/82392
5. J. Gurevitch, J. Koricheva, S. Nakagawa, and G. Stewart. *Nature*, **555**, 7695, 175–182 (2018)
6. S. S. Li, J. J. Zhang, Y. F. Bai, A. A. Degen, T. Wang, Z. H. Shang, L. M Ding, and R. J. Long. *Appl. Anim. Sci*, **36**, 1, 99-106 (2020)
7. M. Khosravi, Y. Rouzbehan, M. Rezaei, and J. Rezaei. *J. Dairy Sci.* **101**, 12, 1–9 (2018)
8. M. T. Harper, J. Oh, F. Giallongo, J. C. Lopes, G. W. Roth, and A. N. Hristov. *J. Dairy Sci.* **100**, 7, 5250–5265 (2017)
9. Y. Yang, G. Ferreira, B. A. Corl, and B. T. Campbell. *J. Dairy Sci.* **102**, 3, 2266–2274 (2019)
10. B. Śliwiński, F. Brzóska, K. Węglarzy, and M. Bereza. *Ann. Anim. Sci.* **12**, 3, 367–379 (2012)
11. H. M. Dann, R. J. Grant, K. W. Cotanch, E. D. Thomas, C. S. Ballard, and R. Rice. *J. Dairy Sci.* **91**, 2, 663–672 (2008)
12. R. Tudisco, V. M. Morittu, N. Musco, M. Grossi, P. Iommelli, B. D'aniello, M. Ferrara, F. Infascelli, P. Lombardi. *Agric.* **11**, 1, 1–11 (2021)
13. P. Zbigniew, and P. Lucyna. *J Cent Eur Agric.* **12**, 2, 294-303 (2011)
14. S. J. Zhang, A. S Chaudhry, D. Ramdani, A. Osman, G. Xue-feng, G. R Edwards, and L. Cheng. *J. Integr. Agric.* **15**, 1, 175–182 (2016)

15. W. G. Carvalho, K. A. D. P. Costa, P. S. Epifanio, R. Perim, D. Teixeira, and L. Medeiros. *Rev. Caatinga*, **29**, 2, 465–472 (2016)
16. E. Sabertanha, Y. Rouzbehan, H. Fazaeli, and J. Rezaei. *J Anim Physiol Anim Nutr*. **00**, 1–12 (2021)
17. N. Nusrathali, M. Mufeeth, A. Ahamed, U. Majeed, and M. Musthafa. *J. Bangladesh Agric. Univ.* **19**, 3, 348 (2021)
18. J. K. Bernard and S. Tao. *Appl. Anim. Sci*. **36**, 2–7 (2020)
19. F. Sarubbi, A. Chiariotti, R. Baculo, G. Contò, and S. A. Huws. *Czech J. Anim. Sci.* **59**, 6, 278–287 (2014)
20. E. Sucu, H. Kalkan, O. Canbolat, and I. Filya. *Rev. Bras. Zootec.* **45**, 10, 596-603 (2016)
21. S. R. Ordoñes, V. M. M. Villalbazo, A. T. Cordoba, J. H. Bautista, and G. V. Villalobos. *Rev. FCA UNCUYO*. **51**, 2, 353-366 (2019)
22. L. Kung, R. D. Shaver, R. J. Grant, and R. J. Schmidt. *J. Dairy Sci.* **101**, 5, 4020–4033 (2018)
23. S. Sandi, Riswandi, S. P. Wijaya, A. I. M. Ali, E. Sahara, A. S. Nurdin, N. Rofiq, and Asmak. *J. Peternak. Sriwij.* **9**, 2, 1–10 (2020).
24. S. Amer and A. F. Mustafa. *J. Dairy Sci.* **93**, 12, 5921–5925 (2010)

The potential of meat and slaughterhouse by-products as sources of bioactive peptides: A literature review

Pramudya Andiana[1*], *Khothibul Umam* Al Awwaly[2], *and Abdul* Manab[2]

[1]Master Student in Animal Products Technology Department, Faculty of Animal Science, Universitas Brawijaya, Malang 65145, Indonesia
[2]Lecturer in Animal Products Technology Department, Faculty of Animal Science, Universitas Brawijaya, Malang 65145, Indonesia

Abstract. The meat and slaughterhouse industry produces two products, namely prime cut and by-products. By-products in some areas are considered waste and thrown away, but there are also some areas that used as animal feed or edible materials that can be consumed directly by humans. By-products from meat and slaughterhouse industry can be in the form of bone, skin, offal, blood, feather, and feet. Many studies show that by-products from meat and slaughterhouse industry also have high protein content and have potential as source of peptides with certain biological activities. Bioactive peptides are molecules composed of 2-20 amino acids in a certain sequence and have biological activity that is not active in the structure of the parent protein. Some of the biological activities of peptides sourced from livestock by-products are antioxidant, antidiabetic, and ACE inhibitor. Bioactive peptides from livestock by-products have also received much attention from researchers, due to their low cost and abundant sources. Several methods have been developed to produce bioactive peptides, such as chemical, enzymatic, and fermentation methods. The aim of this review was to update and summarize several recent studies regarding bioactive peptides from livestock by-products and their potential biological activities.

1 Introduction

The livestock industry plays a vital role in the food sector. The livestock product processing sector comprises several subsectors, including the meat and slaughterhouse industry, dairy industry, and egg industry. Each subsector produces main products and by-products. In general, the term by-product is defined as a part other than carcass, which is edible and non-edible. The meat and slaughterhouse industry produces main products in the form of carcass and by-products in the form of hide/skin, offals, feet, bone, and blood. Some by-products can be utilized as feed and fertilizer, blood being converted into blood meal and used as animal feed has been common. The bone and skin/hide contain collagen which can be used as a raw material for gelatine. The skin/hide is also a valuable raw material in the process of making

* Corresponding author: ramudyaandiana@student.ub.ac.id

shoes and bags [1]. These by-products are known to have a high protein content. Turkey liver, chicken head, goat skin, and beef lungs are reported to contain protein content of 21.9, 12.29, 31.05, and 19.07%, respectively [2-5]. These data show that the study of protein from livestock by-products is interesting and the potential for the development of bioactive peptides from livestock by-products is also still high.

Proteins, the main components of animal tissues and products, are composed of numerous amino acids linked together by peptide bonds [6]. Proteins in meat muscles can be grouped into 3, namely myofibril proteins, sarcoplasm, and stroma. These groups of proteins have their own characteristics based on their solubility properties in water and salt. Stromal proteins are a group of proteins isolated from connective tissue, so they are also called connective tissue proteins. Some examples of stromal proteins are collagen and elastin. Collagen is a type of protein that is abundant in vertebrates and is the most abundant component in livestock by-products. Collagen is a useful biomaterial in the food, pharmaceutical, and biomedical industries. Collagen is also known to be a good precursor of bioactive peptides [7]. Connective tissue proteins can produce hydrophobic amino acids and bioactive peptides that have antioxidant activity through hydrolysis process. The utilization of by-products into bioactive peptides can increase their economic value and potentially can be applied to the nutraceutical and pharmaceutical industries [8].

Bioactive peptides are fragments of a protein that are known to have certain biological functions, such as antioxidant, antimicrobial, antihypertensive, and mineral binding. The composition and sequence of amino acids can influence the activity of a peptide when it is released from its parental protein. Bioactive peptides generally have a residue length of between 2-20 amino acids [9]. Bioactive peptides are a group of biological molecules that are inactive in their parent protein structure and become active when separated from their parental protein [10]. Bioactive peptide can be produced by chemical hydrolysis using acids and bases and enzymatically using protease enzymes. Chemical hydrolysis using acids and bases causes amino acids to denature due to extreme pH. The use of acids and bases during the hydrolysis process is also considered ineffective, because their action is not specific. On the other hand, the advantage of enzymatic hydrolysis is its specific work. The disadvantage of the enzymatic method is that the price of enzymes is quite expensive. Protease enzymes can come from animal, vegetable and microbial sources. Some examples of protease enzymes are papain, bromelain, pepsin, trypsin, and alcalase [11].

By-product source	Method	Biological activities	Reference
Kacang goat (*Capra hircus*) skin collagen	Enzymatic hydrolysis: thermolysin	ACE inhibitor	[12]
Pork (*Sus scrofa*) and chicken (*Gallus gallus*) skin collagen	Enzymatic hydrolysis: collagenase or a protease from *Bacillus licheniformis* (MPRO NX®)	Lipase pancreatic inhibitor	[13]
Chicken (*Gallus gallus*) skin	Enzymatic hydrolysis: alcalase and combination pepsin+pancreatin	Antioxidant	[14]
Bovine (*Bos taurus*) and porcine (*Sus scrofa*) collagen	-	Antioxidant and hypoglycemic activity	[15]

Duck (*Anas platyrhyncos*) liver	Enzymatic hydrolysis: alcalase, neutrase, and papain	Antioxidant	[16]
Porcine (*Sus scrofa*) liver	Enzymatic hydrolysis: alcalase, bromelain, papain, and flavourzyme	Antioxidant	[17]
Chicken (*Gallus gallus*) intestine	Enzymatic hydrolysis: fungal protease	α-amylase inhibitory activity	[18]
Turkey (*Meleagris gallopavo*) head collagen	Enzymatic hydrolysis: alcalase, flavourzyme, and trypsin	Bile acid binding and plasma amine oxidase (PAO) inhibitory activity	[19]
Turkey (*Meleagris gallopavo*) trachea	Enzymatic hydrolysis: Alcalase, flavourzyme, protamex, and papain	Antioxidant and ACE inihibitory activity	[20]
Chicken (*Gallus gallus*) bone	Enzymatic hydrolysis: papain, neutral protease, and trypsin	Antioxidant	[21]
Deer (*Cervus elaphus*), sheep (Perendale), and pig (Duroc) blood plasma	Enzymatic hydrolysis: papain, bromelain, fungal protease concentrate 400, and fungal protease II	Antioxidant	[22]

2 Materials and Methods

Bioactive peptides can be obtained by several methods, such as chemical, enzymatic, and microbial fermentation methods. Each method has its own peculiarities, disadvantages and advantages. The chemical method is cheap and easy to apply, but some amino acids such as tryptophan can be completely destroyed by acid hydrolysis [23]. In addition to the chemical hydrolysis method, enzymatic hydrolysis and microbial fermentation are more preferred. Both of these methods are usually preferred because they work more specifically, making them suitable for producing bioactive peptides and avoiding amino acid denaturation. In the enzymatic hydrolysis method, the enzymes used come from the proteolytic enzyme group. These proteolytic enzymes can be of animal, vegetable, and microbial origin. Some commonly used enzymes are pepsin, trypsin, alcalase, flavourzyme, papain, and bromelain [11]. Chinese pond turtle protein hydrolysate using alcalase and flavourzyme showed the highest ability to inhibit linoleic acid peroxidation compared to trypsin and bromelain. In the other hand, bromelain hydrolysate exhibited the strongest α-amylase inhibitory activity compared to other enzymes. Each enzymes have the specificity to produce bioactive peptides that have certain biological activities [24]. The disadvantage of this method is the price of enzymes that tend to be expensive. Meanwhile, microbial fermentation uses microbes that can produce proteolytic enzymes that will cut long chains of proteins into simple peptides. This method has a high diversity of protease enzymes. Microbial fermentation generally uses lactic acid bacteria (LAB). Some researches showed that LAB can produce bioactive peptides from milk. Based on these three methods above, microbial fermentation method is the most effective method in terms of obtaining bioactive peptide, because microorganism can produce not only one protease, which leads to more bioactive peptide production [11].

Method	Advantage	Disadvantage
Chemical	Cheap and easy to apply	Some amino acids such as tryptophan can be completely destroyed by acid hydrolysis
Enzymatic	Easy to control, specifically producing bioactive peptide, and avoiding amino acids denaturation	Expensive
Microbial fermentation	Producing not only one protease, which leads to more bioactive peptide production	-

3 Results and discussion

Various by-products from the meat industry have high protein content that has the potential to produce bioactive peptides with certain biological activities which can be seen in Table 1.

3.1 Skin/hide

Skin/hide in ruminants reaches approximately 3-12% of live weight, while in poultry it can reach 20%. The skin is mostly composed by a major component referred to as collagen. Collagen is a structural protein in the skin that makes up connective tissue, such as bone and skin [25]. Collagen hydrolysate from kacang goat skin using thermolysin showed an ACE inhibitory activity. Peptides with MW <3 kDa showed the strongest ACE inhibitory activity when compared to fractions 3-5 and >5 kDa. Collagen hydrolysate from kacang goat skin at <3 kDa fraction reported to have an ACE inhibitory activity value of 36.27-91.26% in the range of protein concentration 0.60-1.80 mg/ml and the IC50 value was 82.94 μg/ml [12]. Lipase pancreatic inhibitory activity was reported from pork and chicken skin collagen hydrolysate and peptide fraction. The longer incubation time in the range of 0-24 hours at the time of hydrolysis in chicken and pork skin collagen using collagenase and protease from Bacillus licheniformis (MPRO NX®) showed an increase in the degree of hydrolysis and activity of lipase pancreatic inhibitory. The hydrolysate had a similar or even higher ability to inhibit lipase activity than the ultrafiltrated fractions [13]. Meanwhile, antioxidant activity was reported from chicken skin protein hydrolysate using alcalase and combination pepsin-alcalase. This study have shown that protein hydrolysate from chicken skin and its fractions had the antioxidant activity. Chicken skin protein hydrolysate can scavenge DPPH, hydroxyl, peroxyl, superoxide radicals, and also chelate ferrous ion. Protein hydrolysates can act as antioxidants through the mechanism of electron donation to free radicals [14]. The ability to chelate iron ions is also often associated with the presence of histidine residues that have imidazole rings [26]. Bovine and porcine collagen hydrolysate exhibited antioxidant and hypoglycemic activity through in vitro test. However, α-amylase and α-glucosidase inhibitory activity from bovine and porcine collagen hydrolysate was lower than the acarbose as positive control [15]. The inhibitory activity of α-amylase enzymes by bioactive peptides can come from amino acid residues of Tyr, Trp, Phe, Lys, and cationic residues that tend to bind α-amylase [24].

3.2 Liver

Liver contains high protein, which is around 20%. The liver can also be a good source of several vitamins, such as vitamins B2, B3, B6, B12, ascorbic acid, and vitamin A [16-17, 27]. Duck liver protein hydrolysate by alcalase, neutrase, and papain enzymes showed antioxidant activity. The duck liver protein hydrolysate using the alcalase enzyme had the highest ABTS and hydroxyl scavenging activity, when compared with other enzymes. Six antioxidant peptide sequences (GEHGDSSVPVWSGVN, HLDYYLGK, HLTPWIGK, DTYIRQPW, WDDMEKIWHH, and MYPGIAD) were successfully isolated and identified from duck liver protein hydrolysate using alcalase. These six peptides were also known to increase SOD and CAT activities and decrease MDA production and ROS in H2O2-induced HepG2 [16]. Protein hydrolysate from porcine liver has also been reported to have potential as an antioxidant. Several peptide sequences that can be candidates as antioxidants were APAAIGPYSQAVLVDR, GLNQALVDLHALGSAR, ALFQDVQKPSQDEWGK, LSGPQAGLGEYLFER, and LGEHNIDVLEGNEQFINAAK [17].

3.3 Intestine

Chicken intestine protein hydrolysate showed α-amylase inhibitory activity. The IC50 of the chicken intestine protein hydrolysate, chicken intestine peptide fractions <10 kDa, <3 kDa, and acarbose were 273.04, 959, 362.86, and 158.42 mg/ml, respectively. The highest α-amylase inhibitory activity was in acarbose, however the chicken intestine protein hydrolysate and its fractions still had potential as an antidiabetic [18].

3.4 Head

The protein content of turkey heads was 9.67%, which showed a lower protein content compared to chicken heads, which was 12.29%. Collagen from turkey head was hydrolyzed using single enzyme (alcalase, flavourzyme, and trypsin), dual enzyme (alcalase-flavourzyme, alcalase-trypsin, and flavourzyme-trypsin), and cocktail of all 3 enzymes together. The collagen peptides from all treatments exhibited bile acid binding capacity and PAO inhibitory activity. The increasing of collagen peptide concentration from turkey head increased the percentage of cholic and deoxycholic acids. This shows that the peptide can be an antihypercholesterolemic agent through the mechanism of binding bile acids, thereby preventing reabsorption of bile acids and reducing the amount of cholesterol in the blood. Turkey collagen peptides also inhibited PAO. The ability to inhibit PAO can be used to evaluate the ability of a peptide to act as an anti-inflammatory [3 and 19].

3.5 Trachea

Chicken trachea is a by-product from slaughterhouse. The protein content of chicken trachea was 69.71% on a dry basis. Chicken trachea flakes was reported to have a good solubility at the pH range 6-10 and isoelectric point at pH 4. The soluble protein content in chicken trachea hydrolysate was increasing during 0-6 hours hydrolysis time using papain and alcalase, however the soluble protein content was decreasing after 2-3 hours hydrolysis time in flavourzyme and protamex treatment. Antioxidant and ACE inihibitory activity were reported in chicken trachea hydrolysate. The treatment using alcalase showed the highest antioxidant activity through DPPH and ABTS assay. The lowest antioxidant activity was showed by papain treatment, while protamex and flavourzyme treatment showed the intermediate antioxidant capacity between papain and alcalase. There was a volatility in antioxidant activity among all the treatments during the hydrolysis time. Peptides have the potential to donate hydrogen atoms that can scavenge free radicals. Chicken trachea hydrolysate also exhibited ACE inhibitory activity. The IC50 ACE inhibitor was reported at 0.42, 0.73, 0.55, and 0.72 mg/ml for alcalase, flavourzyme, protamex, and papain, respectively. The greatest ACE inhibitory activity was showed by alcalase. IC50 means the concentration needed to make 50% inhibition. The most abundant amino acids in all chicken trachea hydrolysate were glutamine, glycine, alanine, and proline. Ala and Pro were reported to have antihypertensive activity [20 and 28]. Several ACE inhibitory peptides were reported from chicken parent protein (myosin, creatine kinase, actin, and collagen). Four ACE inhibitory peptides were reported from the hydrolysate of boneless chicken leg meat. The sequences peptide were MNVKHWPWMK, INDNFYDWLP, WDWPY, and VTVNPYKWLP, two peptides (P1 and P4) among the sequences were from myosin [28].

3.6 Bone

Chicken bone protein hydrolysate, its fractions, and their Maillard reaction products showed ability to scavenge DPPH and hydroxyl radical and showed reducing power. Increasing of concentration in all treatments can increase the antioxidant activity. Maillard reaction products derived from peptide fractions had stronger antioxidant activity (DPPH radical scavenging activity and reducing power) than the hydrolysate and its peptides. The Maillard reaction is often occurred between carbonyl and amino groups in food processing. Antioxidant activity from peptides can be increased through the Maillard reaction with reducing sugar. Maillard reaction product derived from protein hydrolysate-sugar has potential for development [21].

3.7 Blood plasma

Blood represents approximately 4% of animal weight live. It's rich in protein and iron. Blood can be used in feed industry in the form of blood meal for feeding livestock and pet [1]. Blood plasma protein hydrolysate from deer, sheep, and pig using papain, bromelain, FP400, and FPII exhibited antioxidant antioxidant activity, but didn't exhibit antimicrobial activity. Enzymatic hydrolysis can change the structure of native protein and open and exposes amino acid residues that can react with oxidants [22]. Porcine blood plasma hydrolysate produced using alcalase also exhibited antioxidant activity through TBARS, DPPH, reducing power, and metal chelating activity test. Increasing of degree of hydrolysis (DH) increased reducing power, DPPH scavenging, and copper and ferrous ion chelating activity dan decreased TBARS value. The highest reducing power and DPPH radical scavenging activity was obtained in fraction <3 kDa compared to fractions 3-6, 6-10, and >10 kDa. The peptide sequence from porcine plasma protein with the highest antioxidant activity was identified as His-Asn-Gly-Asn [29].

4 Conclusion

By-products from meat and slaughterhouse industry is rich in protein. This also shows that by-products from the meat industry and slaughterhouses have the potential to produce bioactive peptides. Production of bioactive peptides from by-products can increase their economic value and support the development of the functional food and nutraceutical industries. Several studies have shown that peptides derived from by-products of the meat industry and slaughterhouses have several biological activities, such as antioxidant , α-amylase inhibitory activity, ACE inhibitory activity, lipase pancreatic inhibitory activity, PAO inhibitory activity, and bile acids binding capacity.

References

1. B. O. Alao, A. B. Falowo , A. Chulayo, V. Muchenje, The potential of animal by-products in food systems: production, prospects and challenges, Sustainability, 9, 1-18, (2017)
2. N. Zouari, N. Fakhfakh, W. B. Amara-Dali, M. Sellami, L. Msaddak, M. A. Ayadi, Turkey liver: Physicochemical characteristics and functional properties of protein fractions, Food and Bioproducts Processing, 89, 142-148, (2011)
3. K. U. Al Awwaly, I. Thohari, M. W. Apriliyani, D. Amertaningtyas, Extraction of chicken head proteins and evaluation of their functional properties, in Proceedings of the International Conference on Environmentally Sustainable Animal Industry, ICESAI, 18-19 November 2020, Malang, Indonesia (2020)
4. T. R. Hakim, A. Pratiwi, Jamhari, N. A. Fitriyanto, Rusman, M. Z. Abidin, D. N. Matulessy, Y. Erwanto, Extraction of collagen from the skin of kacang goat and production of its hydrolysate as an inhibitor of angiotensin converting enzyme, Tropical Animal Science Journal, 44, 222-228, (2021)
5. K. U. Al Awwaly, S. Triatmojo, W. T. Artama, Y. Erwanto, Chemical composition and some functional properties of beef lung protein concentrate extracted by alkaline method, Jurnal Ilmu dan Teknologi Hasil Ternak, 10, 54-62, (2015)
6. [Y. Hou, Z. Wu, Z. Dai, G. Wang, G. Wu, Protein hydrolysates in animal nutrition: Industrial production, bioactive peptides, and functional significance, Journal of Animal Science and Biotechnology, 8, 1-13, (2017)
7. T. Lafarga, M. Hayes, Bioactive peptides from meat muscle and by-products: generation, functionality and application as functional ingredients, Meat Science, 98, 227-239, (2014)
8. E. Susanto, 2019, Bioactive Peptide as Antioxidant: Exploration in Chicken Foot. (Yogyakarta: Deepublish)
9. A. A. Zaky, J. Simal-Gandara, J. B. Eun, J. H. Shim, A. M. A. El-Aty, Bioactivities, applications, safety, and health benefits of bioactive peptides from food and by-products: a review, Frontiers in Nutrition, 8, 1-18, (2022)
10. M. Akbarian, A. Khani, S. Eghbalpour, V. N. Uversky, Bioactive peptides: synthesis, sources, applications, and proposed mechanisms of action, International Journal of Molecular Sciences, 23, 1-30, (2022)
11. D. E. Cruz-Casas, C. N. Aguilar, J. A. Ascacio-Valdés, R. Rodríguez-Herrera, M. L. Chávez-González, A. C. Flores-Gallegos, Enzymatic hydrolysis and microbial fermentation: the most favorable biotechnological methods for the release of bioactive peptides, Food Chemistry: Molecular Sciences, 3, 1-12, (2021)
12. A. Pratiwi, T. R. Hakim, M. Z. Abidin, N. A. Fitriyanto, J. Jamhari, R. Rusman, Y. Erwanto, Angiotensin-converting enzyme inhibitor activity of peptides derived from Kacang goat skin collagen through thermolysin hydrolysis, Veterinary World, 14, 161-167, (2021)

13. J. A. González-Noriega, M. Valenzuela-Melendres, A. Hernández–Mendoza, H. Astiazarán-García, M. Á. Mazorra-Manzano, E. A. Peña-Ramos, Hydrolysates and peptide fractions from pork and chicken skin collagen as pancreatic lipase inhibitors, Food Chemistry: X, 13, 1-7, (2022)
14. J. O. Onuh, A. T. Girgih, R. E. Aluko, M. Aliani, In vitro antioxidant properties of chicken skin enzymatic protein hydrolysates and membrane fractions, Food Chemistry, 150, 366–373, (2014)
15. A. L. A. Gaspardi, D. C. da Silva, L. G. S. Ponte, F. Galland, V. S. N. da Silva, F. M. Simabuco, R. M. N. Bezerra, M. T. B. Pacheco, In vitro inhibition of glucose gastro-intestinal enzymes and antioxidant activity of hydrolyzed collagen peptides from different species, Journal of Food Biochemistry, 46, 1-14, (2022)
16. J. Sun, C. Zhou, J. Cao, J. He, Y. Sun, Y. Dang, D. Pan, Q. Xia, Purification and characterization of novel antioxidative peptides from duck liver protein hydrolysate as well as their cytoprotection against oxidative stress in HepG2 cells, Frontiers in Nutrition, 9, 1-12, (2022)
17. M. López-Pedrouso, P. Borrajo, M. Pateiro, J. M. Lorenzo, D. Franco, Antioxidant activity and peptidomic analysis of porcine liver hydrolysates using alcalase, bromelain, flavourzyme and papain enzymes, Food Research International, 137, 1-9, (2020)
18. E. Vimalraj, R. Ramani, V. A. Rao, M. Parthiban, R. Narendrababu, S. Arulkumar, Antidiabetic activity of peptides extracted from chicken intestine hydrolysate, The Pharma Innovation, 11, 1718-1721, (2022)
19. Z. Khiari, M. Ndagijimana, M. Betti, Low molecular weight bioactive peptides derived from the enzymatic hydrolysis of collagen after isoelectric solubilization/precipitation process of turkey by-products, Poultry Science, 93, 2347-2362, (2014)
20. T. Pramualkijja, T. Pirak, S. R. Euston, Valorization of chicken slaughterhouse by products: Production and properties of chicken trachea hydrolysates using commercial proteases, International Journal of Food Properties, 24, 1642-1657, (2021)
21. X. Nie, D. Xu, L. Zhao, X. Meng, Antioxidant activities of chicken bone peptide fractions and their maillard reaction product: effects of different molecular weight distribution, International Journal of Food Properties, 20, 5457-5466, (2017)
22. C. S. F. Bah, A. E. A. Bekhit, A. Carne, M. A. McConnell, Production of bioactive peptide hydrolysates from deer, sheep and pig plasma using plant and fungal protease preparations, Food Chemistry, 176, 54-63, (2015)
23. C. Alvarez, M. Rendueles, M. Diaz, The yield of peptides and amino acids following acid hydrolysis of haemoglobin from porcine blood, Animal Production Science, 52, 313-320, (2012)
24. Md. S. Islam, W. Hongxin, H. Admassu, A. A. Mahdi, M. Chaoyang, F. A. Wei, In vitro Antioxidant, Cytotoxic and Antidiabetic Activities of Protein Hydrolysates Prepared from Chinese Pond Turtle (Chinemys reevesii), Food Technology & Biotechnology, 59, 360-375, (2021)
25. A. M. E. Matinong, Y. Chisti, K. L. Pickering, R. G. Haverkamp, Collagen extraction from animal skin, Biology, 11, 1-15, (2022)
26. L. L. S. Canabady-Rochelle, C. Harscoat-Schiavo, V. Kessler, A. Aymes, F. Fournier, J. M. Girardet, Determination of reducing power and metal chelating ability of antioxidant peptides: revisited methods, Food Chemistry, 183, 129-135, (2015)
27. M. G. Romero-Garay, E. Montalvo-Gonzalez, C. Hernandez-Gonzalez, A. Soto-Dominguez, E. M. Becerra-Verdin, and M. D. L. Garcia-Magana. Bioactivity of

peptides obtained from poultry by-products: a review, Food Chemistry: X, 13, 1-10, (2022)

28. M. Terashima, T. Baba, N. Ikemoto, M. Katayama, T. Morimoto, S. Matsumura, Novel angiotensin-converting enzyme (ACE) inhibitory peptides derived from boneless chicken leg meat, J. Agric. Food Chem, 58, 7432-7436, (2010)
29. Q. Liu, B. Kong, Y. L. Xiong, X. Xia, Antioxidant activity and functional properties of porcine plasma protein hydrolysate as influenced by the degree of hydrolysis, Food Chemistry, 118, 403-410, (2010)

CSR implementation of palm oil plantations through the integrated farming system program in empowering local farmer groups

Veronika Suryajuanti Arhian[1], *Sigit Heru* Murti B S[2], and *Endang* Baliarti[3*]

[1]Student in Environmental Science Doctoral Program at Universitas Gadjah Mada, Yogyakarta 55281, Indonesia
[2]Faculty of Geography, Universitas Gadjah Mada, Yogyakarta 55281, Indonesia
[3]Graduate School, Universitas Gadjah Mada, Yogyakarta 55281, Indonesia

Abstract. The concept of corporate social responsibility (CSR) emerged in response to the development of sustainable corporations that prioritize sustainable development practices. The Sintang District Government's economic improvement program aims to boost community income by developing the UMKM sector (Micro, Small, Medium Enterprises). To integrate palm oil plantations with livestock, the local government initiated this effort. The research method used is a descriptive study - with qualitative data types through data collection techniques conducted utilizing interviews, observations, and literature studies. The study found that the application of CSR in Indonesia does not fully align with the legal definition of CSR. The implementation of CSR is primarily focused on short-term aid provision to local communities. The integration of livestock farms has not been implemented by all plantation companies, and the ones that do exist are suboptimal. Even though Sintang Regency cannot meet the local beef needs, farmers are only capable of meeting 20-30% of the annual demand. Consequently, to meet the demand, the government must purchase livestock from outside Sintang district and even from outside the province of West Kalimantan.
Keywords: Corporate social responsibility, palm oil plantations, livestock

1 Introduction

The rapid expansion of the palm oil sector in Indonesia has resulted in heightened competition among vegetable oils worldwide. The inclusion of palm oil has altered the vegetable oil market, particularly the four primary vegetable oils: Palm Oil, Soybean Oil, Repseed Oil, and Sunflower Oil. The high demand for vegetable oils and the vast amount of land necessary to meet production requirements have raised concerns. To meet the global demand for vegetable oil in 2025, it would require an additional 50.5 million hectares of land if rapeseed crops were used, 70.4 million hectares for sunflower oil, and only 12.6 hectares for palm oil. This demonstrates the greater efficiency of palm oil compared to rapeseed and sunflower oils. Moreover, palm oil is the sole potentially sustainable

* Corresponding author: bali_arti@ugm.ac.id

vegetable oil and significantly contributes to the welfare of small communities in palm oil-producing countries (Hidayahturrohmah, 2022).

Palm oil plantations have the potential to breed livestock, such as cattle and sheep, in addition to their main production of oil. The integration of livestock and plantations has a positive impact on both the company and the surrounding community. In Sintang District, local farmers are only able to supply around 20-30% of the demand for beef, equating to approximately 6,700 cattle annually. Therefore, the government has to purchase beef from outside the district or the province of West Kalimantan to meet demand. Less than 10% of the plantation companies in Sintang are currently integrating oil palm with cattle.

2 Materials and methods

This study was conducted in two palm oil plantation companies located in Sintang Sub-district and Tebelian River Sub-district. The choice of study locations was based on the fact that one company provides assistance to local farmers while the other company has not yet implemented CSR funds in the livestock sector. Moreover, both companies are situated close to the district city center. This is significant as, during the interviews, the community displayed a relatively heterogeneous and pluralistic composition. Thus, it is assumed to accurately reflect the comprehension of applying CSR through cattle palm integration. The collected data encompasses primary and secondary sources, with primary data being gathered through in-depth interviews with a total of 150 respondents utilizing interview guidelines. Secondary data was collected from technical agencies responsible for CSR and animal husbandry, as well as relevant literature sources.

3 Results and Discussion

3.1 CSR implementation

According to the data, oil palm plantation companies primarily allocate their CSR funds towards enhancing infrastructure and supporting socio-cultural initiatives. However, the allocation towards developing the local economy or aiding local breeder groups is minimal. Over the years, there has been a substantial increase in the realization of CSR funds, which was triggered by the issuance of Minister of Agriculture Regulation Number. No. 38 from 2020 addresses the Implementation of Sustainable Certification for Palm Oil Plantations in Indonesia. Three criteria drive corporate social responsibility and community economic empowerment: community social responsibility, indigenous people's empowerment, and local business development. According to data from a plantation company in Sintang Regency, CSR funds are still predominantly devoted to social, religious, and spiritual aid. CSR funds are provided as social assistance only when a proposal is submitted by the village or community group.

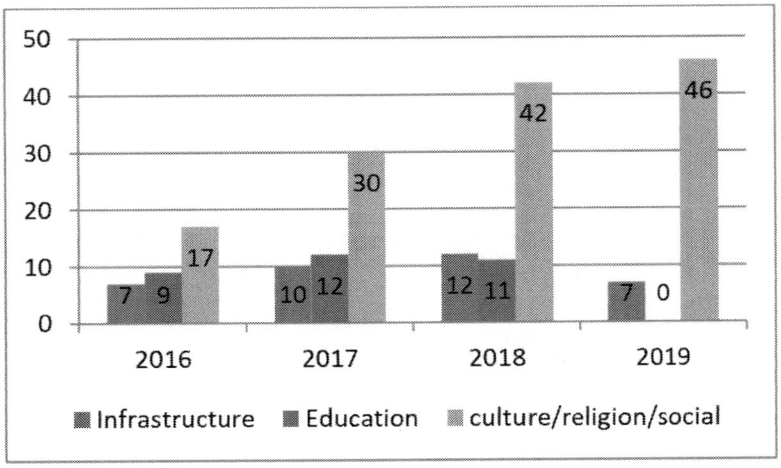

Fig. 1. Realization of the use of CSR funds of PT. SAM

If the CSR realization data reported by companies to the Sintang Regency Government is examined, an increase in the number of CSR realizations can be observed. This is demonstrated in Table 1 and Figure 2.

Table 1. Realization CSR

Year	Realization (Rp)	Year	Realization (Rp)
2013	670,740,000	2018	6.000,000,000
2014	3.775,909,700	2019	7.736,720,632
2015	1.226,441.760	2020	18.340,015,826
2016	2.710,874,274	2021	21.062,865,425
2017	4.097,100,855	2022	33.903,414,774

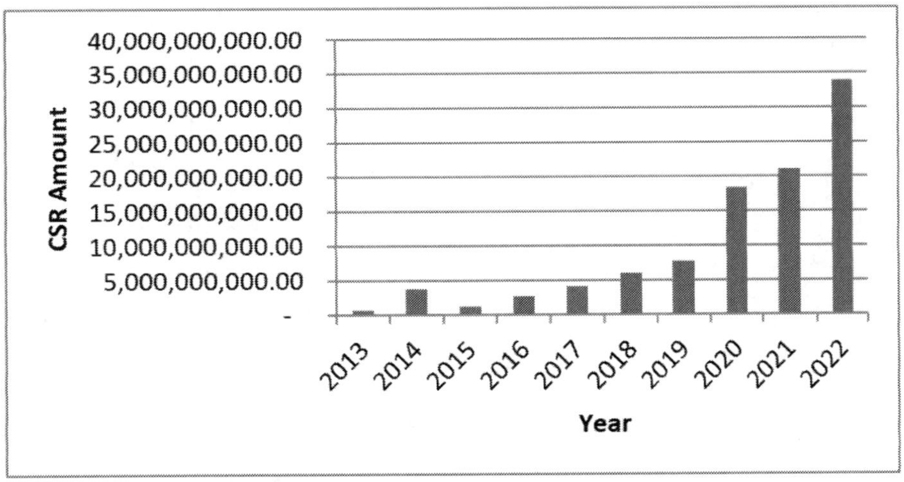

Fig. 2. Graph of CSR implementation in Sintang Regency
Source: Regional Development Planning Agency Sintang Regency

The data illustrated in Figure 3 shows a noticeable uptick in CSR funds over an yearly basis, which is attributable to a rise in the count of companies engaged in CSR programs. The visualization further showcases the range of activities undertaken in CSR implementation.

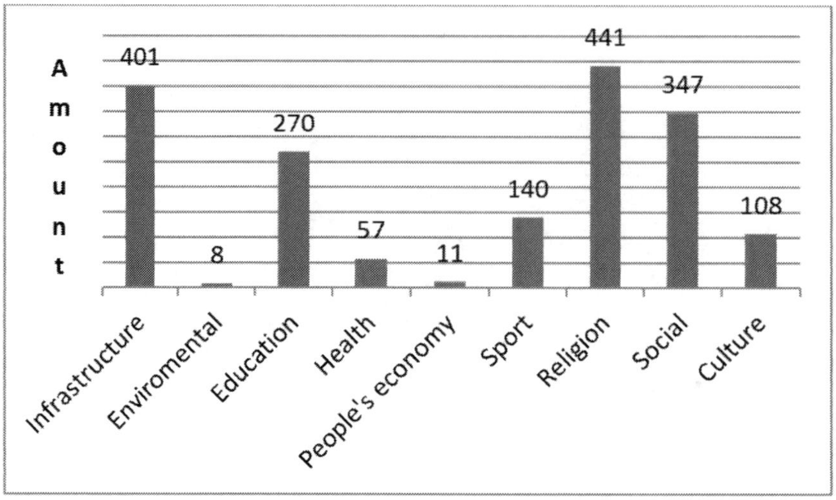

Fig. 3. Field of activity: CSR implementation
Source: Regional Development Planning Agency Sintang Regency

CSR is typically implemented within the realm of religion and encompasses activities such as donations for religious holidays and the provision of facilities for religious institutions. The CSR program should reflect the company's commitment to fostering economic growth in the community surrounding the plantation company. The development of oil palm plantations has the potential to reduce income inequality among community groups, create economic multiplier effects, and enhance community welfare in the proximity of the company, thereby mitigating social conflicts between companies and communities.

Through interviews conducted, the community, in general, did not perceive any significant income changes after the implementation of oil palm plantations. Not all individuals receive direct benefits, such as employment on the plantation, as the community's livelihoods remain largely traditional farming, consistent with pre-existing conditions prior to the oil palm plantations. Figure 4 showcases the results of the interview.

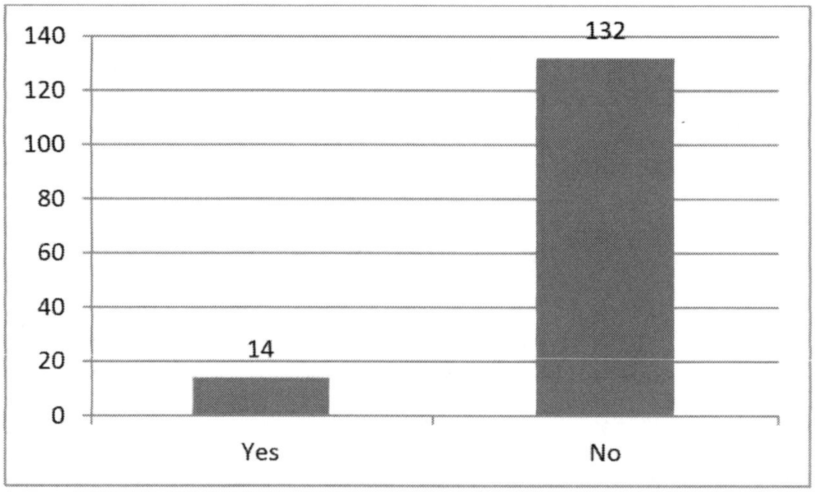

Fig. 4. Change in income after palm oil plantation

The average respondent from the interview results never fully comprehends CSR due to their limited understanding of the concept and the lack of awareness that CSR is a corporate obligation. This results in companies spending their CSR funds only on social assistance when there is a proposal submitted by a group of people and on incidental activities, thus impacting their CSR activities. This aligns with the data gathered by Sintang Regency Bappeda, which indicates that company CSR funds are predominantly utilized for religious and social activities. This is evidenced by the interview findings displayed in Figure 5.

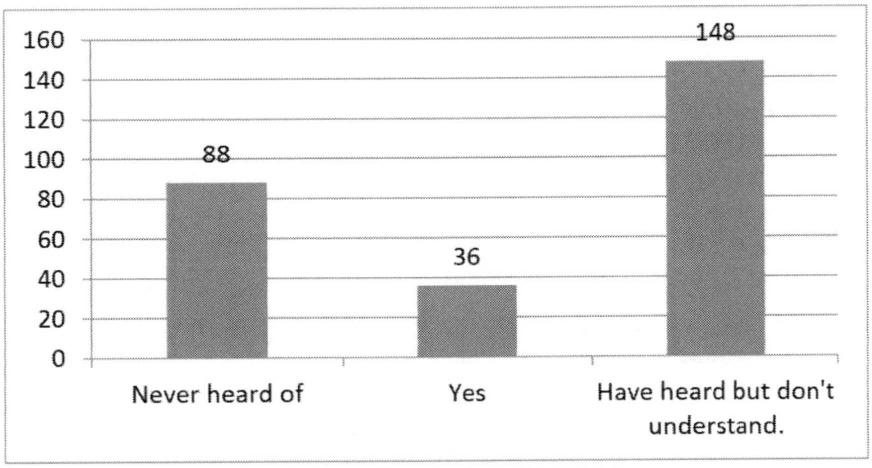

Fig. 5. Community understanding of CSR

Based on the available data, it can be concluded that the utilization of CSR funds in Sintang District has yet to back the livestock sector. Nevertheless, various companies are already offering guidance and aid to nearby farmer groups that are involved in raising livestock within the plantation area. The companies engage in providing cattle seeds, livestock medical personnel, and marketing livestock products. Cattle farming takes place within the oil palm plantation area, depicted in Figure 6.

Fig. 6. Cattle development in palm oil plantations.
Source: Personal

4 Conclusion

1. To meet the local demand for meat in Sintang District, the local government should establish regulations governing the integration of cattle and palm oil using CSR funds.

2. Written cooperation between the local government and plantation companies in Sintang District is necessary to develop local breeder or farmer groups.

References

1. Afifah, N et all . The Impact of Corporate Social Responsibility, Service Experience and Intercultural Competence on Customer Company Identification, Customer Satisfaction and Customer Loyalty (Case Study: PDAM Tirta Khatulistiwa Pontianak West Kalimantan). Social and Behavioral Sciences (2015)
2. Carrol, A. Carrol Pyramid of CSR : Taking Another Look. Carrol International Journal Of Corporate Social Responsibility (2016)
3. Hidayahturrohmah A, N et all. EU behavioral dualism in palm oil discrimination dispute with Indonesia: Environmental commitment or protectionism. Jurnal Inovasi Ilmu Sosial dan Politik, 1-3 (2022)
4. Hinson R, et all. Corporate social responsibility and international business: examining.g the nexus and gaps from a developing economy perspective. International Journal of Corporate Social Responsibility (2017)
5. Morimoto, et all. Coorporate Social responsibility: From Theory to Pracyice, University Of Cambridge, Judge Institute Management Working Paper (2004).
6. Siregar B, et all. Analysis Of Corporate Social Responsibility, Default Risk And Conservatism Effect To Earning Management With Good Corporate Governance As Moderating Variable In Manufacturing Company Whose Shares Incorporated In Indonesia Sharia Stock Indeks. Junior Scientific Researcher Journal (2018).

The use of gambier as a tanner for fur-tanned rabbit skin

Roselini Agustia Andini[1], *Ambar* Pertiwiningrum[1*], and *Mohammad Zainal* Abidin[1]

[1]Faculty of Animal Science, Universitas Gadjah Mada, 55281 Yogyakarta, Indonesia

Abstract. Rabbit fur skin is an exotic skin group and a by-product of slaughtering animal houses. The tanning process is a process for processing. This vegetable tanning can be used as vegetable tanning materials derived from materials obtained from nature. examples of vegetable tanners are gambier and mimosa. The material used for research was rabbit fur 15 pieces and gambier. The research used in this experiment was a completely randomized design with 5 treatments and 3 replications, and there were significant influences would be tested by Duncan's Multiple Range Test Method. The treatment of this research used mimosa 4% (T0), gambier 3% (T1), 4% (T2), 5% (T3) dan 6% (T4). The result showed that using gambier as tanning material has a significant effect on tensile strength, tear strength, thickness, and flexibility. The best result of this research was T4 using 6% of gambier as a tanning agent. The tanned skin had 34,43 N/mm^2 tensile strength, 8,25 Kg/cm, tear strength of 2,33 mm thickness, and 77,09% flexibility. The conclusion of this research is gambier can be added as a tanning agent for rabbit fur and gives good quality.

1 Introduction

Rabbit fur skin is a type of livestock product which is currently used as a trade commodity with a high price. Leather is often used as a material for making jackets and several products are produced by using rabbit skin as a raw material. Leather commodities are classified into raw hides and tanned leather. Raw hides are the raw material for leather that has just been removed from the animal's body until the skin has undergone several preservative processes or is ready to be tanned.

Rabbit fur skin is a group of exotic skins and a by-product of slaughtering animals in slaughterhouses (RPH) and outside slaughterhouses. This by-product can be utilized so that it has high economic value [1]. The process is intended to change the properties of raw hides prone to decay and damage by microbial activity into tanned leather resistant to microbial activity and putrefaction [2].

Gambir or gambier comes from hot water extracted from the leaves and twigs of the gambir plant (*Uncaria gambir* Roxb.) which is then precipitated, drained, molded, and dried. The extraction process is carried out by pressing both traditionally and using

* Corresponding author: artiwi@mail.ugm.ac.id

hydraulic pressed [3]. The main chemical components of Gambir are catechins and tannins. Catechins are monomers of tannins where according to Fengel and Wegener [4] if 3 to 8 catechin molecules form a polymer, the polymer formed belongs to tannins, to be precise tannin condensation. This plant has been ~~widely~~ used in medicine, leather tanning, ink, and dyes. Gambir consists mostly of flavanol monomers such as catechins, epicatechins, and alkaloids [5].

Gambier gives tanning results that are soft, smooth, and light in color. The use of gambier as a tanning agent in recent years has declined rapidly since the early seventies. The tannins produced in the world, including gambier in it, 90% of which are used as leather tanning agents [6]. Mimosa is the essence of acacia bark, which is the result of filtering from the processing of acacia bark (Acacia aureus) which is processed through an evaporation or chemical process. Acacia juice contains several kinds of ingredients, including 63% tanning agent, 16% non-tanning substance, 19.5% water, and 1% dregs (Suparno, Covington and Evans. 2010). Mimosa tanning can affect the quality of the tanned leather. So it is necessary to determine the concentration of the mimosa tanning agent [7]. Gambier is very likely to replace mimosa. This is supported by the availability of gambier which is an agricultural product in the area of West Sumatra [3]. According to Markmann [8], gambier is categorized as a special type of extract for tanning leather as the tannin compounds contained in gambier combine quickly with skin proteins to produce tanning products which are suitable as the basis for belts and bags.

2 Research Materials and Methods

This research is located at the Yogyakarta Leather, Rubber and Plastics Center (BBKKP) and some tests were carried out at the BBKKP and the Leather Waste Development and Processing Laboratory (LP3K).

2.1 Research Materials

The research material used 15 sheets of rex rabbit skin obtained from the RPH in the city of Yogyakarta as well as gambier obtained from the Yogyakarta Leather, Rubber and Plastic Center (BBKKP).

2.2 Research Method

This study used an experimental method with a completely randomized design (CRD) and four treatments, namely the use of mimosa percentages of 4% (P0), the use of gambier percentages of 3% (P1), 4% (P2), 5% (P3) and 6% (P4). Each treatment was repeated three times. The variables tested include tensile strength, tear strength, thickness, and malleability.

3 Results and Discussion

The results of the tensile strength, tear strength, thickness, and malleability tests can be seen in Table 1.

Table 1. The physical characteristics of rabbit skin that tanned with gambier.

Treatment	Tensile strength (N/cm2)	Tear strength (Kg/cm)	Thickness (mm)	Malleability (mm)
T0	28.25±0.96^{a}	7.45±0.31^{b}	2.10±0.05^{b}	82.53±0.99^{b}
T1	30.51±1.13^{a}	5.85±0.46^{a}	1.80±0.1^{a}	86.26±1.63^{cd}
T2	32.31 ± 1^{b}	6.19±0.32^{a}	2.04±0.05^{b}	84.58±0.83^{c}
T3	33.59± 0.77^{b}	7.94±0.39^{b}	2.23±0.05^{c}	81.22±1.34^{b}
T4	34.43±0.61^{b}	8.25±0.30^{b}	2.33±0.15^{b}	77.09±1.38^{a}

3.1 Tensile Strength

The results of analysis of variance showed that between treatments had a very significant effect (p <0.01) on the tensile strength of rabbit skin. The best test result of the tensile strength test of tanned rabbit skin was in the P4 treatment with a value of 34.43 N/cm^2. The test results used a gambier tanning agent with a concentration of 6%. Other results in the study were P0 which was 28.25 N/cm^2, P1 which was 30.51 N/cm^2, P2 which was 32.31 N/cm^2 and P3 which was 33.59 N/cm2. The factor that can affect the tensile strength of tanned rabbit skin is the type of skin used. Purnama [9] states that the high composition of collagen fibers in the skin will affect the high physical strength of the skin, namely the elongation and tensile strength of the skin. The level of tensile strength of the skin is affected by the thickness and thickness of the skin, the density of collagen protein, the angles of the collagen bundles, and the thickness of the corium. The wider the angle of the collagen fiber bundle, the thicker the corium, and the higher the skin fat content, the lower the tensile strength of the skin and the lower the elasticity.

3.2 Tear Strength

The results of analysis of variance showed that between treatments had a very significant effect (P<0.01) on the tear strength of rabbit skin. The tear strength obtained from the results of this study were P0, which was 7.45 kg/cm, P1, which was 5.85 kg/cm, P2 was 6.19 kg/cm, P3 was 7.94 kg/cm and P4 was 8.25 kg/cm when compared to the tear strength of chrome tanned goatskin of at least 17.5 kg/cm. This is because rabbit skin is different from goat skin from the nature of the skin, the thickness, and the bonding of protein fibers to the skin. According to Farid et al. [10], thin skin has loose collagen fibers so it has lower tearing power when compared to thicker skin. Collagen fibers are arranged in interwoven collagen bundles. The angle formed by the webbing and the density of the collagen fibers determines the high and low tensile strength. According to Untari et al. [11] the size of the tear strength is in line with the level of tanner contained in the leather and the physical appearance of the leather will reflect the content of tanning substances in the leather. This means that the tear strength indicates the degree of stability between the tanner and the leather layer.

3.3 Thickness

The results of the analysis of variance showed that between treatments had a very significant effect (P<0.01) on the thickness of the rabbit's skin. The results obtained from testing the skin thickness at P0 was 2.10 mm, P1 was 1.80 mm, P2 was 2.04 mm, P3 was

2.23 mm, P4 was 2.33 mm, the thickness of the tanned skin is very important to note because affect the quality of leather products and the purpose of making the product. Hak [12] states that leather is a potential by-product to be used as a raw material for the tanning industry because it has a specific shape and pattern on the surface of the skin is different from other skins and has a thickness and breadth that allows it to be made into leather products such as wallets, bags, etc. accessories and so on.

3.4 Malleability

The results of the analysis of variance showed that between treatments had a very significant effect ($P<0.01$) on the malleability of the rabbit's skin. There was the highest score in the tanning treatment using gambier P1 with a percentage of gambier 3% of 86.26 mm and the lowest value in treatment P4 with a percentage of gambir 5% of 77.09 mm, and the average in treatment P3 with a percentage of gambir 4% showed a softness value of 81 .22 mm, P2 with a gambier percentage of 3% showed a suppleness value of 84.26 mm while P0 with a mimosa percentage of 4% showed a value of 82.53 mm. The P1 treatment had high laxity due to the inclusion of tannins into the collagen which was less, so that the resulting tannin structure was less dense resulting in the skin becoming softer and weaker.This occurred because there were spaces that were not filled with tannins. Treatments P4, P3, and P0 have lower laxity due to tannins that enter and fill more spaces between collagen fibers so that the resulting tannery structure will become denser and the skin will become more rigid [13].

4 Conclusion

The conclusion from research on tanned rabbit skin using gambier has an effect on tensile strength, tear strength, thickness, and malleability with the best treatment being P4 using gambier 6% showing a tensile strength value of 34.43 N/cm^2, tear strength 8.25 kg/cm, a thickness of 2.33 mm and a softness of 77.09 mm.

Authors wishing to acknowledge assistance or encouragement from RTA *Program* Universitas Gadjah Mada with Grant Number 5075/UN1/P.II/Dit-Lit/PT.01.01/2023. Authors also thanks to all researchers involved in this research.

References

1. Layla Z. and S. Aminah. *Tanning Sheep Leather Using the Chrome Method in An Effort to Utilize Livestock Slaughter By-Product*, in Proceeding of Temu Teknis Nasional Tenaga Fungsional Pertanian, 2007, Bogor, Indonesia (2004)
2. Prayitno. *Research on Making Environmentally Friendly Leather Jackets Using Vegetable Tanners*, in Proceeding of Third National Conference of Leathers, Rubbers, and Plastics, 4 November 2014, Yogyakarta, Indonesia (2014)
3. K. Anwar, D. Novia, S. Mutiar and J Pinem. Leathers, Rubbers, and Plastics Mag. **29**, 1 (2013)
4. F. Fengel and D. Wegener. Wood, Chemistry, Ultrastructure, Reactions (UGM Press, Yogyakarta, 1995)
5. A. Achmad, J. Kassim, T. K. Suan, R. Amat, and T. L. J. Seey. J. of Phy. Sci. **23**, 1 (2012)

6. R. Lehnen. *Test Methods for Determining Formaldehyde Content and Formaldehyde Emissions Of Wood-Based Panels*, in Cost E 49 Training Course of Testing And Evaluation Of Emissions From Wood Based Material, 2006, Hamburg, Germany (2006)
7. S. Triatmojo and M.Z. Abidin. Eco-Friendly Tanning (UGM Press, Yogyakarta, 2009)
8. D.C. Markmann. Eur. Phys. J. E **14**, 7 (2004)
9. A. Mecke, I. Lee, J.R. Baker jr., M.M. Banaszak Holl, B.G. Orr, Eur. Phys. J. E **14**, 7 (2004)
10. J.A. Farid, P.H. Riyadi, and U. Amalia. J. of Sci. Fishery **10**, 2 (2015)
11. S. Untari, T. R. Wiradarya, and B. N. Polii, Sabrina, and D. Mustika. *Various batting materials for tanning lizard skin*, in Proceeding of The Second National Conference of Leathers, Rubbers, and Plastics Industries, 16 October 2023, Yogyakarta, Indonesia (2002)
12. N. Hak. J. of Fishery **2**, 2 (2013)
13. Mustakim, S. W. Aris, U. A. Khotibul and U. Lita. J. of Sci. and Tech. Animal Products **12**, 1 (2017)

The influence of different drying methods on cowhide crackers at PT Java Mandiri's Partner on protein content, expandability, and organoleptic quality

Taufik Rachman Anshori[1], *Dedes* Amertaningtyas[2*], *Djalal* Rosyidi[2], *Imam* Thohari[2], and *Eny* Sri Widyastuti[2]

[1]Student of Animal Product Technology, Faculty of Animal Science, Universitas Brawijaya, Malang 65145, Indonesia
[2]Lecturer of Animal Product Technology, Faculty of Animal Science, Universitas Brawijaya, Malang 65145, Indonesia

Abstract. The aim of this research is to determine the best protein content, expandability, and organoleptic quality among three different drying treatments for cowhide crackers. The material used in this research is cowhide cowhide crackers. Three drying treatments were used in this study, which are drying in an oven for 4 hours (P1), sunlight drying for 2 days (P2), and a combination of both, with sunlight drying for 2 days and oven drying for 30 minutes. The research employs a Completely Randomized Design (CRD) and a Randomized Block Design (RBD) Subsampling for the organoleptic test with three treatments and six replications, and the De Garmo method was used to determine the best treatment. The results of this research were analyzed using Analysis of Variance (ANOVA), and if differences were found, Duncan's test was employed. The findings revealed that the drying treatments significantly ($P<0.01$) affected protein content, expandability, and organoleptic evaluation. The average value of protein content 65.87%-83.20%, expandability 179.53%-530.42%, color rating 3.26-4.01, crunchiness 3.32-4.11, and taste rating 3.03-4.02. The conclusion derived from the De Garmo method is that the best treatment is drying in an oven (P1) with a protein content of 65.87%, expandability of 530.42%, color rating of 4.01, crunchiness rating of 4.11, and taste rating of 4.02.
Keywords: cowhide, cowhide crackers, drying, protein content, expandability, organoleptic

1 Introduction

The cattle slaughter industry plays a crucial role in meeting the protein needs of the Indonesian population. The increasing population growth leads to a higher demand for beef to meet protein requirements. Consequently, many cattle slaughter industry are overwhelmed by the handling of waste generated from animal slaughtering, especially cattle hides. One

* Corresponding author: dedesfptub@ub.ac.id

way to solve this issue is by processing the waste into cowhide crackers, which can be resold at a high market value. This approach aims to harness the diversification of by-products from livestock processing, enhancing the economic value of hides, and creating new job opportunities.

Crackers is a popular snack enjoyed by many Indonesians [1]. It has led to a significant number of Indonesians engaging in crackers processing businesses, which can be found in nearly every region. Ready-to-eat crackers can be found in roadside eateries as well as large restaurants. People of all backgrounds can enjoy this famous snack.

Cowhide crackers is a traditional Indonesian food known to exist since the 9th or 10th century and is native to the island of Java, as recorded in the Batu Pura inscription [2]. Cowhide crackers is typically made from cattle or buffalo hides that undergo various processes to become crispy and popular snack in Indonesia. In processing, industries commonly prefer using cattle hides because they are readily available. The cowhide, as a byproduct of livestock slaughter, has significant potential to be processed into leather crackers [3]. One initiative that can be undertaken to harness this potential is the establishment of a cowhide cracker processing factory.

PT Java Mandiri is a food industry that produces cowhide crackers. The company operates from Monday to Friday, half a day on Saturday, and is closed on Sunday. On a typical day, they can produce approximately 20 sheets of cowhide crackers. However, during peak seasons like approaching Eid al-Adha, the company can produce 30 to 40 sheets of cowhide crackers from cattle hides.

The procedure for handling hides to be processed into cracker is similar to the standard method. However, what sets it apart from others is that lime is not mixed during the boiling process. According to an interview with the company owner, the use of lime can lead to environmental pollution. Another reason is that the company is not yet equipped to handle the waste generated by the cowhide cracker industry. Therefore, this becomes a compelling reason to conduct research by applying different drying methods to observe the resulting protein content, expandability, and organoleptic quality (color, crispiness, and flavor).

2 Material and Methods

This research was conducted from August 9 to October 9, 2023, and was carried out by a group of five researchers. The research took place at three different locations. The sampling and frying of cowhide crackers were performed at PT Java Mandiri, located on Mendalanwetan Street, Wagir District, Malang Regency. Protein content analysis was conducted at the Food Testing and Quality Assurance Laboratory, Faculty of Agricultural Technology, Brawijaya University. Organoleptic testing was carried out in the Livestock Product Technology Reading Room, Faculty of Animal Science, Brawijaya University.

2.1 Research Material

The materials used in this research included cowhide crackers without the addition of lime in the manufacturing process. The cowhide used was from male cattle (with no distinction regarding breed and age) sourced from three different locations: Batu, Surabaya, and Malang. The frying oil used was of the "Kunci Mas" brand, with a total 800 liter for 120 kg of dried cowhide.

2.2 Research Equipment

The equipment used for the production of cowhide crackers included an oven made by PT Java Mandiri, two large gas stoves made by PT Java Mandiri, two large stoves made by PT Java Mandiri, four LPG gas tanks, two large-sized pans, a cowhide cutter made by PT Java Mandiri, two large-sized sieves, an analytical scale of SF-400 brand, a temperature gauge of Benetech brand, and plastic clips. The tools used for sample testing included Kjeldahl flasks, a set of destruction and distillation apparatus, 100 ml measuring flasks, dropper pipets, distillation flasks, measuring glasses, 100 ml Erlenmeyer flasks, Snowman brand pens, Sinar Dunia brand millimeter blocks (for expansion testing), organoleptic assessment questionnaires, and De Garmo test questionnaires.

2.3 Research Method

The method employed in this research involved experiments conducted using a Completely Randomized Design (CRD) with three treatments, each repeated six times, and a Randomized Block Design (RBD) Subsampling for Organoleptic Testing. Samples were collected over a period of 6 days and subjected to different drying methods. The drying treatments were named as follows: P1 (oven drying for 4 hours), P2 (sun drying for 2 days), and P3 (sun drying for 2 days followed by 30 minutes in the oven).

2.4 Research Procedure

The research procedure carried out at PT Java Mandiri, based on previous research and modified accordingly, begins with the cleaning of newly arrived cowhides. Initially, the hides are washed with flowing water, followed by the removal of any remaining fat and meat still adhering to the skin. The cleaned hides are then boiled twice: the first boiling lasts for 10 minutes, and the second for 10-15 minutes, depending on whether the skin is old or young (up to 30 minutes at most) at a temperature 88.8°C. This process makes it easier to remove hair from the cowhide. After boiling, the cowhide is cleaned of hair using a hair scraper knife (if hair remains, burning is done to facilitate scraping). The cleaned cowhide, free from hair, meat, and fat, is then cut according to the order (2x2 cm in this study) to facilitate the drying process. The cut hides are then oven-dried for 4 hours, sun-dried for 2 days, and a combination of sun drying for 2 days and oven drying for 30 minutes. The sun-dried hides enter the degreasing stage or are steamed using oil for 5 hours at a temperature of 225.5°C. If the hides bubble up and rise, turn off the stove and let the hides shrink and go back under the oil. Repeat this process to obtain quality kerupuk rambak. Lift and let the steamed kerupuk cool completely. Then fry the kerupuk until it becomes crispy. The kerupuk is fried twice, with the first frying using low heat at 187°C and the second frying using high heat at 245°C. The aim is to achieve perfectly bloomed cowhide cracker [4].

2.5 Variable Test

The variables used in this research include the protein content, expansibility, and organoleptic quality of cowhide crackers, which are assessed as follows:

1. Protein content testing of cowhide crackers using the Kjeldahl method [5].
2. Expandability testing by calculating the surface area of the crackers by tracing them onto grid paper [6].

3. Organoleptic color using hedonic scale from 1-5 (1 = slightly brownish, 2 = slightly yellowish, 3 = somewhat white, 4 = white, 5 = very white) of the cowhide crackers from the three different drying treatments
4. Organoleptic cunchiness using hedonic scale 1-5 (1 = not crunchy, 2 = somewhat crunchy, 3 = sufficiently crunchy, 4 = crunchy, 5 = very crunchy) of the cowhide crackers from the three different drying treatments
5. Organoleptic taste using hedonic scale 1-5 (1 = unpleasant, 2 = somewhat pleasant, 3 = moderately pleasant, 4 = pleasant, 5 = very pleasant) of the cowhide crackers from the three different drying treatments. This hedonic test is performed by 20 untrained panelists [7].

2.6 Data Analysis

The data obtained from the three tests mentioned above are then analyzed using Analysis of Variance (ANOVA) with the Microsoft Excel 2019 application. If there are differences, the analysis is further continued with the Duncan's Multiple Range Test (DMRT). The data obtained is then analyzed to find the best treatment using the De Garmo method [8]

3 Results and Discussion

3.1 Protein Content

The analysis of variance results shows that the drying methods significantly affect the protein content of cowhide crackers with a very significant impact ($P<0.01$). The average protein content of cowhide crackers can be seen in Table 1. The table shows that the protein content of cowhide crackers ranges from 65.87% to 83.20%. The highest protein content is found in treatment P2, which is 83.20% with sun drying, while the lowest average is obtained from treatment P1, which is 65.87% with oven drying. The protein content in this study is relatively high when compared to the other research which reported a protein content of 6.10% for cowhide crackers, water content of 0.11%, calcium content of 1.88%, expansion power of 372.12%, crispness score of 5.38, and taste score of 6.89 [9].

The average protein content of cowhide crackers is influenced by several factors. The first factor is the use of raw materials from cowhide. Cowhide crackers contain 83 grams of protein, 4 grams of fat, 1 gram of minerals, 5 milligrams of calcium, 10 milligrams of phosphorus, 12 grams of water, and an energy content of 268 calories per 100 grams. The high protein content is affected by the drying and frying processes [10]. There are differences between treatment P1, P2, and P3. P1 has the lowest value due to the drying method that uses an oven with a higher temperature than sun drying. Higher temperatures and longer processing times lead to increased protein damage in the material [11]. Higher cooking temperatures lead to a decrease in protein content in food products [12].

Table 1. The average protein content, expandability, organoleptic quality, and de garmo on cowhide cracers with different drying method.

Treatment	Protein Content (%)	Expandability (%)	Organoleptic		
			Color	Crunchiness	Taste
P1	$65.87^{a}\pm3.22$	$530.42^{b}\pm89.24$	4.01±0.85	4.11±0.78	4.02±0.82
P2	$83.20^{b}\pm1.00$	$179.53^{a}\pm35.10$	3.26±0.82	3.32±1.22	3.03±1.11
P3	$81.51^{b}\pm2.00$	$223.95^{a}\pm112.08$	3.44±0.99	3.56±0.99	3.23±1.06

Note: Superscripts $^{a\text{-}b}$ in the same column indicate highly significant differences ($P < 0.01$)

3.2 Expandability

The analysis of variance results show that the different drying methods have a highly significant effect (P<0.01) on the expandability of cowhide crackers. The average expandability of cowhide crackers can be seen in Table 1. The data demonstrates that the expandability of cowhide crackers ranges from 179.53% to 530.42%. The highest expandability is observed in treatment P1, which is 530.42% with oven drying, while the lowest expandability is found in treatment P2, which is 179.53% with sun drying. The expandability of all three treatments is notably high compared to the other research with the expandability of commercial crackers ranged from 38% to 145% [13]. The higher the expandability of the crackers, the better the quality of the final product [14].

The expandability of cowhide crackers is related to the protein content contained within them. The higher the protein content in cowhide crackers, the lower the expandability, and conversely, the lower the protein content, the higher the expandability. The higher the protein content in cowhide crackers, the lower the expandability [15]. High protein content can reduce the expandability of crackers, leading to smaller air pockets due to the density of air pockets filled with protein [16]. In addition to protein content, there are other factors influencing the expandability of crackers, such as moisture content. The moisture content of raw crackers significantly affects the bloom value produced when the crackers are fried, with an optimal moisture content ranging from 9-10% [13].

3.3 Organoleptic Quality

The sensory evaluation of crackers was carried out to determine the effect of different drying treatments on the sensory characteristics of the product, including color, crunchiness, and taste. The method used to determine the sensory quality was the hedonic test. The analysis of variance results shows that the different drying methods have a highly significant effect (P<0.01) on the sensory quality of crackers. The average values of the sensory evaluation of cowhide crackers can be found in Table 1.

Color. The analysis of variance results shows that the different drying treatments for cowhide crackers have a highly significant effect (P<0.01) on color evaluation. Based on the data presented in Table 1, the organoleptic color values range from 3.26 to 4.01 (somewhat white to white). This indicates that the color of the tested cowhide crackers is acceptable to the panelists. The lowest value for color in the sensory evaluation is 3.26 for P2, which involves sun drying, while the highest value is 4.01 for P1, which involves oven drying. However, both P2 and P3 have lower average color evaluation scores compared to P1. Based on this data, the panelists tend to prefer cowhide crackers dried using the oven method (P1) due to the color being less yellowish or brownish, which can occur with excessive frying.

The color of cowhide crackers is influenced by the boiling and frying process. The color of crackers is influenced by the original ingredients of the crackers, such as protein and carbohydrates, and is affected by the frying process [17]. The desirable color for crackers is a natural, unobtrusive color that arises from the composition of the ingredients used. The resulting color depends on the duration and temperature of frying and the chemical composition of the surface of the food, while the type of oil used has a minimal effect on the color change of the food [18].

Crunchiness. The analysis of variance results shows that different drying methods for cowhide crackers have a highly significant effect (P<0.01) on crunchiness evaluation. According to the data presented in Table 1, the organoleptic crunchiness values range from 3.32 to 4.11 (somewhat crispy to crispy). This suggests that the crunchiness produced by the tested cowhide crackers is acceptable to the panelists. The lowest value for crunchiness in the sensory evaluation is 3.32 for P2, involving sun drying, while the highest value is 4.11 for P1, involving oven drying. However, both P2 and P3 have lower average crunchiness evaluation scores compared to P1. According to this data, the panelists tend to prefer cowhide crackers dried using the oven method (P1). This occurs because cowhide crackers dried using the oven are easier to break apart compared to the other two treatments.

Crunchiness is closely related to the expandability of cowhide crackers. Crunchiness is influenced by the expandability; the higher the expandability, the higher the preference for crunchiness [19]. The crunchiness of cowhide crackers is also influenced by the protein content. High protein content can reduce crunchiness and the expandability of crackers [20].

Taste. The analysis of variance results shows that the different drying treatments for cowhide crackers have a highly significant effect ($P<0.01$) on the taste evaluation. Based on the data presented in the table, the organoleptic taste values range from 3.03 to 4.02 (fairly delicious to delicious). This suggests that the taste produced by the tested cowhide crackers is acceptable to the panelists. The lowest value for taste in the sensory evaluation is 3.03 for P2, involving sun drying, while the highest value is 4.02 for P1, involving oven drying. However, both P2 and P3 have lower average taste evaluation scores compared to P1.

Based on this data, the panelists tend to prefer cowhide crackers dried using the oven method (P1). This is because cowhide crackers dried using the oven have a milder skin flavor compared to the other two drying methods. The preference level of the panelists, ranging from 3.03 to 4.02, is influenced by several factors. The first factor is related to cowhide crackers from PT Java Mandiri, which underwent three different drying treatments and did not use flavorings, allowing the panelists to experience the natural taste of cowhide. The quality of crackers, including their taste, can be influenced by the quantity of raw materials, the type of raw materials, and the amount of seasoning used [21]. The second factor is related to the oven treatment, which imparts a better taste than the other two treatments. This occurs because the protein contained in the crackers undergoes hydrolysis during the cooking process, producing amino acids and glutamic acid, which enhance the umami taste of the crackers [22].

3.4 The Best Treatment

The selection of the best treatment aims to make informed decisions by comparing the research variables with the treatments that have been implemented. The best treatment can be determined using the De Garmo method. Parameters used include protein content, crispiness, color, taste, and crunchiness. The results of the best treatment analysis using the De Garmo method are selected based on the highest productivity value. The calculation data using the De Garmo method can be seen in Table 2.

Table 2. The productivity value using De Garmo

Treatment	Productivity Value
P1	0.74
P2	0.26
P3	0.33

Note: * Indicates the best treatment, which is cowhide crackers dried using the oven drying method for 4 hours

Based on the calculation of treatment values using the Effective Index method, the best treatment is found in P1 with a protein content of 65.87%, crispiness of 530.42%, color rating of 4.01, crunchiness rating of 4.11, and taste rating of 4.02.

4 Conclusions

Based on the research results, it can be concluded that the different drying treatments significantly affect the protein content, expandability, and organoleptic qualities of cowhide crackers. Drying the cracker in an oven for 4 hours yielded the best results with a protein content of 65.87%, expandability of 530.42%, color score of 4.01, crunchiness score of 4.11, and a taste score of 4.02.

Acknowledgements for the Brawijaya University Research and Community Service Institute through Dana Penerimaan Negara Bukan Pajak (PNBP) following the DIPA of Brawijaya University Number: DIPA-FPt-251101/2023-1 on 13 January 2023.

References

1. R.T, Indrawati, F.T. Putri, R.A. Rochmatika, H. Prawibowo, *Increased Production Capacity Through the Design and Construction of a Semi-Automatic Machine for Cutting Crackers Dough*, Journal of Mechanical Engineering, **16**, 437-446 (2021)
2. A. Setiawan, R. Arifin, S. Asiyah, *The Influence of Price, Product Quality, Distribution Channels, Brand Image, Brand Trust, and Promotion on the Purchase Decision of Lestari Eco Sembung Tulungagung Rambak Crackers*, Electronic Journal of Management Research, **9**, 138-155 (2020)
3. Arianti and Waluyati, *The Analysis of Added Value and Development Strategy of Red Sugar Agroindustry in Madiun Regency*, Jurnal Ekonomi Pertanian dan Agribisnis (JEPA), **3**, 20-25 (2019)
4. F.B. Lilir, C.K.M. Palar, N.N. Lontaan, *The Effect of Drying Duration on the Processing of Cowhide Cracker*, Zootec, **41**, 214-222 (2021)
5. M. Pratama, M. Baits, N.A.A.R. Saman, *Analysis of Protein and Fat Levels in Smoked Halfbeak Fish (Hemiramphus far) in Kayoa Subdistrict, North Maluku, Using the Kjeldahl and Gravimetric Methods*, As-Syifaa, **6**, 178-186 (2014)
6. R. Chaniago, D. Lamusu, L. Samaduri, *Combination of Wheat Flour and Tapioca Flour on the Expansion and Organoleptic Properties of Terubuk Crackers (Saccharum edule Hasskarl)*, Food Processing Journal, **4**, 1-8 (2019)
7. I.H.L. Lay, B. Sabtu, H. Armadianto, *The Effect of Boiling Time on the Physical, Chemical, and Organoleptic Properties of Pork Skin Crackers*, Journal of Tropical Animal Science and Technology, **3**, 10-20 (2021)
8. W. Setiaboma, D. Desnilasari, A.C. Iwansyah, et al, *Chemical Characterization and Organoleptic Testing of Catfish (Arius thalassinus, Ruppell) Meatballs with Fresh and Steamed Moringa Leaves*, BIOPROPAL Industry Journal, **12**, 9-18 (2021)
9. M. Sari, Ernawati, R. Wilis, *Study of the Production of Cowhide Crackers in Kapalo Koto Village, Tanjung Barulak Subdistrict, Tanah Datar Regency*, Buana Journal. **2**, 167-179 (2018)
10. D. Rahayu, *Local Food Technology: Petos Chips, Milk Tofu, Rambak Crackers, Sagon Tungku, Crang Madu, Tiwul, Wajik Cake & Asinan*, Central Java: Pustaka Rumah Cinta (2021)
11. R.S. Widia, A.B. Amhar, Dzarnisa, *The Effect of Soaking with Bromelin Enzyme on the Quality and Organoleptic Test of Cowhide Crackers*, Scientific Journal of Agricultural Students. **7**, 270-278 (2022)
12. D. Sundari, Almasyhuri, A. Lamid, *The Effect of Cooking Process on the Nutrient Composition of Protein Sources*, Media Litbangkes, **25**, 235-242 (2015)

13. D. Zulisyanto, P.H. Riyadi, U. Amalia, *The Effect of Steaming Dough on the Physical and Chemical Quality of Catfish (Clarias gariepinus) Crackers*, Food Processing and Resulting Biotechnology Journal, **5**, 26-33 (2016)
14. T.S. Nugroho, U. Sukmawati, *The Effect of Drying Method on the Expansion and Organoleptic Value of King Prawn Crackers (Paneaus Monodon)*, Manfish Journal, **1**, 111-112 (2020)
15. N. Wahyuningtyas, B. Basito, W. Atmaka, *Study on the Physicochemical Characteristics and Sensory Crackers A Material Wheat Flour, Tapioca Flour, and Kepok Yellow Banana Flour*, Food Science Journal, **3**, 76-85 (2014)
16. D.W. Setiawan, T.D. Sulistiyati, E. Supriyatno, *Utilization of Snakehead Fish (Ophiocephalus striatus) Meat Residue in Making Albumin Fish Crackers*, THPi STUDENT JOURNAL, **1**, 21-32 (2013)
17. V.K. Dinda, A. Faridah, R. Holinesti, *Analysis of the Quality of Shrimp Skin Crackers*, HOME ECONOMICS AND TOURISM, **15**, 1-14 (2017)
18. E. Susanti, B. Saragih, Yuliani, *The Effect of Composite Wheat Flour and Banana Flour Formula on the Sensory, Expansion Power, Color Intensity, and Carotenoid Content of Pumpkin Donuts*, Journal of Tropical Agrifood, **3**, 79-85 (2021)
19. T.D. Kusuma, T.I.P. Suseno, S. Surjoseputro, *The Effect of Proportion of Tapioca and Wheat Flour on the Physicochemical and Organoleptic Properties of Celery Crackers*, Food Technology and Nutrition Journal, **12**, 17-28 (2013)
20. S. Listyarini, Asrani, J. Santoso, *Lele Dumbo Catfish (Clarias gariepenus) Protein Concentrate Waste in Melarat Crackers to Achieve Sustainable Development Goals*, Mathematics, Science, and Technology Journal. **19**, 106-113 (2018)
21. Hastian, E. Basir, *The Influence of Flour Type on Carbohydrate Characteristics and Organoleptic Properties of Tempeh Stick Crackers*, Sultra Journal of Economic and Business, **3**, 79-92 (2022)
22. F.L. Widyawati, D.K. Suwardiah. N. Purwidiani, I.F. Romadhoni, *The Effect of the Proportion of Milkfish (Chanos chanos) and Catfish (Clarias sp) on the Organoleptic Properties of Amplang Crackers*, Tata Boga Journal, **11**, 94-104 (2022)

Customer satisfaction of goat kids using Customer Satisfaction Index and Importance Performance Analysis (case study in goat production farm of East Java Livestock Agency)

Ahmad Zarkasi Efendi[1*], *Priyo Sugeng* Winarto[1], and *Kuswati* Kuswati[2]

[1]Department of Livestock Agribusiness, Faculty of Animal Science, Universitas Brawijaya, Jl. Veteran, Malang 65145, Indonesia
[2]Department of Animal Production, Faculty of Animal Science, Universitas Brawijaya, Jl. Veteran, Malang 65145, Indonesia

Abstract. This study aims to analyze customer satisfaction of goat kid and service quality attributes that are a priority to increase customer satisfaction. This research was conducted at Goat Production Farm of East Java Livestock Agency in Toyomarto Village, Singosari District, Malang, East Java, which is a producer of Etawa Crossbreed and Senduro goats. Respondents of this research were carried out using the total sampling method of buyers of goat kid from 2021 to 2023 with a total of 16 respondents. The research variables represent the elements of the marketing mix known as the 7P (product, price, place, promotion, process, physical evidence, and people) which are divided into 17 attributes. Primary data in this study were obtained through surveys and interviews assisted with instruments in the form of questionnaires. At the same time, the secondary data in this study were performance reports and literature. The results showed a customer satisfaction index of 74.02% in the good category. The main priority for performance improvement can be made to improve the genetic quality of parents in order to produce good quality offspring, completeness of sales supporting documents, adequacy of sales promotion information, after-sales service in the form of customer technical assistance. Furthermore, the mechanism determines the price of goat kid, the ease of the payment process through receipts, the availability of non-cash payments, and the existence of proof of payment accompanied by agency identity.

1 Introduction

The interest of the people of East Java in developing goat livestock agribusiness is very high. Livestock and Animal Health Statistics data shows that the goat population in East Java is recorded at 3.89 million heads, which is one level below Central Java Province which has the largest national goat population, namely 3.96 million heads [1]. The national goat population at the same time was 19.39 million heads, which means that the contribution of East Java

* Corresponding author: zarkasi@student.ub.ac.id

Province to the national population was 20.07 percent, while that of Central Java Province was 20.45 percent.

The high interest of the community in goat agribusiness needs support to guarantee the provision of quality and standardized goat kids. This is the mandate of Law Number 41 of 2014 concerning Animal Husbandry and Health as stated in article 13 (5) [2]. One of the goat kid producers in East Java is the Goat Production Farm of East Java Livestock Agency which is under the hierarchy of the East Java Province Livestock Service. As a producer, Goat Production Farm produces breeds of the Etawa breed of goats and Senduro lines. These two commodities are dual-purpose type goats as both meat and milk producers.

As a production unit with the core business of breeding local goats, Goat Production Farm of East Java Livestock Agency is expected to be able to produce effectively and efficiently through the use of its resources. As a service unit, Goat Production Farm has distributed goat kid to the community. Based on identification at the pre-research stage, there were no repeat purchases of goat kid from customers. Repeat purchase behavior is the activity of purchasing product carried out more than once or several times by customers. Repeat purchases are consumer behavior that is influenced by at least three things, namely promotion, price and product [3]. These three things are closely related to customer satisfaction so they are important to use to evaluate service performance and increase competitiveness. Customer satisfaction is an aftersale evaluation through a comparison between customer expectations and satisfaction with the product. Meanwhile, dissatisfaction arises if the results of the purchase do not meet expectations.

The marketing mix is a marketing strategy that has several elements in marketing. Initially, the marketing mix only had four elements, namely product, price, place and promotion. This concept can also be known as 4P which is only product oriented. Now, the marketing mix concept is not limited to 4P, but is even more complex so that it becomes 7P which is not focused on products, but rather accommodates relationships with the market or target customers. The concept of seven elements which is the basis for designing a marketing strategy consists of product, price, promotion, place, people, process and physical evidence [4].

Products are basic elements as objects offered to customers. The quality factor is the main thing for a product because the higher the quality of a product, the easier it is to sell and produce a positive experience for customers. Price is the economic value of a product. Price also means the amount of money that customers have to pay when they want to get the product. Price is a form of consumer sacrifice to obtain the desired product. Prices are determined through capital and production costs. Meanwhile, the selling price takes into account a number of profits targeted by the seller as the product owner. Place is a marketing mix element related to business location. A strategic location can improve the image of the seller and product and make it easier for consumers to reach the product. Suitability of location to target consumers can maximize sales and profits. Promotion is an important part that connects business people as product owners with consumers in the market. Promotion aims to attract attention and convince potential consumers to buy and use the product. People is an element of the marketing mix concept in the form of human resources. Good quality human resources will have a good impact on a company because they have the potential to have good performance. Process is an element related to the handling of business actors in serving consumers. This section starts from ordering until the product reaches the consumer. Physical evidence or physical appearance is related to the image or appearance of the company which is represented by the design of the business premises, product packaging and the image displayed is part of the physical appearance. Marketing strategy through the marketing mix concept is important for creating and maintaining customer loyalty. Customer loyalty can be created through good product quality, appropriate prices, good service and being superior to competitors [5].

There is a positive and significant relationship between the marketing mix and consumer purchase intentions for dairy products. The research results can be a basis for marketing companies to develop better products, reasonable prices, promotional strategies and appropriate sales locations to attract the interest of dairy product consumers [6]. Marketing is an effective way to stimulate consumers interests, desires and actions. The most efficient marketing actions could be mentioned as the marketing mix paradigm [7]. These actions consider 7Ps to influence the consumers: place, price, product, promotion, people, physical evidence, and process. The most influential marketing mix variable was price. The best strategy to increase purchasing decision of Malang Apples at modern market is odd price.

As a producer that produces quality and standardized goat breeds, Goat Production Farm is expected to provide a wide range of benefits to customers and breeders. The expected benefits are in the form of increasing the genetic quality of goat kids circulating in the community through dissemination or sales. Based on goat production data, it is known that the production capacity in 2021 was 89 heads, in 2022 was 87 heads and in 2023 (temporary data) was 26 heads. Unfortunately, this production capability is not matched by good marketing capabilities so that in 2021 only 17 are marketed, in 2022 34 are marketed, and in 2023 19 are marketed. One of the factors causing the low ability to market goat kids can be indicated because of the unsatisfactory quality of service.

Considering what is known from previous research, this research seeks answers to the following questions: How do customer satisfaction and service quality affect the 7P marketing mix at Goat Production Farm? What marketing variables need to be improved? This study aims to analyze customer satisfaction of Etawa Crossbreed and Senduro goat breeds as well as service quality attributes that are a priority to increase customer satisfaction of goat breeds in the future.

2 Research Methodology

2.1 Research Location and Time

This research was carried out at Goat Production Farm of East Java Livestock Agency located in Toyomarto Village, Singosari District, Malang Regency. The choice of location was made deliberately with the consideration that Goat Production Farm is a producer of Etawa Crossbreed and Senduro goat breeds which manages goats with a dynamic population of around 450 heads per month [8]. Data collection was carried out from July to August 2023.

2.2 Research Respondents

Respondents in this study consisted of sixteen goat kid customers. Determining the number of respondents in this study was carried out by the total sampling method. Total sampling is a sampling technique according to the population size because the population number is less than one hundred [9].

2.3 Research variable

This study aims to measure the level of Goat Production Farm performance and the level of customer importance in goat kid sales service. The variables in this study include the marketing mix with the 7P approach to identify overall customer satisfaction consisting of product, price, place, promotion, process, physical evidence, and people [10].

In this study, the level of satisfaction and importance of each variable was measured using a questionnaire containing seventeen questions with a scale range of 1-4 with the lowest value

(1) meaning very dissatisfied or very unimportant to the highest (4) meaning very satisfied or very important. The variables measured in this study are translated into measurable indicators as shown in Table 1.

Table 1. Variables and indicators for measuring the level of performance and interests of Goat Production Farm goat kid customers

Variable	Attribute code	Indicators
Product	1	Goats Parents quality (performance of production and reproduction)
	2	Goats kids quality (performance of production and reproduction)
	3	Completeness of supporting documents
Place	4	Affordability of office and stable locations
	5	Time (hours and working days) of service
Promotion	6	Sufficiency of information
	7	Availability of dedicated presentation sessions
	8	After-sales service
Price	9	Conformity of price and product quality of goat kids
	10	Pricing negotiation mechanism
People	11	Politeness and friendliness of sales service officers
	12	Willingness to share complete information on goat kids
	13	Officer honesty
Process	14	Ease of payment process
	15	Speed of service process
	16	Goat seed delivery service
Physical Evidence	17	Cleanliness and comfort of the environment and production facilities

2.4 Data source

The data collected in this study are primary and secondary data. Primary data is data or information obtained directly from the object under study, which can be in the form of direct interviews, surveys and questionnaires. While secondary data is data obtained indirectly from the object under study can be in the form of organizational reports, documents, or literature [9].

Primary data collected in this research is related to the 7P marketing mix variables through interviews with sixteen selected respondents. At the same time, secondary data in this research is in the form of organizational profiles, livestock populations and performance reports.

2.5 Data analysis

2.5.1 Importance and Performance Analysis

Importance Performance Analysis or can be abbreviated as IPA was first introduced by Martilla and James in 1997. This method measures the relationship between consumer perceptions and priorities for improving the quality of production or service. This method is also known as quadrant analysis. Calculation of this analysis obtained the level of compatibility between importance and performance that will be applied to develop management strategies to increase customer satisfaction [11].

Measuring the level of importance and level of performance of a service quality attribute is carried out using a scale with a score range of 1-4 for each attribute. The lowest value in the form of a score of 1 indicates the level of performance or importance is very unimportant or very dissatisfied. While the highest value in the form of a score of 4 indicates the level of performance or importance is very important or very satisfied.

The first step in measuring IPA is to calculate the total score for the level of importance and level of performance for each attribute for all respondents. The next step is to calculate the average level of importance and level of performance for each attribute to obtain the level of suitability (in %) for each attribute. The next stage is to create a Cartesian diagram consisting of four quadrants bounded by 2 vertical cutting lines at points X and Y. Mapping of the 7P attributes in each diagram is carried out based on the average value of satisfaction and importance. This is done to show an overview of the attributes that need to be improved and maintained in order to obtain maximum satisfaction from consumers.

The four quadrant matrix has the following meaning [12]:

- Quadrant I is the Concentrate Here. In this quadrant, an attribute is considered very important by customers but the service is not satisfactory, so the company's service quality must be improved;
- Quadrant II is Keep Up the Good Work. In this quadrant, an attribute is considered very important by customers and the service is very satisfying, so the company's service quality must be maintained;
- Quadrant III is Low Priority. In this quadrant, an attribute is considered unimportant by customers and the service is less than satisfactory. Thus, attributes in this quadrant receive low priority in improving service quality and have very little influence on customer satisfaction;
- Quadrant IV is Possible Overkill. In this quadrant, an attribute is considered unimportant by customers but the service is satisfactory. One of the treatments in this quadrant is to reduce expenditure costs for this type of attribute service so that it is more economical.

Table 2. Quadrants in Importance and Performance Analysis (IPA)

Quadrant I Concentrate Here	Quadrant II Keep Up the Good Work
Quadrant III Low Priority	Quadrant IV Possible Overkill

2.5.2 Customer Satisfaction Index

Customer Satisfaction Index is one of the main measurement tools in integrating company performance because it is closely related to company profits. Basically, customer satisfaction is an aggregate number that describes customer assessments of several elements of company performance [13]. According to [14] stated that the CSI method has several advantages, namely efficient, easy to use, and uses a scale that has high sensitivity and reliability. This measurement is useful for knowing customer satisfaction and can determine future improvement targets.

According to [15] the customer satisfaction index value can be obtained by calculating the total average of the importance column (I) to obtain Y. The next step is to multiply the importance items (I) with performance (P) in the score column (S). The results of the multiplication are then added together to form T. The customer satisfaction index is obtained from the calculation of (T/4Y) x 100%. The value 4 on 4Y is the maximum value on the measurement scale.

Table 3. Example of calculating the customer satisfaction index

Attribute	**Importance (I)**	**Performance (P)**	**Score (S)**
	1-4	**1-4**	**S = (I) x (P)**
1			
2			
3			
Etc			
Total score	Total (I) = (Y)		Total (S) = (T)

$$Customer\ Satisfaction\ Index = \frac{T}{4Y} x100\ \%$$

Notes:

T : The total importance score multiplied by the performance score

4 : Measurement scale

Y : Total importance score

Determining the satisfaction category value is carried out through several steps. The first is to determine the lower limit value of the interval according to formula 1 [12].

$$The\ lower\ limit\ value\ of\ the\ interval = \frac{Number\ of\ respondents\ x\ smallest\ score}{Number\ of\ respondents\ x\ biggest\ score} x\ 100\%$$

$$\frac{16\ x\ 1}{16\ x\ 4} x\ 100\% = 25\%$$

The next step is to determine the difference in range values using formula 2 [12].

$$difference\ in\ range\ values = \frac{100\% - lower\ limit}{number\ classes}$$

So, the satisfaction category values can be arranged as in the table

Table 4. Satisfaction category value

Interval	**Performance quality**
25 % - 44 %	Very Not Good
45 % - 64 %	Not good
65 % - 84 %	Good
> 85 %	Very good

3 Results and Discussion

3.1 Respondent Profile

The number of respondents in this study are 16 people. The profile of the respondents in this study is shown in Table 5.

Table 5. Respondent profile

Profile	Total (person)	Percentage (%)
Education		
- Elementary school	6	37
- Junior high school	1	6
- Senior high school	6	37
- Bachelor	3	20
Occupation		
- Paramedic veteriner service	1	7
- Farmer	3	20
- Teacher	2	13
- Animal Farmer	9	60
Livestock ownership		
- < 10 ekor	5	31
- 11-25 ekor	6	37
- 26-50 ekor	3	19
- >50 ekor	2	13
Experience		
- < 1 tahun	1	6
- 1-5 tahun	12	75
- 5-10 tahun	1	6
- >10 tahun	2	13
Motivation		
- Primary business	2	12
- Non primary business	7	44
- Saving	7	44
Purpose of livestock production business		
- Kids production	11	69
- Milk production	5	31

Based on the table, it is known that the education level of the most respondents is elementary and high school with six people each. The majority of respondents work as farmers as many as nine people. However, only two people out of all respondents made livestock farming their main business, which was especially closely related to the number of livestock owners of more than fifty heads. While each of the seven people make livestock as a nonprimary business and savings. The majority of respondents owned less than 25 livestock, namely eleven people. The majority of respondents claimed to have 1-5 years of farming experience as many as twelve people. While the purpose of the goat livestock production business, as many as eleven respondents answered raising goats for kid production and five respondents answered for milk production. Farming experience and number of livestock ownership influences motivation to farming. At the same time, age, education level and number of family have no effect on motivation to livestock farming [16]. A study of goat farmer in Kuwarasan Village, Kebumen Regency, shows that the farmers are moderately motivated. The characteristics of farmers represented in age, education, farming experience, number of families and number of livestock ownership do not show any influence on motivation to raise goats [17].

3.2 Customer Satisfaction Index

Based on the description above, the customer satisfaction index can be used to determine the level of customer satisfaction from a service as a whole through analyzing the level of importance of product attributes. After analyzing the customer satisfaction index, it is known that the value produced in this research is 74.02%, which means the level of customer satisfaction is in the range of 65% - 84% with the service performance predicate in the good category. This means that overall customer satisfaction with the product is classified as satisfied. However, service performance can still be improved to obtain an excellent title by improving the quality of several attributes so that customers feel very satisfied. Finding result show the same case happened to South Bandung Livestock Cooperative. As the producer of milk product, the consumers of liquid milk products are satisfied with a satisfaction index of 78.8%. The cooperative must improve promotion, packaging design and nutritional content need to be improved and its performance improved [18].

3.3 Importance and Performance Analysis

To determine the importance and performance of all attributes, it is necessary to analyze the level of importance and performance. The output of this analysis is operational suggestions and improvements to attributes that have low performance but have high importance. This analysis is used to calculate the value of X (average performance score) and Y (average importance score) and Tki (level of conformity). The level of suitability is a comparison between the performance score and the importance score which can be used to determine the priority scale for service improvement. The results of calculating performance scores, interests, averages and suitability levels are shown in Table 6.

Based on the calculation results, the results show that thirteen attributes, namely attributes 1, 2, 4, 5, 7, 9, 10, 11, 12, 13, 15, 16 and 17 have a 100% suitability level. While attributes 6 and 14 have a suitability level of 97.78% and 95.83%. In addition, there are two attributes that have low scores, namely attribute 3 with a suitability level of 35.8% and attribute 8 with a suitability level of 42.19%. Attribute 3 describes the completeness of supporting documents which shows that most of the respondents expressed interest in these supporting documents, but the producers had not fully complied with them. Meanwhile, attribute 8 describes after-sales service which shows that most customers express an interest in after-sales service, but manufacturers do not provide this service. Finally, in general, the level of suitability of performance and the importance of each attribute is 93.73% which means that the performance of service quality is very good.

The next stage is creating a performance (x) and importance (y) quadrant matrix. The horizontal line on the y-axis is formed from the mean of importance which is 2.26 which extends on the x1(0) and x2(4) axes. While the vertical line on the x-axis is formed from the average performance of 3.17 which extends on the y1 (0) and y2 (4) axes. The results of mapping on this quadrant matrix can be used to determine the position of attributes that need to be repaired or need to be maintained so that customer satisfaction can be achieved. The results of the mapping of the performance and interest quadrant matrix are shown in Fig 1.

Table 6. The results of the calculation of the level of importance and performance analysis

Attribute	Performance (X_i)	Importance (Y_i)	Average (P)	Average (I)	Conformity (Tk_i)	Score (I x P)
Product						
1	51.00	43.00	3.19	2.69	100.00	8.57
2	48.00	41.00	3.00	2.56	100.00	7.68
3	19.00	53.00	1.19	3.31	57.58	3.93
Place						
4	64.00	18.00	4.00	1.13	100.00	4.50
5	48.00	19.00	3.00	1.19	100.00	3.56
Promotion						
6	44.00	45.00	2.75	2.81	97.78	7.73
7	60.00	34.00	3.75	2.13	100.00	7.97
8	27.00	64.00	1.69	4.00	42.19	6.75
Price						
9	47.00	34.00	2.94	2.13	100.00	6.24
10	45.00	39.00	2.81	2.44	100.00	6.86
Producer						
11	64.00	34.00	4.00	2.13	100.00	8.50
12	52.00	41.00	3.25	2.56	100.00	8.32
13	64.00	31.00	4.00	1.94	100.00	7.75
Proceses of sales						
14	46.00	48.00	2.88	3.00	95.83	8.63
15	64.00	29.00	4.00	1.81	100.00	7.25
16	54.00	26.00	3.38	1.63	100.00	5.48
Physical evidence						
17	64.00	17.00	4.00	1.06	100.00	4.25
Total				38.5		113.98
Average			3.17	2.26		

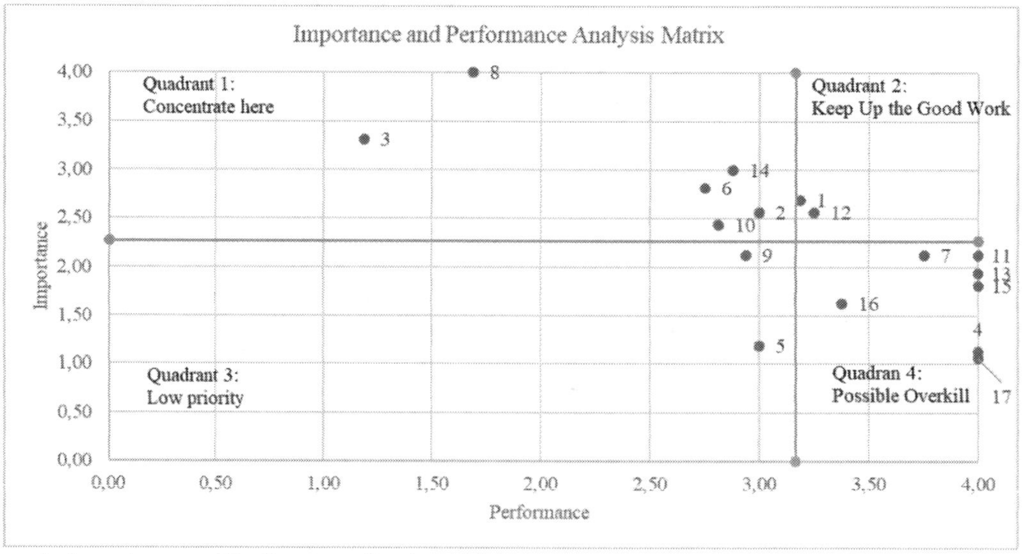

Figure 1. Importance and Performance Quadrant Matrix
Source: Data processed, 2023

Based on Figure 1, there are four quadrants resulting from the intersection of the average interest and performance. In quadrant 1 there are six attributes that need to be given top priority attention, namely attributes 2, 3, 6, 8, 10 and 14. Quadrant 1 is the result of mapping which shows a high level of customer interest, but the level of performance or customer satisfaction is low so it needs to be done repair. In quadrant 2 there are two attributes that fall into the category of maintaining performance, namely attributes 1 and 12. Quadrant 2 represents the same high level of importance and performance for customers and must still be maintained. In quadrant 3 there are two attributes, namely 5 and 9. which need to be given low priority or can be ignored temporarily because they have very little effect on customer satisfaction. In quadrant 4 there are seven attributes, namely 4, 7, 11, 13, 15, 16, 17. Quadrant 4 represents excessive importance and performance, which means it is not very important but has good satisfaction.

After mapping based on quadrants, it is necessary to develop improvement efforts to increase the efficiency and effectiveness of performance and service, especially on the attributes that are in quadrant 1. Some efforts that can be made on the attributes of quadrant 1 include:

1. Attributes 2 and 3 represent product variables. Attribute 2 is goat kid quality (performance of production and reproduction) as well as the number of kid at birth. The proposed improvement on this attribute is the need to replace the parents, both male and female parents, who have entered the age of culled. The older the etawa crossbreed goat, the lower the quality of the semen. The best quality semen is found in etawa crossbreed goats aged 3-4 years. So, the use of males in natural mating is recommended at the age range of 3-4 years to obtain good quantity and quality of semen [19]. Attribute 3 is the completeness of sales supporting documents consisting of a travel document, a health certificate from an authorized veterinarian, and a certificate of proper breed. The proposed improvement to this attribute is the inclusion of supporting documents in every sale to all customers or buyers. The age of the doe determines productivity in the form of litter size. In addition, the older the parent, the ability of the ovaries to produce ova decreases in quality and quantity, which has an impact on reducing the quality of the kids [20];
2. Attributes 6 and 8 represent promotions. Attribute 6 is the adequacy of information during sales promotion consisting of goat breed or strain, sex, age or date of birth, pedigree, growth potential/production of meat and milk. The proposed improvement on this attribute is the need for a system and information materials that can be presented in the form of a brief presentation to customers of goat seeds. These information systems and materials are related to production data in the form of growth in body weight or milk production as well as other production data related to stock availability of ready-to-sell goat breeds, especially related to transparency of data on age and pedigree. Attribute 8 is after-sales service which consists of technical guidance on maintenance management and recording and quality control of goat kids at least once after purchase. The proposed improvement on this attribute is that there is a need for an after-sales service program for customers, especially after lambing and the kids are more than 3 months old until they enter the productive breeding age. Promotions influence repeat purchases. Promotion can build communication and can have a good influence on both parties by building trust and avoiding suspicion. As a business actor, it is necessary to increase attractive sales promotions in a business unit so that it can influence consumers and make repeat purchases in the future [21];
3. Attribute 10 represents price. Attribute 10 is a price negotiation mechanism which consists of price negotiation, the price requested by the customer is considered by the producer and the price accommodates additional needs proposed by the customer. The proposed improvement to this attribute is the existence of regulations or rules that can accommodate producers to sell goat kids at more flexible auction prices. In this case, flexible means that in determining prices there is a lower limit on offers and customers

can determine the reasonableness of the price of the individual goat kids offered. The price set by a company is one of the factors that influences consumers repurchase interest. Price is the amount of money charged for a product or service [22].

4. Attribute 14 is the ease of the payment process which consists of a commitment to complete or pay-off, cash or non-cash payment, payment via the receipt treasurer or direct deposit via a virtual account, as well as proof of payment and settlement containing the identity of the agency and the official name of the officer. The proposed improvement to this attribute is that there needs to be a transparent and accountable mechanism in regulating payment transactions so that there is no potential for fraud or losses incurred in the sales process. The process of sales, as well as the producers, are the most important marketing mix elements and receive strong indications of customer satisfaction for short food supply chains. So that emphasis should be placed both on the processes of sales and producers [5].

While the attributes in quadrant 2 are attributes that need to be maintained. This quadrant means that there are factors that are considered important and are expected to be one of the factors that support customer satisfaction so that it is mandatory to maintain this performance achievement [23]. The attributes in this quadrant are attributes 1 and 12. Attribute 1 is an assessment of the production and reproductive performance of goat kids, consisting of quantitative and qualitative performance, as well as male and female reproductive characteristics. At the same time, attribute 12 is an assessment of the willingness to share information about goat kid products to be purchased, which consists of conveying the stock of the number of goat kids available, conveying information on the productive age of goats, conveying information on the advantages and disadvantages of Etawa Crossbreed or Senduro goat kid products.

In quadrant 3 there are two attributes, namely attributes 5 and 9. This quadrant shows that the attributes in this position need to receive low priority. Attribute 5 is the days and hours of service and the frequency of visits to buying goat kid. While attribute 9 is the suitability of the price with the product quality of goat seeds. This quadrant indicates that satisfaction with this service attribute is low and is considered not too important for customers. Thus, the organization does not need to pay attention to these attributes.

In quadrant 4 there are seven attributes, namely 4, 7, 11, 13, 15, 16, 17. Quadrant 4 can be redundant which means that in this quadrant there are factors that are not too important and are not really expected by customers, so that the focus of the organization can be shifted on other attributes that have a higher priority level. Attribute 4 is good access to Goat Production Farm locations and is affordable by private vehicles. Attribute 7 is the availability of special presentation sessions for goat breed customers on the dimensions of place, time, use of teaching aids and discussions and questions and answers. Attribute 11 is the politeness and friendliness of the sales service staff. Attribute 13 is the officer's honesty. Attribute 15 is the speed of the service process in selecting goat seeds less than 2 hours from arriving at the location. Attribute 16 is the availability of goat seed delivery service and the agreement about it. Attribute 17 is the cleanliness and comfort of the environment and production facilities in the form of odor-free, piles of garbage and dirt, and free of noise that disturbs comfort.

4 Conclusion

4.1 Conclusion

1. Based on the customer satisfaction index, the level of customer satisfaction is 74.02% or the predicate is "good". The satisfaction index is measured through marketing mix elements commonly known as the 7Ps, namely product, price, place, promotion, process,

physical evidence, and people. However, there are still opportunities to improve service quality to the very good category by improving the service quality of several attributes.
2. Based on the results importance and performance analysis, there are six performance attributes that must be improved. First, is the quality of offspring through production performance, reproduction and the chance of multiple births. Second, completeness of sales supporting documents. Third, the adequacy of sales promotion information. Fourth, after-sales service in the form of customer technical assistance. Fifth, the mechanism for determining the price of goat seeds. Sixth, the ease of processing payments through receipts, the availability of non-cash payments, as well as the existence of proof of payment accompanied by the agency's identity.

4.2 Recommendation

Increasing the customer satisfaction index can be done through improvements to service quality as follows:
1. Evaluate the use of male and female parents so that production and reproductive performance as well as the chance of multiple births can increase according to customer expectations.
2. Include sales supporting documents in each kid sales transaction process.
3. Provide a complete explanation of the available goat kids stock including breed/line of goat kids, gender, age or date of birth, pedigree, as well as meat growth potential or milk production potential.
4. Provide technical assistance to customers at least once in terms of good farming practices management and recording.
5. Improve the pricing mechanism through a negotiation process so that it can accommodate customer needs.
6. Increase the convenience of payment methods through agreement completion, availability of non-cash payments, as well as transparent and accountable proof of transactions.

References

1. Direktorat Jenderal Peternakan dan Kesehatan Hewan Kementerian Pertanian RI. Statistik Peternakan dan Kesehatan Hewan (Kementerian Pertanian RI, Jakarta, 2022)
2. Law No. 41 of 2014 concerning Livestock and Animal Health
3. E. Permatasari, H. Luthfiana, N. A. Pratama dan H. Ali. J Ilmu Manaj Ter **3** (5): 469 (2022)
4. S. G. Dhita, Mukson, A. Setiadi. JEPA **6** (1), 175 (2022)
5. A. Tiganis, E. Grigoroudis and P. Chrysochou. Food Qual and Pref, 104: 1 (2023)
6. M. S. Farid, A. Cavicchi, M. M. Rahman, S. Barua, D. Z. Ethen, F. A. Happy, M. Rasheduzzaman, S. Sharma, M. J. Alam. Heliyon. **9** (2023)
7. C. N. Leonidou, C. S. Katsikeas, N. A. Morgan, J. Acad. Marketing Sci **41**, 151 (2012)
8. Dinas Peternakan Provinsi Jawa Timur. Laporan Kinerja Tahun 2022 (Dinas Peternakan Provinsi Jawa Timur, Surabaya, 2022)
9. Sugiyono. Metode Penelitian Kuantitatif. Kualitatif dan R&D (Alfabeta, Bandung, 2017)
10. E. Constantinides, J. of Market Manag **22** (3): 407 (2006)

11. M. Pranata, A. Hartiati and C. Sadyasmara, J. Rekay dan Manaj Agroindustri 7(4): 594 (2019)
12. H. Syahputra, A. Ramadhanu, R. B. Putra, J. Eko dan Manaj Sist Info. **1**(4): 334 (2020)
13. M. Goic, C. Levenier. and R. Montoya, J of Retailing and Consumer Services. **60** (2021)
14. L. D. Anggraini, P. Deoranto, dan D. M. Ikasari. J. Fak Tek Industri Pertanian Universitas Brawijaya. **4**(2): 74 (2015)
15. I. Sa'adah, Mukson, dan Y. S. Ondho, J. Eko Pert dan Agri (JEPA). **3**(3): 557 (2019)
16. Nurdayati, NI Fidin and Supriyanto. J. Agri Ext Devt, **17**(32): 121 (2020)
17. E. T. W. Utami, N. Nuraeni, WS Ashar, I. Faelasuf. J. Saintek, **5**(1): 305 (2023)
18. M. Yulita, DAH Lestari, D. Haryono. JIIA **2**(2): 158 (2014)
19. E. Heriyanta, M. N Ihsan and N. Isnaini, J. Tropical Livestock **14** (2): 1 (2013)
20. A. Sudrajat, IGS Budisatria, S. Bintara, ERV Rahayu, N. Hidayat and RF Christi, J. Animal Science, **21**(1): 27, 2021
21. M. Khomilah, JEMS, **1**(2):202 (2020)
22. P. Kotler, LC Harris, G. Armstrong, N. Piercy, Principles of Marketing (Pearson Publisher, UK, 2017)
23. Y. Siyamto, JIEI, **03** (1): 63–76 (2017)

Successful artificial insemination of sexed semen using albumin sedimentation on Friesian Holstein cows

Habib Asshidiq Syah[1], *Nurul* Isnaini[2], *Aulia* Puspita Anugra Yekti[2], *Muhaimin* Rifai'i[3], and *Trinil* Susilawati[2*]

[1] Post Graduate Program, Faculty of Animal Science, Universitas Brawijaya, Malang 65145, Indonesia
[2] Faculty of Animal Science, Universitas Brawijaya, Malang 65145, Indonesia
[3] Faculty of Mathematics and Natural Sciences, Universitas Brawijaya, Malang 65145, Indonesia

Abstract. This study aimed to determine the success of artificial insemination (AI) using unsexed and sexed frozen semen with albumin sedimentation. This research was conducted from August to December 2022 at a farmer community in Pandesari Village, Pujon District, Malang Regency. The material used in this study was 78 heads of Friesian Holstein cattle, with the criteria of having a minimum body condition score of 2.5 (scale of 1–5), having normal reproductive organs, and showing signs of estrus. 38 cows were used as acceptors in artificial insemination (AI) with unsexed frozen semen, and 38 cows in AI with albumin sedimentation frozen semen. The variables used are non-return rate 1, non-return rate 2, and conception rate. The data obtained were then analyzed descriptively. The results showed that the percentage values of NRR1, NRR2, and CR in AI using frozen semen unsexed were 76.3%, 63.1%, and 47.3%, respectively, while the percentage values of NRR1, NRR2, and CR in AI using frozen semen sexed were 94.7%, 86.15%, and 63.15%. Based on the study's results, insemination success using albumin sedimentation frozen semen sexed showed better results than AI using unsexed semen.

1 Introduction

The demand for cow's milk in the country has increased along with the increase in population. Still, the consumption level of cow's milk in Indonesia is not supported by high milk production. Efforts can be made to increase cow's milk production by increasing the population of female dairy cows with reproductive technology, Artificial Insemination (AI). Artificial insemination has become one of the most essential techniques developed for the genetic improvement of farm animals [1]. The AI program can be improved by spermatozoa sexed, which aims to produce superior cattle with the appropriate sex for maintenance [2]. Spermatozoa sexed technology that can be used to increase the value of AI is sexed using the albumin sedimentation method. One method of sexed spermatozoa

* Corresponding author: tsusilawati@ub.ac.id

that is easy to use is Bovine Serum Albumin (BSA). Sexed spermatozoa using albumin is a method of separating spermatozoa that is easy to do and can produce spermatozoa with X and Y chromosomes between 71-76% [3]. Assessment of the reproductive success of AI cattle can be conducted by observing the parameters of Non-Return Rate (NRR) and Conception Rate (CR) [4]. This study aims to evaluate the success of AI using frozen semen sexed results with albumin sedimentation method in Friesian Holstein cows based on NRR1, NRR2, and CR, which are the evaluation parameters of artificial insemination success.

2 Materials and methods

This study was conducted in the working area of inseminator KOP SAE Pujon Pandesari Village, Pujon Subdistrict, Malang, East Java, and was conducted from August 13 to December 13, 2022. The materials used in this study were 76 female PFH cows. Samples were selected by purposive sampling with the requirements of BCS at least 2.5-3.5 (scale 1-5), having normal organs and showing signs of estrus. The cows were divided into 2 treatments, where 38 females were inseminated with non-sexed frozen semen, and 38 females were inseminated with sexed frozen semen using albumin sedimentation. AI was conducted by an inseminator in the local village and was done when there were reports from farmers that cattle showed signs of estrus. The inseminator inseminated using the single dose method at the 8th hour, by positioning the semen on the corpus utery. After AI, the cows were injected with 20 ml of ADE vitamin. Semen deposition treatment in position 4 in this study is expected to increase the birth of female calves in Friesian Holstein cows. Research conducted by Azmi *et al* [5] stated that AI with a single dose of frozen semen in position 4 or corpus utery obtained sex identification using the ultrasound method of 72.89% of female calves and 27.11% of male calves, while in double dose AI treatment received identification of 69.33% female calves and 30.67% male calves.

3 Variable

3.1. Non-Return Rate

Non-Return Rate 1 (NRR_1) is the percentage of cows that do not return to estrus on days 19-22 and 39-43 after AI. Cows that do not show signs of estrus are assumed to be pregnant; if the cow shows signs of estrus, it is assumed that the female is not pregnant. According to Iswoyo *et al* [6], NRR_1 can be calculated with the formula:

$$NRR_1 = \frac{\text{Total AI in cows} - \text{number of estrous cows}}{\text{Total cows in AI}} x\ 100\%$$

$$NRR_2 = \frac{\text{Total AI in cows} - \text{number of estrous cows (2nd cycle)}}{\text{Total cows in AI}} x\ 100\%$$

3.2. Conception Rate

Conception Rate (CR) is the percentage of mothers who become pregnant at the first insemination. The CR value is obtained from the results of acceptors who do not show signs of estrus in NRR_1 and NRR_2 observations and have carried out pregnancy checks using the rectal palpation method, CR calculation can be obtained by the formula [4]:

$$CR = \frac{\textit{Total cows pregnant at first insemination}}{\textit{Total cows in AI}} x\ 100\%$$

3.3. Body Condition Score

Body Condition Score (BCS) is a subjective assessment method to determine the energy reserves of cattle, this assessment is based on fat reserves in the back and flanks, the scale commonly used is 1-5 [7].

3.4. Age

Age determination can be done in two ways, based on birth records/interviews with breeders and permanent tooth replacement. The age determination method based on permanent teeth is one pair of permanent teeth = 18-24 months, two pairs of permanent teeth = 24-36 months, and 3 pairs = 38-44 months.

4 Results and discussion

4.1. Artificial Insemination Success Based on Non-Return Rate

Table 1. Observation Results of Non-Return Rate 1 and Non-Return Rate 2 of Friesian Holstein Cows Inseminated with Sexed and Non Sexed Frozen Semen

Treatment	Number of acceptors (head)	NRR_1		NRR_2	
		Number of acceptors (head)	%	Number of acceptors (head)	%
T1 (*nonsexed*)	38	29	76.3	25	63.1
T2 (*sexed* SDA)	38	36	94.7	33	86.15

Evaluation of the success of pregnant cattle can be seen from the appearance of signs of estrus at a certain period, the NRR value is used to measure the ability of the cow, when the cow does not show signs of estrus in the estrous cycle, then the acceptor is assumed to be pregnant [8]. Other factors include the ability of the inseminator to thaw [9], the ability of the farmer to recognize signs of estrus, and other external factors that affect the outcome of artificial insemination [10]. According to Susilawati [11], NRR observations can be observed on days 0-30 and 30-60 after AI, while in this study, NRR observations were observed on days 19-22 for NRR1 and days 39-43 for NNR2. Acceptors who showed signs of estrus in NRR1 observation were considered failed, while acceptors that didn't show signs of estrus will be observed in NRR2, if the acceptor showed signs of estrus, then the acceptor was considered failed, if it didn't show signs of estrus, then a pregnancy test will be conducted on day 60 after AI. The percentage of NRR in this study can be seen in table 1.

Based on table 1, the results of NRR observations in T1 (AI unsexed) obtained the percentage of NRR_1 and NRR_2 of 76.3% and 63.1%. Then, for the observation of NRR T2 (AI sexed), the rate of NRR_1 and NRR_2 was 94.7% and 86.15%. The lower NRR value in T1 than T2 was due to the number of old cows in T1. The increase in the number of parities causes the age of the cows to get older, leading to a decrease in the condition and reproduction of the cows [12]. Another factor causing the lower NRR at T1 was probably due to observation error [11]. According to Iswoyo *et al* [6], the NRR value can be categorized as good if it reaches 79.53%. The percentage of NRR of unsexed and sexed frozen semen in the study was higher than the research conducted by Rosita *et al* [13], namely the NRR_1 and NRR_2 values of unsexed frozen semen of 74% and 74%, while the percentage of NRR_1 and NRR_2 of sexed frozen semen was 70% and 66%. Based on table 1. the rate of NRR values at T1 and T2 decreased. According to Yekti, *et al* [14], the decrease in NRR value is caused by silent heat and early embryonic death. Silent heat is a condition where cattle do not show signs of estrus, but have a normal reproductive cycle, thus increasing the possibility that the estrus cycle will be missed from observation [15]. This is due to low estrogen, leading to no signs of estrus [14]. Silent heat can also be caused by heat stress [16].

4.2. Artificial Insemination Success Based on Conception Rate

Table 2. Conception Rate in Friesian Holstein Cows Artificially Inseminated with Albumin Sedimentation Sexed and Unsexed Frozen Semen

Treatment	Number of receptors (head)	CR	
		Number of pregnant cows	%
T1 (*nonsexed*)	38	18	46,15
T2 (*sexed* SDA)	38	24	63,15

Conception rate (CR) is the percentage of acceptors who successfully conceive in the first AI known after rectal palpation. Costa *et al* [17] stated that CR can be used as an indicator of female cattle fertility, CR results can be measured by rectal palpation 60 days after AI. The calculation of CR value in this study is based on pregnancy examination using the rectal palpation method on days 60-90 after AI.

Table 2 shows that T2 has a better CR value than T1, Priyanto *et al* [18] stated that the CR standard ranges from 60%-70%, so T2 has an ideal value, and T1 is still below the ideal value. The CR value in this study had a higher percentage than the CR value in a study conducted by Rosita *et al* [13] for AI using sexed frozen semen, which was 59%, while for AI using non-sexed semen in this study had a lower value of 74%. The decrease in NRR2 value to CR value can be caused by factors such as acceptor age, BCS, animal parity, estrus detection management, and cow physiology [19]. In addition, the success of AI is also influenced by the quality of spermatozoa used [20].

The interesting data in this study is that the CR value of acceptors AI using sexed frozen semen is higher than acceptors AI using non-sexed frozen semen. The low CR in T1 (unsexed) was also due to the large number of old acceptors, there were eight acceptors over six years old in T1, and 7 of them suffered pregnancy failure. According to Jainudeen *et al* [4], the older the cow, the lower the pregnancy rate. Susilawati [21] added that cows aged 6-7 years have lower fertility.

4.3. Artificial Insemination Success Based on Body Weight

Based on Table 3, acceptors with the highest percentage of pregnancy are acceptors with a BW of 300-400 for T1 and 401-500 for T2, this is because the BW is the ideal weight for cattle to be mated. According to Moran [22] cows with a body weight of 300 are the ideal weight for mating. Cows that are mated when the weight is not ideal will lead to difficulties during parturition. In the first mating, it is crucial to see that the cattle are mature and have reproductive maturity so that the resulting offspring have good performance. Furthermore, cattle that are mated when the weight is less than ideal will lead to difficulties during parturition. Mammary gland development is influenced by the hormone estrogen and progesterone produced by the ovaries under the influence of FSH and LH hormones [12].

Table 3. Percentage of Artificial Insemination Success Based on Conception Rate at Various Body Weights

Body Weight (kg)	T1 (%)	T2 (%)
300-400	39	12.5
401-500	33	58.3
501-600	22	20.8
<600	6	8.3

4.4. Artificial Insemination Success Based on Body Condition Score

Table 4. Percentage of Artificial Insemination Success Based on Conception Rate at Various Body Condition Scores

BCS	T1 (%)	T2 (%)
2,5	44	29,17
2,75	33	37,5
3,0	17	33,33
3,5	6	0

Based on Table 4, the BCS with the highest pregnancy rate is BCS 2.5 for T1 and 2.75 for T2. The high pregnancy rate of acceptors with BCS 2.5-3 is because the ideal BCS for cows to become pregnant is cows with BCS 2.5-3. Roche *et al* [23] reported that the ideal BCS of a cow is between 2.5-3.5. Cows with a BCS of less than 2.5 or more than 3.5 have a higher chance of reproductive disorders.

4.5. Artificial Insemination Success Based on Age

Table 5. Percentage of Artificial Insemination Success Based on Conception Rate at Various Ages

Age (years)	T1 (%)	T2 (%)
1,5	38,9	12,5
2,5	5,6	0
3	5,6	20,8
3,5	5,6	8,3
4	16,7	20,8
4,5	16,7	8,3
5	5,6	20,8
<6	5.6	8,3

Table 5 shows that in this study, the age of cows that experienced the most pregnancies was at the age of 1.5 years. This could be due to the ideal age of AI implementation, according to Jainudeen *et al* [4], the ideal age for FH heifers to be mated for the first time is at the age of 15 months. The age of old cows will reduce the percentage of pregnancy, in this study, seven acceptors were more than seven years old, and all experienced pregnancy failure. Susilawati [20] states cows older than six years have lower fertility than younger cows.

5 Conclusion

The success of artificial insemination using SDA-sexed frozen semen showed better results than artificial insemination using non-sexed semen in NRR_1, NRR_2, and CR values. Artificial insemination using unsexed frozen semen obtained NRR_1, NRR_2, and CR values of 76.3%, 63.1%, and 46.15%, respectively, while the percentage of NRR_1, NRR_2 and CR values in AI using SDA sexed frozen semen was 94.7%, 86.15% and 63.15%.

References

1. A. M. Abdula, Z. M. Bilal, Vet. Med. – Open J, **7** (2022)
2. D. L. Garner, K. M. Evans, G. E. Seidel, Spermatogenesis: Methods and protocols, **927** (2013)
3. M. Gunawan, E. M. Kaiin. S. Said, Pros Sem Nas Masy Biodiv Indon. **1** (2015)
4. M. R. Jainudeen, E. S. E. Hafez, Cattle and Buffalo in Reproduction in Farm Animals ed By E. S. E. Hafez and B. Hafez, (2000)
5. D. U. Azmi, A. P. A. Yekti, T. Susilawati, Ternak Trop. J. Trop. Anim. Prod, **23** (2022)
6. Iswoyo, P. Widiyaningrum, J. Ilm. Ilmu-Ilmu Peternak, **11** (2008)
7. J. E. Pryce, M. P. Coffey, G. Simm, J. Dairy Sci, **84** (2001)
8. S. Dash, A. K. Chakravarty, A. Singh, A. Upadhyay, M. Singh, S. Yousuf, Vet. World, **9** (2016)
9. D. C. Nisa, A. Rachmawati, T. Susilawati, A. P. A. Yekti, J. Ilmu-Ilmu Peternak., **32** (2022)
10. A. P. A. Yekti, R. Prafitri, Kuswati, A. N. Huda, Kusmartono, T. Susilawati, Am. J. Anim. Vet. Sci, **17** (2022)
11. T. Susilawati, J. Ternak Trop, **12** (2011)
12. M. Zainudin, M. N. Ihsan, and D. Suyadi. *J. Ilmu-Ilmu Peternak.* **24** (2014)
13. A. Rosita, T. Susilawati, S. Wahyuningsih, J. Ilmu Ilmu Peternak, **24** (2014)
14. A. P. A. Yekti, T. Susilawati, M. N. Ihsan, S. Wahjuningsih, *Fisiologi Reproduksi Ternak. Malang* (2017)
15. T. K. S. Rao, N. Kumar, P. Kumar, S. Chaurasia, N. B. Patel, Vet. World, **6** (2013)
16. A. Sammad, S. Umer, R. Shi, H. Zhu, X. Zhao, Y. Wang, J. Anim. Physiol. Anim. Nutr. (Berl), **104** (2020)
17. N. Da Costa, T. Susilawati, M. N. Ihsan, IOSR J. Pharm, **6** (2016)
18. L. Priyanto, H. Herdis, S. Santoso, R. I. Anwar, T. P. Priyatno, P. I. Sitaresmi, F. Azhari, M. Gunawan, O. D. Putranti, Vet. World, **16** (2023)
19. S. Yotov, I. Fasulkov, A. Atanasov, E. Kistanova, B. Sinapov, B. Ivanova, D. Yarkov, D. Zaimova, Animals, **13** (2023)
20. S. A. Aldini, N. Isnaini, A. P. A. Yekti, T. Susilawati, J. Ilmu-Ilmu Peternak, **32** (2022)
21. T. Susilawati, *Sexed Spermatozoa* (2014)
22. J. Moran, *Managing High Grade Dairy Cows in the Tropics* (2012)
23. J. R. Roche, D. P. Berry, J. Dairy Sci, **89** (2006)

Conception rate of artificial insemination using sexed frozen semen with albumin sedimentation method on Friesian Holstein cow

Ismail Hasan[1], *Rika* Amalia[1], *Aulia* Puspita Anugra Yekti[2], *Sri* Wahjuningsih[2], and *Trinil* Susilawati[2*]

[1]Post Graduate Program, Faculty of Animal Science, Universitas Brawijaya, Malang 65145, Indonesia
[2] Faculty of Animal Science, Universitas Brawijaya, Malang 65145, Indonesia

Abstract. This study aimed to determine the success of Artificial Insemination (AI) using unsexed and sexed semen of albumin sedimentation method based on Non-Return Rate, Conception Rate, reproduction performance, estrus characters, and feed nutrition. This research was conducted from August to December 2022 in Bendosari Village, Pujon District, Malang Regency. The material used in this study was 76 Friesian Holstein (PFH) cows with the criteria of having a minimum Body Score Condition (BCS) of 2.5 (scale of 1-5), normal reproductive organs and showing signs of heat. Thirty-eight cows were inseminated with unsexed frozen semen and 38 with albumin sedimentation sexed semen. The results showed that the percentage value of CR in AI using unsexed frozen semen was 55.23%, while CR in AI using albumin sedimentation sexed was 60.53%. In conclusion, AI's success using frozen albumin sedimentation sexed semen shows better results than AI using unsexed frozen semen.

1 Introduction

Dairy farms in Indonesia are still small-scale and traditional, which has led to low livestock productivity. One step to increase livestock productivity is to improve reproductive performance. Efforts are made to increase livestock productivity by using Artificial Insemination (AI) [1]. Artificial insemination is one of the reproductive technologies compatible with and has succeeded in improving the genetic quality of livestock so that in the future, it can produce good quality offspring in large numbers by using superior males as much as possible [2]. The utilization of sexing or separation of X and Y spermatozoa is the technology that supports AI in increasing the efficiency of livestock breeding. Various sexing methods have been carried out, including sedimentation, albumin column, percoll density gradient centrifugation, electrophoresis, H-Y antigen, flow cytometry, and filtration with sephadex column [3]. The success of pregnancy and the accuracy of the genital sex of the calves born are the final results of the application of AI with sexed sperm. In the

* Corresponding author: tsusilawati@ub.ac.id

research conducted by [4] at the AI center of Lembang, the sex suitability of X sperm reached 87% and Y sperm reached 89.5%. This study aims to determine the success of AI in friesian holstein cow dairy cows in terms of Non-Return Rate (NRR) and Conception Rate (CR) using sexed frozen semen through the albumin sedimentation method.

2 Materials and methods

This research was conducted from August to December 2022. This research was conducted in Bendosari Village, Pujon District, Malang Regency, East Java. The materials used in this study were 76 female friesian holstein females with at least 1,5 years old criteria, BCS value of at least 2.5 (scale 1-5), normal reproductive organs and showing signs of estrus. Samples were selected by purposive sampling with the criteria of female cows, vulva color, swollen vulva, vagina temperature, mucus cervix. The cows were divided into 38 cows inseminated with unsexed frozen semen (T0) produced by the AI center of Singosari Malang, then compared with 38 cows inseminated with sexed frozen semen using the albumin sedimentation method (T1) produced by the AI center of Lembang.

3 Variable

3.1 Non-Return Rate

The NRR method is based on the assumption that cows that have been inseminated and no longer in heat are considered pregnant [5]. NRR-1 observations are made by looking at signs of heat, including vulva condition (vulva swelling), vulva colour, and riding or being ridden by cows on the 19 to 22 days after AI. Cow that are not in heat will continue to NRR-2 observation, which includes vulva condition (vulva swelling), vulva color, and climbing and riding other cows on the 39 to 42 days after AI [6] with the following formula:

$$\text{NRR} = \frac{\text{Total inseminated cows} - \text{total cows in heat}}{\text{Total inseminated cows}} \text{ X } 100\%$$

3.2 Conception Rate

The number of pregnant cows can obtain the CR value at the first AI [7]. [8] states that CR is the percentage of cows that become pregnant at the first IB. The CR formula is:

$$\text{CR} = \frac{\text{Number of cows pregnant at first insemination}}{\text{Number of acceptors}} \text{ X } 100\%$$

4 Results and discussion

4.1. Body Performance of Friesian Holstein Breeders

The reproductive physiology of female cows is influenced by various factors, including body weight, Body Condition Score (BCS), and age. Body weight measurements are used to estimate BCS so that phenotypes are better in cows because BCS directly affects dairy cows' nutritional, health, and reproductive status [9]. Underweight body weight will have reproductive disorders commonly referred to as subfertile [10], and BCS that is <2.5 experiences nutritional deficiencies resulting in the reproductive system not functioning optimally and if livestock are mated and too young, will cause suboptimal body weight [11]. Management in optimizing BCS can affect reproductive performance [12]. Low body weight is caused by nutritional deficiencies. The composition of the feed ration is low and does not meet the daily feed needs and nutrient deficiencies in the body of livestock resulting in the reproductive system will not run normally [10]. [13] added that when feed consumption cannot meet the daily needs of livestock, it will cause a decrease in BCS and will interfere with their reproductive status.

One of the factors that affect the low percentage of pregnancy in livestock is the age factor. This is reinforced by the opinion of [3] which states that the older the age of the livestock, the lower the pregnancy rate. According to [14] female cattle fertility will increase continuously until the age of 4 years and will flatten at the age of 6 years until finally it will decrease gradually when the old mother is older.

Table 1. Body Weight, Body Condition Score, Age of Friesian Holstein Cattle

Treatment	Age	Weight	BCS
T0	3.89±1.34 years	336.94 ± 34.16 kg	3.52±18.67
T1	4.07±1.82 years	323.44 ± 38.07 kg	2.66±0.20

The acceptors used in this study are 76 acceptors divided into three groups, namely acceptors weighing 300, 350, and 400. The average weight of acceptors in this study for T0 is 336.94 ± 34.16 kg and T1 is 323.44 ± 38.07 kg. The BCS range used in this study was BCS 2.5-3.5. This study's mean BCS of acceptors was 3.52±18.67 for T0 and 2.66±0.20 for T1. The age of acceptors in this study ranged from 1.5-7 years. The average age in this study was 3.89±1.34 years in T0 and 4.07±1.82 years in T1.

4.2. Estrus Performance

Table 2. Estrus Performance on Artificial Insemination Success Rate Using Unsexed Semen and Albumin Sedimentation Sexed Semen Before First Artificial Insemination

Treatment	Vulva Colors			Cervical Mucus	
	Evenly Red (%)	Red Unevenly (%)	Pale (%)	Much Mucus (%)	No Mucus (%)
	Head (%)	Head (%)	Head (%)	Head (%)	Head (%)
T0	17 (80,95 %)	1 (4,76 %)	3 (14,29 %)	20 (95,2 %)	1 (4,8 %)
T1	19 (82,61 %)	0 (0,0 %)	4 (17,39 %)	23 (100 %)	0 (0,0 %)

Estrus or heat is a condition where female cow have shown heat, namely swelling of the vulva, reddening of the vulva color, warmth, and willingness to ride. Estrus in livestock is characterized by mucus discharge and changes in the physical appearance of the vulva, which is influenced by estrogen hormone levels [15]. [16] added that cows show signs of estrus, including increased restlessness and activity, mucus discharge from the vagina, willingness to ride other cows, or standing heat. Table 2 shows the percentage of estrus appearance in the condition of evenly red, unevenly red, and pale vulva, as well as the quality of mucus with mucus and traces of mucus. Differences in vulval color can be caused by the hormonal activity of each individual cow. Vulvar color is used as an indicator because an increase in the hormone estrogen in cattle increases will cause blood to flow to the vulva, so the vulva will turn red. The color of the vulva turning reddish is one of the signs of heat because the blood flow to the reproductive tract will be higher, so it will have an impact on the condition of the vulva to be swollen, evenly red and the temperature will increase [17]. The amount of blood supply to the vulva can cause swelling of the vulva and vestibule area, so that it will become bright red due to congestion of blood vessels, while the vulva that shows an uneven red color indicates cattle in estrus cycle conditions towards metestrus and at the end of estrus cattle will show a pink or pale color [18].

The results of the research showed that acceptors who were positive for post-AI pregnancy with red mucus color were evenly distributed in the T0 and T1 treatments as many as 17 heads and 19 heads with a percentage of 80.95% and 82.61%, for uneven red vulva color only in the T0 treatment as many as one heads with a percentage of 4.76%, for pale vulva color in the T0 and T1 treatments as many as three heads and four heads with a percentage of 14.29% and 17.39%. In addition to the color of the vulva, the appearance of estrus observed in this study is cervical Mucus, where the results showed that acceptors who were positive for post-Al pregnancy with a description of cervical Mucus had Mucus in the T0 and T1 treatments as many as 20 heads and 23 heads with a percentage of 95.2% and 100 % and for the description of former Mucus there was only one head in the T0 treatment with a percentage of 4.8%. Based on Table 1, it was found that the CR value in cows with evenly distributed red vulva color was higher than uneven and pale red, which was 80.95% in P0 and 82.61% in T1, while the CR value in cows with cervical Mucus was higher than mucus marks, which was 95.2% in T0 and 100 % in T1.

The quality of heat shown by each acceptor at the time of AI is closely related to the success of pregnancy, which includes vulvar reddening and transparent (clear) mucus [19]. Cows in normal and healthy heat will indirectly secrete cervical mucus with a large volume, thick and watery during estrus. The success of pregnancy is positively correlated with transparent mucus secreted from the cow's vulva [20]. According to [21] stated that cervical mucus secreted is influenced by estrogen which makes the hormone adrenaline and the hormone oxytocin secreted, the hormone oxytocin will make blood vessel endothelial cells become permeable which increases the activity of goblet cells so that water accumulation occurs so that high goblet cell pressure results in goblet cells rupturing and cervical mucus will come out. [22] say the hormone estrogen stimulates thickening of the vaginal wall, increased vascularization, causing the external genitals to swell, reddish in color, and increased vaginal secretions characterized by mucus hanging on the vulva or sticking around it.

4.3. Feed Quantity and Quality

Table 3. Observation results of Non-Return Rate-1 and Non-Return Rate 2 of filial Friesian Holstein Artificially Inseminated Using Unsexed Semen and Albumin Sedimentation Sexed Semen.

Treatment	Number of Acceptor	$NRR_{0\text{-}22}$		$NRR_{23\text{-}43}$	
		Head	%	Head	%
T0	38	4	89.47%	9	76.32%
T1	38	2	94.74%	4	89.47%

Farmers play a significant role in detecting heat in female cow, and maintenance management also determines the success of AI. One of the things that is considered is the feed given to livestock. The types of feed given in this study are *Pennisetum purpureum* and corn stover. According to [23], the type of feed and nutritional content will affect the physiology of livestock. Lack of feed intake can result in less visible conditions because of inhibited production of reproductive hormones. [24] stated that nutrition is essential in cow reproductive performance that affects fertility, especially at various stages of the reproductive cycle. [25] added that feed nutrition received by cows before and after calving also affects CR because nutritional deficiencies can cause delays in the estrus cycle.

Adult cows that are deficient in nutrients for a long time can cause the ovaries to become inactive, resulting in irregular cycles and anestrus. Low quality feed such as lack of fat and carbohydrates can affect ovarian activity, suppressing follicle growth rate and encouraging anestrus [26]. According to [5] the lack of feed nutrition will cause the basic needs not to be fulfilled, so that milk production will decrease and reproductive efficiency is low. The protein intake of cattle given irregularly will result in disrupted fat metabolism, decreased LH, cholesterol enzyme activity and estradiol which will affect follicular development, so that it can reduce reproductive performance.

4.4. Pregnancy Rate Based on Non-Return Rate

Non-return rate is the first step to determine the success rate of Al with the understanding that 78 acceptor cows are routinely observed on the 22 and 43 days. Susilawati (2011) [5] explained that NRR is the percentage of female AI acceptors that are no longer in heat for 20-60 days or 60-90 days after the implementation of Al.

Based on Table 3, the percentage of T0 unsexed semen NRR-1 and NRR-2 is 89.47% and 76.32%. The NRR value in the T1 sexing treatment using the albumin method obtained the observation values of NRR-1 and NRR-2 of 94.74% and 89.47%. The decrease in NRR value may be caused by silent heat in female cattle. Cattle affected by silent heat are unable to show signs of heat due to low levels of the hormone estrogen, so farmers will have difficulty in detecting and eventually will be considered not in heat [27]. Silent heat is a condition where cows do not show clear signs of heat but when checked using per rectal palpation there is ovarian activity in the form of follicular development or development of the corpus luteum which is characterized by ovulation [3]. Follicle Stimulating Hormone (FSH) during silent heat is able to encourage the development of follicles in the ovaries until ovulation occurs, but cannot encourage granulosa cells to be able to synthesize the hormone estrogen, so that signs of heat cannot appear [28]. According to [29] one of the causes of silent heat is due to disturbances from ectoparasites and endoparasites that can cause stress on acceptors.

The decrease in NRR value can also be caused by embryonic death. It is in accordance with the opinion of [30] that early embryonic death is caused by livestock disturbed by disturbances from ectoparasites and endoparasites so that livestock will be exposed to stress which will cause neurotransmitters, namely inhibition of the release of hormones caused by the influence of the five senses or stress [10]. [29] explained that early embryonic death is closely related to the implantation process that occurs in the endometrium, the process of implantation will be disrupted if there is a lack of steroid hormones. Acceptors who experience early embryonic death after 9 days of age, the next estrous cycle will experience a setback or longer than the normal cycle (long cycle), this is because the body needs time to destroy the conceptus and be absorbed by the endometrium, then the endometrium secretes the hormone PGF2α and estrus occurs again [10].

4.5. Artificial Insemination Success Based on Conception Rate

Table 4. Conception Rate Results of Filial Friesian Holstein Cows Inseminated Using Unsexed Semen and Albumin Sedimentation Sexed Semen. first AI

Treatment	Number of Acceptor	Pregnant	
		Head	%
T0	38	21	55,26%
T1	38	23	60,53%

The high level of reproductive efficiency can be measured using the conception rate method. [5] stated that CR can be measured by knowing based on pregnancy between 60-90 days after the first insemination. The value of CR is caused by many factors, one of which is detection in heat. Accuracy at the time of heat detection significantly affects the CR value. The CR value is influenced by the accuracy of heat detection and the time of implementation of AI. Silent heat can occur due to errors during heat detection. [31] added that the high and low CR and PR values are influenced by several factors, including estrus condition when the cow is inseminated and the time of AI. There are several factors that cause cows to go into heat, one of them is the lack of attention from breeders in the detection of heat and the delay in reports to the Inseminator, as a result, the implementation of AI was delayed. The low CR value can be caused by the fact that during the implementation of unsexed AI, the animals were still recovering after the Foot and Mouth Disease (FMD) outbreak. In contrast, the implementation of sexing AI was carried out after FMD recovery. [32] stated that foot and mouth disease has a negative effect, which can interfere with several reproductive hormones, which can harm cow productivity.

The ideal CR value is 65-75% [33]. The CR percentage in this study can be said to be below the normal value. The low CR value can be caused by silent heat that occurs due to ovary hypofunction and ovary cystic follicles and feed consumption that cannot meet the daily needs of livestock. [34] explained that ovarian hypofunction is a condition where ovarian function decreases, so that there is no optimal follicular development caused by impaired secretion of FSH and LH hormones, while cystic follicles are young follicles that grow on the ovary for 10 days or more caused by a lack of LH hormones so that ovulation or anovular does not occur. [35] added that the ovum produced by ovary hypofunction has low fertility, making it difficult to fertilize even though the spermatozoa are of good quality.

5 Conclusion

The success rate of AI using unsexed semen is lower than albumin sedimentation sexed semen. The non-return rate value in unsexed semen treatment was 89.47% and 76.32%, while for albumin sedimentation, sexed semen was 94.74% and 89.47%. The conception rate in unsexed and sexed semen was 55.26% and 60.53%, respectively.

The authors would like to thank the Faculty of Animal Science, Brawijaya University through the Hibah Guru Besar 2023 with contract number: 2044.15/UN10.F05/PN/2023

References

1. K. Kastalani, H. Torang, A. Kurniawan, Jurnal Ilmu Hewani Tropika (Journal of Tropical Animal Science), **8** (2020)
2. T. Susilawati, *Pedoman Inseminasi Buatan pada Ternak*, (2013)
3. E.S.E. Hafez, B. Hafez, *X and Y Chromosome-Bearing Spermatozoa in Reproduction in Farm Animals ed By E. S. E. Hafez and B. Hafez*, (2000)
4. M. Gunawan, E. M. Kaiin. S. Said, Pros Sem Nas Masy Biodiv Indon. **1** (2015)
5. T. Susilawati, J. Ternak Trop, **12** (2011)
6. A. P. A. Yekti, E. A. Octaviani, K. Kuswati, T. Susilawati, J. Ternak Trop, **20** (2019)
7. I. F. Puspitasari, N. Isnaini, A. P. A. Yekti, T. Susilawati, J. Ternak Trop, **19** (2018)
8. C. A. Yulyanti, T. Susilawati, M. N. Ihsan, Indonesian Journal of Animal Science, **24** (2014)
9. B. M. Martins, A. L. C. Mendes, L. F. Silva, T. R. Moreira, J. H. C. Costa, P. P. Rotta,

M. L. Chizzotti, M. I. Marcondes, Livestock Science, **236** (2020)
10. A. P. A. Yekti, T. Susilawati, M. N. Ihsan, S. Wahjuningsih, *Fisiologi Reproduksi Ternak*, (2017)
11. Iskandar, Jurnal Ilmiah Ilmu-Ilmu Peternakan, **14** (2011)
12. A. K. Singh, Journal of Applied Animal Science, **12** (2022)
13. Y. R. Bindari, S. Shrestha, N. Shrestha, T. N. Gaire, Bangladesh Journal of Animal Science, **4** (2013)
14. F. Febrianthoro, M. Hartono, Suharyati, Jurnal Ilmiah Peternakan Terpadu, **3** (2015)
15. N. Ducha, D. Hariani, W. Budijastutik, T. Susilawati, Iranian journal of veterinary Sciance and Technology, **15** (2023)
16. Irfan, S. Wahjuningsih, T. Susilawati, Jurnal Ternak Tropika, **18** (2017)
17. C. Venvers, M. V. Z. Langhout, M. Govaere, Soom, Vlaams Diergeneeskundig Tijdschrift, **84** (2015)
18. Z. Meratti, A. Towhidi, Iranian Journal of Applied Animal Science, **12** (2022)
19. S. Sudarmaji, A. Malik, A. Gunawan, Majalah Ilmiah Peternakan, **10** (2007)
20. Y. R. Bernandi, A. Rinaudo, P. Marini, Iranian Journal of Veterinary Research, **17** (2013)
21. E. Anisa, Y. S. Ondho, D. Samsudewa, Jurnal Sains Peternakan Indonesia, **12** (2017)
22. D. R. Frandson, W. L. Wilke, A. D. Fails, *Anatomy and Physiology of Farm Animal*, (2003)
23. Z. Abidin, Y. S. Odho, B. Sutiyono, Animal Agriculture Journal, **1** (2012)
24. M. Luthfi, S. Wahjuningsih, G. Ciptadi, T. Susilawati, Indian Journal of Animal Sciences, **93** (2023)
25. N. Nuryadi, S. Wahjuningsih, Jurnal Ternak Trop, **12** (2011)
26. F. F. Dirgahayu, M. Hartono, P. E. Santosa, Jurnal Ilmiah Peternakan Terpadu, **3** (2015)
27. T. K. S. Rao, N. Kumar, P. Kumar, S. Chaurasia, N. B. Patel, Veterinary World, **6** (2013)
28. T. G. O. Pemayun, I G. N. B. Trilaksana, M. K. Budias, Jurnal Veteriner, **15** (2014)
29. L. Wahyudi, T. Susilawati, N. Isnaini, Jurnal Ternak Tropika, **14** (2014)
30. M. Saifudin, N. Isnaini, A. P. A. Yekti, T. Susilawati, Jurnal Ternak Tropika, **19** (2018)
31. Kuswati, D. Prasetyo, A. P. A. Yekti, T. Susilawati, Jurnal Ilmu-Ilmu Peternakan, **32** (2022)
32. A. K. Shaban, R. H. Mohammed, A. M. Zakaria, E. M. Baheeg, Veterinary World, **15** (2022)
33. T. B. Kaufmann, M. Drillich, B. A. Tenhagen, D. Forderung, W. Heuwieser, Theriogenology, **71** (2009)
34. Y. Malda, N. Layla, A. P. A. Yekti, A. N. Huda, Kusmartono, T. Susilawati, Livestock and Animal Reasearch, **20** (2022)
35. Sutiyono, D. Samsudewa, A. Suryawujaya, Jurnal Veteriner, **18** (2017)

The physical quality of cream cheese influenced by the addition of encapsulated *Lactococcus lactis*

Feronika Mei Figaliah[1], *Abdul* Manab[2*], *Manik* Eirry Sawitri[2], and *Ria* Dewi Andriani[2]

[1]Student of Faculty of Animal Science, Universitas Brawijaya, Malang 65145, Indonesia
[2]Lecturer of Faculty of Animal Science, Universitas Brawijaya, Malang 65145, Indonesia

Abstract. This study aimed to determine the best treatment of *Lactococcus lactis* encapsulated with gelatin and Na-Alginat at different concentrations of : T_0 0% had (w/v); T_1 1% had (w/v); T_2 2% had (w/v), and T_3 3% had (w/v). The research was conducted in Animal Product Technology Laboratory. The material used was cream cheese made from pasteurized milk and added *Lactococcus lactis* starter culture encapsulated. *Lactococcus lactis* starter was obtained from the Center for Food and Nutrition Studies, Gadjah Mada University, encapsulated with Haqiqi brand beef gelatin and Food Grade brand Na-alginate. The method of this research was an experimentally designed completely randomized design (CRD) with four treatments and three replications. The parameter observed in this research were water content, water binding power, syneresis, free oil, emulsion stability, emulsion activity, emulsion index, and turbidity of cream cheese. The result showed that the addition of *Lactococcus lactis* encapsulated did not have a significant effect on water content, water binding power, syneresis, free oil, emulsion stability, emulsion activity, emulsion index, and turbidity of cream cheese (P>0.05). It can be concluded that the best cream cheese was achieved with the addition of 3% *Lactococcus lactis* encapsulated with the 55.11% water content, 76.80% water holding capacity, 19.03% syneresis, 45.28% free oil respectively, 64.88% emulsion stability, 1.54 m^2/g emulsion activity, 3,68% emulsion index, and 2.19% turbidity.
Keywords : Cream Cheese. *Lactococcus lactis*. Encapsulated

1 Introduction

Cream cheese is a type of soft cheese rich in nutrients that has a soft, smooth, white and creamy texture and a sour taste [1]. The process of making cream cheese uses the main ingredient, namely skim milk, which is pasteurized with the addition of lactic acid bacteria [2]. The lactic acid bacteria used in this research were *Lactococcus lactis*. The purpose of adding *Lactococcus lactis* to cream cheese is as a fermentation starter [3], improving the distinctive taste and aroma, appearance, color, and texture [4]. *Lactococcus lactis* is a lactic

* Corresponding author: manabfpt@ub.ac.id

acid bacterium that is sensitive to extreme environments [5], so it can cause the loss of the functional life of the bacteria. In order for *Lactococcus lactis* to work well, it requires the right growth temperature, so it needs a coating treatment using encapsulation technology with the technique of entrapping the core material in a certain coating material. Encapsulation is an immobilization technique that protects active ingredients in a matrix from environmental influences. Microencapsulation of probiotic bacteria often uses a matrix in the form of Na alginate and gelatin. Na alginate has been tested to increase probiotic survival by 80-95%. Immobilization of *Lactococcus lactis* bacteria in alginate and gelatin makes the bacteria surrounded by a gel network, which limits the movement of the bacteria. When growth occurs, bacteria press on the gel network, and bacterial colonies are formed [5]. As the colony grows, bacteria can reach the surface of the alginate beads. This can lead to a colony eruption, resulting in the colony being released into the environment.

Cream cheese has a standard water content for fresh cheese, namely a maximum of 80% [6]. In the process of making cream cheese, there is a process of increasing the total amount of acid, which triggers a decrease in pH, resulting in the cream cheese experiencing an increase in water holding capacity and low syneresis. The quality of cream cheese seen during storage can affect free oil. Free oil formation or fat leakage is the tendency of liquid fat to separate from the melted cheese and accumulate as pockets of oil, especially on the surface [7]. The use of encapsulated *Lactococcus lactis* resulted in decreased emulsion stability in cream cheese, increased emulsion activity, and increased emulsion index and turbidity values. The decrease in emulsion stability is caused by protein denaturation along with decreasing pH during storage, which causes damage to protein bonds as an emulsifying agent [8]. The addition of encapsulated *Lactococcus lactis* to cream cheese is a solution to produce quality cream cheese because its viability is stable in the external environment, thereby reducing the risk of damage to food ingredients [9].

2 Materials and Methods

2.1 Materials

The material used in this research was cream cheese made from pasteurized milk and added *Lactococcus lactis* starter culture. The ingredients used in this research were powdered skim milk (Super) and added with *Lactococcus lactis* starter (obtained from the Center for Food and Nutrition Studies, Gadjah Mada University) encapsulated with Haqiqi brand beef gelatin and Food Grade Na-alginate, rennet enzymes (QSO), 0.1 M Catalog brand, and table salt (NaCl) (Ship Stamp).

2.2 Methods

The research method used was a laboratory experimental method using a completely randomized design with 4 treatments and 3 replications which added an encapsulated *Lactococcus lactis* starter culture with the variables observed being water content, water holding capacity, syneresis, free oil, emulsion stability, emulsion activity, emulsion index, and turbidity. Data were analyzed using Analysis of Variance (ANNOVA) with the Completely Randomized Design (CRD) method. If there are differences then they are analyzed further using Duncan's Multiple Distance Test (UJBD).

3 Results and discussion

3.1 Effect of Using Encapsulated *Lactococcus lactis* in Cream Cheese on Water Content

Data and analysis of variations in the average water content of cream cheese with different treatments using encapsulated *Lactococcus lactis* bacteria can be seen in Table 1.

Table 1. Average water content (%) with different encapsulated *Lactococcus lactis* addition treatments.

Treatments	Water Content (%)
P0	53.50 ±0.05^{a}
P1	54.50±0.13^{a}
P2	55.11±0.09^{b}
P3	55.98±0.47^{c}

Note: a, b, and c different superscripts in the same column show very significant differences in water content (P<0.01).

The results of the analysis of variance showed that the treatment using encapsulated *Lactococcus lactis* had a very significant effect (P<0.01) on the water content of cream cheese. Table 1 shows the average water content between 53.50% and 55.98%. The water content of cream cheese tends to increase with an increasing concentration of encapsulated Lactococcus as a starter in making cream cheese. The control sample (P0) with the addition of *Lactococcus lactis* without encapsulation had a low water content among all treatments, namely 53.50%. This was because the lactic acid produced was low due to the decreased viability of *Lactococcus lactis* due to its lack of bacterial resistance in vulnerable and extreme environments. protective layer because *Lactococcus lactis* is a non-sporulative gram-positive bacteria [10]. In P1, P2, and P3, there was an increase due to the increasing ratio of the addition of encapsulated *Lactococcus lactis*. The highest average water content was found in the P3 treatment, namely 55.98%. The higher the concentration, the more the addition of encapsulated *Lactococcus lactis* tends to increase the water content. Water content is free water and adsorbed water, where the adsorbed water is water bound in hydrocolloid networks [11].

The greater the amount of *Lactococcus lactis* added, the higher the lactic acid produced, resulting in milk proteins clumping and releasing water. This event is also related to syneresis. The release of water increases the water content of the cream cheese. Stated that water separation occurs due to LAB activity, which produces proteolytic enzymes in an acidic atmosphere, and clumping occurs. This is due to the presence of hydroxyl in the encapsulant material (Na alginate and gelatin), which has a tendency to bind water so that the water content becomes high [12]. The water content of all treatments in cream cheese is 53–55.98%, so the resulting cheese is included in the soft cheese group. [13] states that the water content in soft cheese ranges from 40–75%. [14] the water content of cream cheese is ≤ 55%, so the best treatment is P1 and P2 because they are included in the cream cheese standard. Soft cheeses are generally consumed fresh and have a limited shelf life. High water content can weaken the casein network so that cheese particles can easily separate [15].

3.2 Effect of Using Encapsulated *Lactococcus lactis* in Cream Cheese on Water Holding Capacity

Data and analysis of the average holding capacity of cream cheese water with treatment using encapsulated *Lactococcus lactis* bacteria can be seen in Table 2.

Table 2. Average water holding capacity with different encapsulated *Lactococcus lactis* addition treatments.

Treatments	Water Holding Capacity (%)
P0	75.19±0.74
P1	75.32±1.22
P2	76.80±0.56
P3	76.37±1.32

The results of the analysis of variance showed that the treatment using encapsulated *Lactococcus lactis* did not make a significant difference ($P > 0.05$) to the water holding capacity of cream cheese. Table 2 shows that the treatment of adding encapsulated *Lactococcus lactis* to cream cheese has an average water holding capacity ranging from 75.32% to 76.80%. In P1 and P2, there was an increase in water holding capacity. This was due to the addition of a large number of encapsulated *Lactococcus lactis*, where the encapsulant in *Lactococcus lactis* contained gelatin and Na-alginate as stabilizers, which affected the water holding capacity of cream cheese. Casein micelles have the ability to bind large amounts of water. Mechanical water trapping plays a role in increasing water holding capacity [16]. There was a slight decrease in water holding capacity in P3 due to the influence of the lack of protein binding in gelatin as an encapsulant compared to Na-alginate in comparison to *Lactococcus lactis* encapsulation. Decreased protein-water interactions will result in aggregation and precipitation due to the release of water from the protein network [16]. The addition of gelatin can increase water holding capacity. This is because gelatin functions as a stabilizer, inhibiting hydrogen bonds and increasing the hydrophilic properties of proteins [17]. Giving 2% encapsulated *Lactococcus lactis* (P2) to cream cheese namely, 76.80%, is the best treatment because it is able to increase water holding capacity optimally.

The protein contained in the encapsulant will increase the amount of protein contained in the cream cheese so that more water is bound because the casein molecule in the cream cheese experiences a change in negative ion charge to positive. This event will result in an interaction inhibiting hydrogen bonds between the casein molecule and the lactic acid molecule. Manab [16] stated that the water-holding capacity of yogurt is influenced by casein micelles. The casein molecule in cream cheese experiences a change in negative ion charge to positive; this event will result in an interaction inhibiting the hydrogen bond between the casein molecule and the lactic acid molecule. [17] argue that casein and lactic acid do not bond with each other because the ion charge is the same, which can inhibit hydrogen bonds between casein molecules and lactic acid molecules and maintain the binding of water molecules by protein molecules. Water molecules are suspended in the semi-solid phase due to hydrogen bonds (C == H) between water molecules and protein amide groups (N -- H), and interactions between water molecules and ionic groups to form a hydration structure consisting of hydrogen bonds between water molecules.

3.3 Effect of Using Encapsulated *Lactococcus lactis* in Cream Cheese on Syneresis

Data and analysis of the average variance of cream cheese syneresis with different treatments using encapsulated *Lactococcus lactis* bacteria can be seen in Table 3.

Table 3. Average syneresis with different encapsulated *Lactococcus lactis* addition treatments.

Treatments	Syneresis (%)
P0	26.09±1.14^{b}
P1	21.58±2.84^{ab}
P2	19.03±0.71^{a}
P3	21.97±0.35^{ab}

Note: a and b different superscripts in the same column show very significant differences in syneresis (P<0.01).

Based on Table 3, it can be seen that the amount of syneresis ranges from 19.03% to 26.09%. The results of the analysis of variance showed that the addition of encapsulated *Lactococcus lactis* with different treatments had a very significant effect (P<0.01) on cream cheese syneresis.

One of the main determining factors for the quality of cream cheese is syneresis. Syneresis is a gel shrinkage event that occurs naturally without external force or pressure. This is related to unstable gel bonds, which result in a loss of whey binding ability [18]. The breaking and formation of casein bonds affect the syneresis behavior of the product, namely the release of whey. According to [16] the occurrence syneresis is caused by changes in casein solubility and the shrinkage of casein particles. Curd cutting and pressing can also influence syneresis [19].

Based on the results of research, cream cheese treated with encapsulated *Lactococcus lactis* (P1, P2, and P3) experienced a decrease in syneresis compared to the control treatment with the addition of non-encapsulated *Lactococcus lactis* (P0). At P0, it has a higher level of syneresis because it is influenced by a higher serum release factor due to the reduced viscosity in the water phase of the cream cheese. P2 with an added concentration of 2% encapsulated *Lactococcus lactis* showed the lowest syneresis, namely 19.03%, so treatment P2 was the best treatment among the other treatments because it was close to the syneresis of commercial cream cheese. According to [18], 15 samples of commercial cream cheese had an average syneresis value of 14.4%. The increase in P3 again is probably due to the large amount of lactic acid metabolism products coming out of the encapsulant pores due to the lack of binding to the protein from gelatin, which can completely encapsulate the pores of the microencapsulation membrane due to the small ratio between gelatin and Na-alginate (1:3). Low protein solubility is easily released so that it comes into direct contact with the environment, causing protein-protein interactions from lactic acid acidification, so the tendency for syneresis to occur cannot be inhibited [16].

The syneresis value in cream cheese is reduced because *Lactococcus lactis*, which is used as a starter in making cream cheese, is coated with an encapsulant coating containing gelatin and sodium alginate, which is able to form peptide bonds with casein and prevent hydrogen bonds between casein and lactic acid in acidic conditions, which can result in increasing water holding capacity and reducing syneresis. [20] reported that yogurt syneresis was influenced by the addition of stabilizer concentration, caused by a highwater holding capacity, which could reduce syneresis.

3.4 The Effect of Using Encapsulated *Lactococcus lactis* in Cream Cheese on Free Oil

Data and analysis of the average variance of free-oil cream cheese with different treatments using encapsulated *Lactococcus lactis* bacteria can be seen in Table 4.

Table 4. Average free oil with different encapsulated *Lactococcus lactis* addition treatments.

Treatments	Free Oil (%)
P0	45.61±0.34
P1	46.79±2.22
P2	45.28±0.13
P3	44.58±0.42

Based on Table 4, it can be seen that the amount of free oil ranges from 44.58% to 46.79%. The results of the analysis of variance showed that the treatment using encapsulated *Lactococcus lactis* did not provide a significant difference (P > 0.05) to the free oil in cream cheese. Free oil is also known as oiling off or fat leakage. Free oil is expressed as a percentage of cheese or cheese fat. In each treatment, it was seen that each addition of encapsulated *Lactococcus lactis* to cream cheese did not reduce the free oil much. In P1, P2, and P3, there was a decrease in free oil due to the formation of a complex bond between gelatin and na-alginate, thus preventing fat leakage. Free oil occurs when the casein matrix collapses during heating, allowing fat globules to coalesce and flow to the surface [21]. The formation of free oil in cream cheese is also influenced by fat globules. The smaller the size of the fat globule, the smaller the free oil produced, and the formation of free oil depends on varying fat mobility [22].

According to [7], the low quantity of oil released causes the formation of emulsions in fat, resulting in reduced fat content in cheese. The more free oil that comes out, the more it will affect the quality of the cream cheese. The longer storage can also affect the increase in free oil. [23] believes that the oil released comes from larger fat globules, which are unstable as emulsion droplets in cream cheese.

3.5 Effect of Using Encapsulated *Lactococcus lactis* Cultures in Cream Cheese on Emulsion Stability

Data and results of statistical analysis of the average value of emulsion stability using encapsulated *Lactococcus lactis* culture in cream cheese can be seen in Table 5.

Table 5. Average emulsion stability (%) using culture *Lactococcus lactis* encapsulated in cream cheese.

Treatments	Emulsion Stability (%) ± SD
P0	88.39±1.27
P1	87.96±0.96
P2	82.99±8.64
P3	64.88±3.47

In Table 5, the results of the ANOVA test show that the treatments given showed that there was no significant difference (P > 0.05) in the stability of the emulsion. The use of encapsulated *Lactococcus lactis* cultures resulted in decreased emulsion stability. According to [24], emulsion stability is the ability of the emulsion to remain stable and unchanged against coalescence and flocculation and to form cream quickly. Factors that influence emulsion stability are temperature during the emulsification process, pH, emulsification viscosity, and the amount and type of soluble protein [25].

The average value of emulsion stability when using *Lactococcus lactis* culture encapsulated in cream cheese ranges from <64.88% to >88.39%. The research results are comparable to those of [26], who found that the best treatment was the emulsion stability value of guava juice probiotic drink samples on various variations of sucrose and skim milk with the addition of Lactobacillus casei starter, namely with an average value ranging from 71% to 100% and a decrease in the value of emulsion stability. In this study, the higher the use of encapsulated *Lactococcus lactis* culture, the lower the emulsion stability value. According to [8], the decrease in emulsion stability is caused by protein denaturation, as the pH decreases during storage and causes damage to protein bonds as an emulsifying agent. The capacity of encapsulated *Lactococcus lactis* to stabilize emulsions depends on the molecular composition of the cell wall and the resulting surface properties. *Lactococcus lactis* can be dispersed in water but not in oil, but when high-cell surface hydrophobic molecules are mixed with hydrocarbons, they will disappear from the water phase and find the oil-in-water interface of the emulsion formed [27].

[28] explains that the fermentation process using skim milk will produce high levels of lactic acid due to the high protein content in skim milk. Increasing lactic acid levels cause the pH to drop so that casein becomes unstable and coagulates to form a gel. Acid coagulation causes the stability of casein to be disturbed and can modify the internal bonds between proteins, causing protein aggregation and denaturation, which will also increase the hydrophobic properties and form a gel-like texture [29]. The ability of proteins to form and stabilize emulsions is a critical factor in food processing. The decrease in emulsion stability is caused by the process of replacing casein with phospholipids that are adsorbed on the surface of oil globules [30]. Low emulsion stability will cause the emulsion to separate quickly and also be damaged.

3.6 Effect of Using Encapsulated *Lactococcus lactis* Cultures in Cream Cheese on Emulsion Activity

Data and results of statistical analysis of the average value of emulsion activity using encapsulated *Lactococcus lactis* culture in cream cheese can be seen in Table 6.

Table 6. Average emulsion activity (m^2/g) using culture *Lactococcus lactis* encapsulated in cream cheese.

Treatments	Emulsion Activity (%) ± SD
P0	1.22±0.15
P1	1.36±0.27
P2	1.45±0.40
P3	1.54±0.38

In Table 6, the results of the ANOVA test show that the treatments given showed that there was no significant difference ($P > 0.05$) in the emulsion activity. The use of encapsulated *Lactococcus lactis* culture resulted in increased emulsion activity. [31] explain that emulsion activity is the area between the surfaces of oil globules with protein stabilization, which can show the emulsion index by calculating the area between the surfaces stabilized per unit mass of protein (m^2/g). Factors that influence activity are the amount of pectin and protein bound to the structure [32].

The average value of emulsion activity when using *Lactococcus lactis* culture encapsulated in cream cheese ranges between <1.22 $m2$/g - >1.54 $m2$/g. This is comparable to [33] that the value of emulsion activity in the fortification of synbiotic yoghurt with red dragon fruit peel extract (Hylocereus polyrhizus) and stevia on emulsion and color properties is aroun7.68–9.4040 and has increased value on emulsion activity. In this study, the higher the use of encapsulated *Lactococcus lactis* culture, the greater the emulsion activity value. [31] explained that the increase in the emulsion activity value was caused by the hydrophobic and hydrophilic groups moving more easily to the surface. The higher the emulsion activity value shown, the higher the protein emulsification ability. The higher the emulsion activity, the smaller the size of the oil globules in the emulsion.

The addition of encapsulated *Lactococcus lactis* also causes the pH of cream cheese to be low. A low pH and a value far from its isoelectric point can cause the protein to release and contain exopolysaccharides. Exopolysaccharides are able to bind water during storage, thereby maintaining a soft texture in gel products [34]. [35] stated that a low pH and a value far from the isoelectric point can also increase solubility so that it can increase the interaction of electron charges in the encapsulation molecules with oil and water molecules in the emulsion system. This causes the emulsion formed from the use of encapsulation to have high emulsion activity.

The casein contained in cream cheese is an amphoteric compound that can react with acids and bases. Casein molecules have positive and negative charges; when proteins and polysaccharides carry opposite charges, they will form complex agglomerations that are soluble or insoluble [36]. According to [37], high total dissolved solids can trigger an increase in emulsion activity.

3.7 Effect of Using Encapsulated *Lactococcus lactis* Cultures in Cream Cheese on the Emulsion Index

Data and results of statistical analysis of the average value of the emulsion index using encapsulated *Lactococcus lactis* culture in cream cheese can be seen in Table 7.

Table 7. Average emulsion index (%) using encapsulated *Lactococcus lactis* culture in cream cheese.

Treatments	Emulsion index (%) ± SD
P0	2.46±0.60
P1	3.13±0.005
P2	3.17±0.015
P3	3.68±0.35

In Table 7, the results of the ANOVA test show that the treatments given show that there is no significant difference ($P > 0.05$) in the results of the emulsion index. The use of encapsulated *Lactococcus lactis* culture resulted in an increase in the emulsion index. According to [38], the quality of uniformity of a dispersion is an indication of the emulsion index. A small emulsion index value indicates a narrow particle size distribution, which means the particle size is more uniform. Factors that influence the emulsion index are seen in the average volume and diameter of the fat globules. Smaller fat globules can contribute to the overall continuity of the matrix by acting as fillers, resulting in a product with a stronger structure [39].

Average emulsion index value when using culture *Lactococcus lactis* in cream cheese ranges from < 2.46% - >3.68%. This is comparable to [40] that the emulsion index value for the potential of Bacillus cereus in biosurfactant production is 44.83%–51.5%, and the value of the emulsion index has increased. The use of encapsulated *Lactococcus lactis* at higher concentrations plays a significant role in increasing the viscosity of the emulsion so that the total dissolved solids or viscosity also increases [41]. The addition of encapsulated *Lactococcus lactis* will cause an increase in milk acidity, which can cause a low pH, and the release of protein, which is replaced by exopolysaccharide compounds, which can cause the gel texture to become denser so that the absorbance value decreases [42]. The more stable the emulsion, the oil droplets will spread relatively more evenly with a more uniform size and not experience as many droplets coalescing [38].

3.8 Effect of Using Encapsulated *Lactococcus lactis* Cultures in Cream Cheese on Turbidity

Data and results of statistical analysis the average value of turbidity using encapsulated *Lactococcus lactis* culture in cream cheese can be seen in Table 8.

Table 8. Average turbidity (NTU) using encapsulated *Lactococcus lactis* culture in cream cheese.

Treatments	Turbidity (NTU) ± SD
P0	1.68±0.74
P1	1.72±0.20
P2	1.74±0.98
P3	2.19±0.61

In Table 8, the results of the ANOVA test show that the treatments given showed that there was no significant difference ($P > 0.05$) in turbidity. The use of encapsulated *Lactococcus lactis* cultures resulted in increased turbidity. [38] explain that turbidity is turbidity that occurs due to the presence of suspended substances. Factors that influence turbidity are the content of cellulose, hemicellulose, lignin, starch, and water-soluble carbohydrates [43].

The average turbidity value of using encapsulated *Lactococcus lactis* in cream cheese ranges from <1.68 NTU to >2.19 NTU. This is comparable to [33] that the turbidity value in the fortification of synbiotic yoghurt with red dragon fruit peel extract (Hylocereus polyrhizus) and stevia on emulsion properties and color ranges from 0.43 to 0.53 and there is an increase in the value of turbidity. In this research, the higher the use of encapsulated *Lactococcus lactis* in cream cheese, the higher the turbidity value. The use of encapsulated *Lactococcus lactis* in cream cheese is thought to increase the activity of lactic acid bacteria in producing organic acids, thereby increasing turbidity. [44] explained that turbidity can increase, which is influenced by the time the electrocoagulation process takes place. The faster the oxidation and reduction process takes place, it makes it difficult to remove particulates, and the more turbid the liquid waste becomes, plus the increase in voltage applied during the electrocoagulation process will also increase the potential. chemical energy produced. The chemical energy produced is in the form of $Al3+$ ions at the anode, which acts as a coagulant. The more the amount of coagulant increases, the more flocs will be deposited. Gelatin as a coating material in encapsulated *Lactococcus lactis* cultures can cause a rough and lumpy texture. Gelatin can increase water binding capacity by means of electrostatic interactions with casein molecules so that hydrophobic interactions can be inhibited in non-polar molecules or molecules that have non-polar groups fused together due to the acidification process [17].

4 Conclusion

In conclusion, P2 adding 2% encapsulated *Lactococcus lactis* starter to cream cheese can increase the water content and water holding capacity and reduce syneresis and free oil. The addition of 2% encapsulated *Lactococcus lactis* produces cream cheese with a water content of 55.11%, a water holding capacity of 76.80%, syneresis of 19.03%, and free oil of 45.28%. Treatment P3 (3% encapsulated *Lactococcus lactis*) is the best treatment for emulsion stability, emulsion activity, emulsion index, and turbidity, with an average value of emulsion stability of 64.88% and emulsion activity of 1.54%. m^2/g, emulsion index 3.68%, and turbidity 2.19 NTU.

References

1. M. Fadhlillah, A. Safari, S.D. Rachman, F.F. Isnanisafitri, S. Ishmayana, Chimica et Natura Acta. **8** (3), 98-103 (2020)
2. M.R. Sainani, K. Vyas, P. S. Tong, J. Dairy Sci. **87**, 2854-2863 (2004)
3. E. Delavenne, J. Mounier, F. Déniel, G. Barbier, G. Le Blay, International Journal of Food Microbiology, **155** (3),185-190 (2012)
4. P. Khemariya, S. Singh, G. Nath, A.K. Gulati, Turkish Journal of Agriculture-Food Science and Technology, **5** (6), 556-562 (2017)
5. K. Cahyono, E. Nurcahyani, S. Wahyuningsih, B. Irawan, Sumardi, Jurnal Ilmiah Farmasi Farmasyifa, **4** (1), 33-40 (2021)
6. Codex Alimentarius Commission, Codex Alimentarius Milk and Milk Products Volume 12, 2nd Edition (Food and Agriculture Organization of The United Nations, World Health Organization, 2011)
7. N. Novianti, H. Ardiansyah, F. Melli, Jurnal Media Laboran, **8** (1), 1-4 (2018)
8. L. Yuniarti, Murhadi, Suharyono, S. Hidayati, Jurnal Agroindustri Berkelanjutan, **1** (1), 130-142 (2022)
9. A.Z. Mustofa, Hasim, S. Amelia, Jurnal Biologi Indonesia, **14** (1), 61-71 (2018)
10. K. Cahyono, E. Nurcahyani S. Wahyuningsih., B. Irawan, Sumardi, Jurnal Ilmiah Famasi Farmasyifa, **4** (1), 33-40 (2021)
11. I.R. Putri, B. Basito, E. Widowati, Jurnal Teknosains Pangan, **2** (3), 112-120 (2013)
12. J. Sahib, Miskiyah, A. Kusuma, Jurnal Penelitian Pascapanen Pertanian, **16** (2), 56-63 (2020)
13. D. Rahmawati, S. Juni, W. Kusuma, Jurnal Ilmu Ternak, **1** (9), 46-51 (2014)
14. USDA, USDA Specifications for cream cheese, cream cheese with other foods, and related products, Dairy Division (Agricultural Marketing Service United States Department of Agriculture, 1994)
15. McSweeney PLH, Cheese Problems Solved (England: CRS Pr, 2007)
16. A. Manab, Jurnal Ilmu dan Teknologi Hasil Ternak, **3** (1), 52-58 (2008)
17. M.E. Sawitri, A. Manab, T.W.L. Palupi. Jurnal Ilmu dan Teknologi Hasil Ternak, **3** (1), 35-42 (2008)
18. A.F. Wolfschoon-Pombo, B.P. Dang, B.C. Chiriboga, International Dairy Journal, **85**, 129-136 (2018)
19. C.D. Everard, D.J. O'Callaghan, M.J. Mateo, M. Castillo, F.A. Payne, C.P. O'Donnell, J. Dairy Sci. **94**, 2673-2679 (2011)
20. M.N. Radi, S. Amiri, Journal of Applied Sciences, **9** (11), 2194-2197 (2009)
21. P.S. Kindstedt, P.F. Fox, Journal of Food Science, 56 (4), 1115-1116 (1991)

22. P. Schenkel, R. Samudrala, J.Hinrichs, International Dairy Journal, **30** (2), 79-87 (2013)
23. E. Fitasari, Jurnal Ilmu dan Teknologi Hasil Ternak, **4** (2), 17-29 (2009)
24. T. Estiasih, J. Tek. Pert. **4** (3), 141-154 (2003)
25. D.F. Rosida, U. Sarofa, R.C. Dewi, Jurnal Rekapangan, **9** (1), 19-27 (2015)
26. F. Nurainy, S. Rizal, S. Suharyono, U. Ekarisa, Jurnal Aplikasi Teknologi Pangan, 7 (2), 47-54 (2018)
27. M. Tarazanova, T. Huppertz, M. Sttarenburg, T. Todt, S. van Hijum, J. Kok, H. Bachmann, Plos one, **14** (7), 1-17 (2019)
28. A. Triyono, Seminar Rekayasa Kimia dan Proses, ISSN : 1411-4216 (2010)
29. S. Melia, I.M. Sugitha, J.Indon.Trop.Anim.Agric. **32** (2), 86-90 (2007)
30. F. Rohmah, T. Estiasih, Jurnal Pangan dan Agroindustri, **6** (3), 30-36 (2018)
31. K.G.S. Ahmadi, T. Estiasih, Agritech, **31** (3) (2012)
32. N. Bayar, M. Friji, R. Kammoun, Food Chemistry, **241**, 127-134 (2018)
33. T.M.K. Ningrum, M.E. Sawitri, A. Manab, Jurnal Agripet, 22 (1), 120-126 (2022)
34. N. Suryawanshi, S. Naik, J.S. Eswari, Food Science and Applied Biotechnology, **5** (1), 22-44 (2022)
35. F. Ricardo, D. Pradilla, J.C. Cruz, O. Alvarez, International Journal of Molecular Sciences, 22, 1-27 (2021)
36. Wusigale, X. Fu, X. Yin, C. Ji, H. Cheng, L. Liang, Journal of Agricultural and Food Chemistry, **69** (42), 12505-12516 (2021)
37. T. Aewsiri, S. Benjakul, W. Visessanguan, P.A. Wierenga, H. Gruppen, (2013). Food and Bioprocess Technology, **6**, 870-881. (2013)
38. S.H. Anwar, D. Hasni, S. Rohaya, M. Antasari, C. Winarti, Heliyon, **6** (7), 1-11 (2020)
39. W. Hoffmann, K. Schrader, International Journal of Food Science & Technology, **50** (4), 950-957 (2015)
40. N.A. Setiani, N. Agustina, I. Mardiah, S. Hamdani, D.Astriany, Jurnal Biologi, **24** (2), 135-141 (2020)
41. W.E. Hanzen, U.S. Hastuti, B. Lukiati, Jurnal Pendidikan: Teori, Penelitian, dan Pengembangan, **1** (11), 2140-2144 (2016)
42. J.B. Divya, K.M. Nampoothiri, Bioresource Technology, **188**, 226-230 (2015)
43. D.Y. Ali, S.S. Yuwono, N. Istianah, J. Ilmu Pangan dan Has. Pertan, **2** (1), 63-71 (2018)
44. B. Chezeau, L. Boudriche, C. Vial, A. Boudjemaa, Separation Science and Technology, **55** (14), 2510-2527 (2020)

The effects of dried betel (*Piper betle* Linn) leaf extract and betel leaf meal on the production of quail egg

Deden Sudrajat[1*], *Dewi* Wahyuni[1], and *Singgih* Singgih[1]

[1]Department of Animal Sciences. Faculty of Agriculture, Universitas Djuanda, Bogor 16720, Indonesia

Abstract. One of the purposes of the inclusion of antibiotics in the ration is to improve production efficiency. Betel leaves are medicinal herbs containing essential oil which can be used as natural antibiotics. This research aimed to examine the effect of giving betel leaf extract and betel leaf flour on quail egg production. A completely randomized design with 4 treatments and 4 replicates was used. Treatments consisted of basal ration (P0), the inclusion of betel leaf meal (BLM) in ration by 1% (P1), the inclusion of dried betel leaf extract (DBLE) in ration by 0.5% (P2) and 1% (P3). The results showed that the inclusion of betel leaf extract and flour did not affect egg production, feed intake, and egg mass. But it improved the feed conversion ratio.

1 Introduction

In the livestock industry, antibiotics in feed additives are used as a growth promoter, increasing efficiency in feed use and livestock production [1]. The use of antibiotics, drugs, or hormones that stimulate animal growth, which have the effect of not complying with the prescribed dosage recommendations, can cause residues in the livestock products produced, which can cause health problems in livestock. Adding antibiotics to animal feed is starting to cause serious problems, with residues or pathogenic bacteria found in livestock carcasses and other livestock products. Residues contained in consumed meat will increase bacterial resistance to antibiotics, so antibiotic replacement feed ingredients are needed [2].

Several studies have been carried out to replace the role of antibiotics as antibacterials, by using feed additives derived from medicinal plants (herbal medicine). Herbal rations are given to livestock, one of which is quail. Feed ingredients derived from medicinal plants can provide additional benefits and advantages and can be an alternative to replacing antibiotics. One of the traditional medicinal plants that can be used as an herbal feed ration and can replace antibiotics is green betel leaf (*Piper betle* Linn).

As an alternative feed ingredient, betel leaves can be used as a ration ingredient to increase metabolism in the body, stimulate the intestines to absorb important nutrients and vitamins for the poultry body, and be used as a strong antimicrobial and anti-fungal agent. In

* Corresponding author: deden.sudrajat@unida.ac.id

general, betel leaves contain as many essential oils as possible 4.2% which is antimicrobial because it can inhibit the growth of several types of bacteria such as *Escherichia coli* and *Salmonella* sp. [3]. However, according to Sudrajat et al. [4], medicinal plants given to animals will have better results if they are an extract solution. Carrasco et al. [5] stated that plant extracts can stimulate growth and efficient feed use. The research results of Sudrajat and Kardaya [2], showed that the boiled betel leaf solution has antibacterial activity, especially against *E. coli* bacteria, and has the potential to increase nitrogen retention and metabolic energy in broiler chicken rations. This research was conducted to test the effect of betel leaf flour and betel leaf extract on quail egg production.

2 Materials and methods

2.1 Birds and diets

This research used 128 *Coturnix coturnix japonica* quail with an average weight of 173.00 ± 17.98 grams/head, aged 7 weeks. The quail is in good health, has good body posture, is not deformed, has a light brown face with white under the jaw, has black or brown spots on the chest area, and has no lumps in the cloaca area. The experimental feed was prepared independently according to nutrient requirements (Table 1).

Table 1. Effect of feeding media composition on Black Soldier Fly larvae development

Feeds (%)	**P1**	**P2**	**P3**	**P4**
Corn Meal	60.4	60	59.4	58.85
CGM	7	6.8	6.6	7.8
Sorbean Meal	13.05	13.05	12.95	12.00
Fish Meal	8.8	8.5	8.5	8
Premix	1	1	1	1
$CaCO_3$	5.6	6.5	7.2	8
CPO	1.5	1.5	1.7	1.7
DCP	2	2	2	2
Cr-O	0.15	0.15	0.15	0.15
BLE	0	0	0.5	1
BLF	0	1	0	0
Nutrient				
Metabolizable Energy (Kcal/kg)	2932	2952	2932	2935
Crude Protein (%)	20.4	20.4	20.4	20.4
Calcium (%)	2.72	2.72	2.72	2.72
Posfor (%)	0.6	0.6	0.6	0.6

Notes : Corn Gluten Meal (CGM), Dicalsium Phosphate (DCP), Calsium Carbonat (CaCO3), Crude Palm Oil (CPO), Cromium Organik (Cr-O), Betel Leaf Extract (BLE), Betel leaf meal (BLM)

2.2 Experimental design and parameter

This research used a completely randomized design (CRD) consisting of four treatments and four replications, each containing 8 quail. The treatment for adding betel leaf extract to the ration in this study was: P0: Ration Without Betel Leaf Meal; P1: Ration Contains 1% Betel Leaf Flour; P2: Ration Contains 0.5% Betel Leaf Extract; P3: Ration Contains 1% Betel Leaf Extract.

Rations are given twice a day, namely in the morning and evening ad libitum. Drinking water was given ad libitum. There has been no treatment with medication or vitamins for quail since treatment began. The variables observed were feed consumption, egg production, egg mass, and feed conversion. Variable measurements are carried out every week. Data analysis was carried out using the ANOVA test. If it shows a significant effect ($P<0.05$) or very significant ($P>0.01$), then proceed with the Duncan test. Data analysis was assisted using the SPSS application.

3 Results and Discussion

The feed intake of quails that were given feed containing betel leaf extract and flour was not significantly different compared to quails that were not given betel leaves (control) (Table 2). This showed that the rations' palatability level was the same in all experimental diets. Feed palatability refers to the sensory characteristics of animal feed that influence an animal's willingness to consume it. According to Mcdonald et al. [6], palatability refers to the taste and texture of the feed and how the animal responds to it. A palatable feed is readily eaten by the animal. It may also refer to the aesthetic appeal of a feed, which might relate to the appearance or smell of the feed. Providing betel leaves in the feed showed that it did not affect the quail's sensory abilities so it did not affect the palatability of the feed. Apart from betel leaf extract, betel leaf flour can also be used as a feed additive in feed. The results of research by Herlina and Ibrahim [7] stated that the use of 0.2% betel leaf flour in commercial feed had a very significant effect on feed consumption but had no effect on body weight and feed conversion.

Table 2. Effect of feeding betel leaf on performances of quail-phase laying

Variables	P1	P2	P3	P4
Feed intake (g)	18.90±2.04	17.39±2.09	20.19±1.44	17.98±2.00
Egg production (%)	40.43±7.08	34.47±8.54	38.42±8.55	30.38±7.88
Mass production (g)	3.82±0.70	3.19±0.84	3.45±0.95	2.64±0.86
Feed conversion	4.32±0.91	3.67±0.91	4.37±1.30	3.31±0.59

Notes: The mean values of all variables show that they are not significantly different ($P > 0.05$)

According to the analysis of variance, feeding quail flour and betel leaf extract had no impact on the birds' ability to lay eggs. These findings are consistent with those of Sudrajat et al. [4] study, which found that the production and egg-mass production of quail eggs was not significantly affected by the administration of betel leaf solution to quail. Studies on the use of betel (*Piper betel* L.) leaf meal in chickens increased the population of non-pathogenic bacteria and reduced the population of pathogenic bacteria in intestines [8], and increased body weight gain [9]. This research is in line with research by Sudrajat et al. [4] on laying quail who were given betel leaf solution to drink, which did not affect egg production.

Providing betel leaf extract in feed improved feed conversion (Table 2). The average conversion rate for quail-given betel leaf extract was 3.31, better than that for quail without betel leaf extract, which was 4.32. In rations given betel leaf extract, there was an increase in ration efficiency by 23%. Plant extracts in meals may impact hens' functioning, intestinal health, and endogenous secretions [10]. This is due to the substance content contained in betel leaf extract can suppress the growth of microorganisms in the gastrointestinal tract [11].

4 Conclusion

The consumption and production of eggs were unaffected by feeding extract and flour of betel leaf. The ration conversion of laying quail can be improved by administering betel leaf extract in ration of laying quail.

References

1. S. Bahri, E. Masbulan, A. Kusumaningsih, Jurnal Litbang Pertanian, **24**, 27-35 (2010)
2. D. Sudrajat, D. Kardaya, Jurnal Peternakan Nusantara, **1**, 33–40 (2017)
3. J. Reveny, Jurnal Ilmu Dasar, **12**, 6-12 (2011)
4. D. Sudrajat, D. Kardaya, S. Sahroji, Jurnal Peternakan Nusantara, **1**, 159–166 (2015)
5. J.M.D. Carrasco, L.M. Redondo, E.A. Redondo, J.E. Dominguez, A.P. Chacana, M.E.F. Miyakawa, BioMed Res. Int. **2016**, 3278359 (2016)
6. P. Mcdonald, R.A. Edwards, J.F.D. Greenhalgh, C.A. Morgan, L.A. Sinclair, R.G. Wilkinson, M.E. Greenhalgh, M.S. Wilkinson, Animal nutrition, seventh edition (Pearson, Harlow, 2010)
7. B. Herlina, W. Ibrahim, Jurnal Peternakan, **6**, 60-64 (2022)
8. B.A. Pradikdo, E. Widodo, A. S. Wardhani, E. Sudjarwo, Int. Res. J. Adv. Eng. Sci. **4**, 310–312 (2019)
9. T.G. Patagao, J.S. Gannapao, Int. J. Ecol. Conserv. **32**, 102–109 (2006)
10. D.E. Cross, R.M. McDevitt, K. Hillman, T. Acamovic, Br. Poult. Sci. **48**, 496–506 (2007)
11. E.M.F.R. Lodang, G.A.M.K. Dewi, I.M. Nuriyasa, Int. J. Life Sci. **4**, 19–25 (2020)

Identification and nutritive value of forages in grazing land consumed by Kuantan cattle in Benai District, Kuantan Singingi Regency

Arsyadi Ali[1*], *Restu* Misrianti[1], *Elfawati* Elfawati[1], *and Tri Regina* Putri[2]

[1]Animal Science Departement, Faculty of Animal and Agriculture Science, Universitas Islam Negeri Sultan Syarif Kasim Riau, Pekanbaru 28298, Indonesia
[2]Student in Animal Science Departement, Faculty of Animal and Agriculture Science, Universitas Islam Negeri Sultan Syarif Kasim Riau, Pekanbaru 28298, Indonesia

Abstract. This study investigated the effects of egg storage duration prior to incubation on performances of Arab Chickens during the first week post-hatch. A total of 100 day-old-chicks were distributed among 20 experimental units, each representing a different egg storage duration before incubation (1, 4, 7, and 10 days) with five replicates. There were five birds of each replicate totaling 25 birds per treatment. The variables examined in this study included the growth performance of Arab Chickens during the first week post-hatch, specifically the initial body weight, final body weight, body weight gain, feed intake, water intake, feed conversion ratio, and water conversion ratio. The findings of the study revealed that egg storage duration prior to incubation did not significantly affect ($P>0.05$) feed intake, water intake, and water conversion ratio. Storing eggs for up to 4 days did not result in any adverse effects ($P>0.05$) on initial body weight, final body weight, body weight gain, and feed conversion ratio. However, eggs stored for 7 days or longer exhibited a significant reduction ($P<0.05$) in initial body weight, final body weight, and body weight gain. Additionally, a notable increased ($P<0.05$) in feed conversion ratio observed for eggs stored for 10 days. In conclusion, the hatching eggs of Arab Chickens can be safely stored for up to 4 days without any detrimental effect on their performance during the first week post-hatch.

1 Introduction

Kuantan Singingi Regency is one of the regions in Riau Province which has an indigenous livestock, namely Kuantan Cattle. This cattle is widely raised by the local community. Kuantan cattle have been designated in the Republic of Indonesia Ministry of Agriculture Decree Number 1052/Kpts/SR.120/10/2014 which states that Kuantan cattle are a wealth of genetic resources for local Indonesian livestock and have a lineage from Bos Indicius cattle [1].

* Corresponding author: arsyadi.ali@uin-suska.ac.id

In general, the management of livestock in Kuantan Singingi Regency is based on traditional system, which is the extensive system where all cattle are in the grazing fields from morning to evening [2]. The grazing areas include land along the Kuantan river and open pastures where the type of forage is not known. Under this management system the productivity of cattle is relatively low.

Forages growing on the pasture land for livestock consists of a number of grass species (Gramineae), legumes (Leguminoseae) and several types of trees with leaves that can be consumed by livestock such as jackfruit leaves, hibiscus leaves and Leucaena [3]. The foragea are grouped into grasses, legumes and other plants. Grass is a type of weed plant that has small stems, many varieties, segmented stems, long narrow leaves, spike-shaped flowers and grain-shaped fruit [4]. Legumes are feed plants that have higher nutritional value, especially crude protein (CP) content compared to grass. Legume plants can provide additional protein because their leaves have a much higher protein content than grass. Legume leaves also provide essential minerals and vitamins for livestock growth [5].

Most of the forage growing in pastures are native grasses and a few legume plants [6]. The lack of legume plants in natural pastures results in low forage quality, especially during the dry season when legumes do not grow.During the dry season native grass becomes the only source of feed and its quality becomes very low. Other local forage problems include low productivity, nutritional content, and limited development [7], so it is necessary to optimize regional potential in providing sustainable forage throughout the year. To determine the availability of types and nutritional potential of forage in an area, an inventory is needed. Inventory is a recording activity to create a data list for an item. This is very important to build an information system for the development of ruminant livestock, especially Kuantan cattle. Base on this information system, efforts are being made to optimize the use of locally available feed ingredients so that they can meet the standard needs of livestock and can prevent the extinction or extinction of Kuantan cattle. This research aims to determine the type and nutritional content of forage consumed by Kuantan cattle in Benai District, Kuantan Singingi Regency.

2 Materials and Methods

This research was carried out in Benai District, Kuantan Singingi Regency where forage samples were taken. Analysis of sample for their nutritional contents was carried out at the Nutrition and Feed Technology Laboratory, Faculty of Agriculture and Animal Science, Universitas Islam Negeri Sultan Syarif Kasim Riau.

The research method was carried out by observing and analyzing forage with the following steps: 1) Kuantan cattle were followed as they graze on the pasture from the time the cattle was released until when the cattle was penned; 2) Observations were made on the types of forage selected by Kuantan cattle during grazing time; 3) About 300 g of fresh samples of all species of forages found were taken and later analysed for nutritional content 4) Forages selected and consumed by cattle were sampled and analysed for the nutritional contents.

The parameters observed were, 1) Kind of forage that Kuantan cattle grab and eat, 2) Nutrient content, namely: Dried Matter (DM), Crude Protein (CP), Crude Fiber (CF), Ether Extract (EE), Ash, and Nitrogen Free Extract (NFE).

3 Results and discussion

Results of field observations shown that there were several types of forage eaten by Kuantan cattle during grazing period, namely grass, legume and other plant. Kind of forage and name of forage shown in Table 1. There were grasess (n=7), legume (n=3) and other plant (n=12).

Table 1. Kind and name of forage that eaten by Kuantan cattle during grazing period time

No.	Kind of Forage	Name of Forage
1	Grass	*Paspalum conjugatum*
		Eleusine Indica
		Chrysopogon aciculatus
		Kyllinga brevifolia
		Imperata cylindrica
		Leersia hexandra
		Brachiaria mutica
2	Legume	*Gliricidia sepium*
		Centrosema pubescens
		Arachis hypogaea L
3	Other palnt	Banana leaves (*Musa paradisiaca*)
		Areca leaves (*Areca catechu* L.)
		Palm oil leaves (*Elaeis guineensis* Jacq)
		Sago leaves (*Metroxylon sagu*)
		Mango leaves (*Mangifera indica* L)
		Sweet potato leaves (*Ipomoea batatas*)
		Cassava leaves (*Manihot esculenta*)
		Sugarcane leaves (*Saccharum officinarum* L)
		Fern leaves (*Pteridophyta*)
		Jackfruit leaves (*Artocarpus heterophyllus* L)
		Cocoa leaves (*Theobroma cacao* L)
		Caesarweed leaves (*Urena Lobata*)

Observation results showed that grasses were the dominant species of forage eaten by Kuantan cattle during the grazing period. The composition of forages consumed by Kuantan cattle were relatively similar to the botanical composition of forage for cattle in Lembang West Java. The variety of forage consumed is known to affect livestock productivity, due to difference in nutritional value [8]

Composition of forage type shown in Figure 1. High compositions of forage were other plant (54,55%) followed by grass (31,82%) and legume (13,63%). The composition of forage eaten by Kuantan cattle during grazing time was less than ideal. This due to the portion of legume was only 13,63%. According to [9] the proportion of grass and legume in pasture should be 60%:40%. This condition causes Kuantan cattle to consume forage that is low in nutrient so that their productivity becomes low. The availability of legumes eaten by Kuantan cattle must be sufficient. Legumes is a forage source of protein.

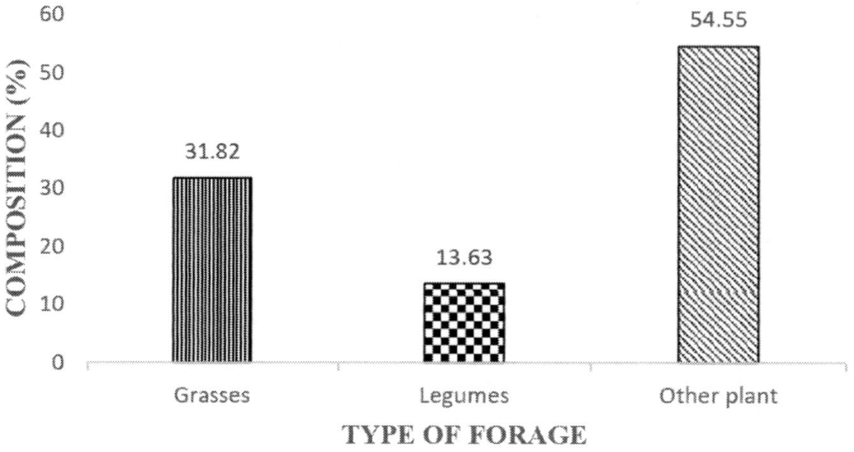

Fig. 1. Composition of forage types

The macro-nutrient content (DM, CP, EE, CF, ash and NFE) of grasses eaten by Kuantan cattle during grazing period is shown in Table 2. The value range of each nutrient were DM (62,56-94,26%), CP (4,33-11,07%), CF (21,78-33,66%), EE (1,49-2,00%), Ash (6,98-16,50%) and NFE (47,42-58,60%). The average CP content of grasses in this study was 9,08%. This CP value has exceeded the CP content of natural pasture grass which is only 3,99% [10], but still lower than CP content of field grass reported by [11] which range 9,22-13,56% on average 11,55%.

Table 2. Nutrient content of grasses that eaten by Kuantan cattle during grazing period time

Kind of Grass	Nutrient Content (%)					
	DM	CP	CF	EE	Ash	NFE
Paspalum conjugatum	24.40	10.94	24.51	1.50	12.97	50.08
Eleusine indica	23.56	11.07	23.53	1.99	7.95	55.46
Chrysopogon aciculatus	28.61	7.89	28.43	1.49	6.98	55.21
Kyllinga brevifolia	19.0k	9.96	21.78	2.00	7.67	58.60
Imperata cylindrica	40.77	4.33	33.66	1.98	6.98	53.04
Leersia hexandra	32.58	9.10	25.49	1.49	16.50	47.42
Brachiaria mutica	13.57	10.30	28.00	1.49	8.87	51.34

Macro nutrient contents of legume was eaten by Kuantan cattle during grazing period is shown in Table 3. The highest CP in *Gliricidia sepium* (22,24%) and lower in *Arachis hypogaea* (14,13%). The CP content of *Gliricidia sepium* eaten by Kuantan cattle during the grazing period was camparable to the CP content of *Gliricidia sepium* reported by [12], namely 22,6%. Although the CP content of *Gliricidia sepium* is high, the NFE content is low. The NFE content of *Gliricidia sepium* eaten by Kuantan cattle during the grazing period was slighty lower than the result of research reported by [13] of 39.50%. This differences is due to differences in climate and soil type. The CP content of *Arachis hypogaea* and *Centrosema pubescens* was relatively lower than legumes in general. This lead to the low quality of forage eaten by Kuantan cattle during the grazing period. Moreover, the proportion of legumes was only 13,53%.

Table 3. Nutrient content of légumes that eaten by Kuantan Cattle during grazing period time

Kind of other plant	Nutrient Content (%)					
	DM	CP	CF	FAT	ASH	NFE
Banana leaves (*Musa paradisiaca*)	21.99	9.27	24.00	5.45	9.18	52.10
Areca leaves (Areca catechu L)	26.21	12.96	25.00	3.00	8.80	50.24
Palm oil leaves (*Elaeis guineensis* Jacq)	40.51	7.35	22.55	2.48	12.15	55.47
Sago leaves (*Metroxylon sagu*)	46.08	7.80	26.73	0.50	7.78	57.18
Mango leaves (*Mangifera indica* L)	31.07	9.10	23.76	2.49	5.58	59.07
Sweet potato leaves (*Ipomoea batatas*)	38.37	22.76	14.85	1.99	12.59	47.80
Cassava leaves (*Manihot esculenta*)	26.68	17.77	16.00	6.50	10.54	49.19
Sugarcane leaves (*Saccharum officinarum* L)	29.94	7.28	27.45	2.00	23.13	40.13
Fern leaves (*Pteridophyta*)	14.48	14.48	16.67	1.99	12.92	53.95
Jackfruit leaves (*Artocarpus heterophyllus* L)	33.09	11.18	18.63	3.96	16.77	49.46
Cocoa leaves (*Theobroma cacao* L)	29.76	7.09	38.24	1.00	6.23	47.44
Caesarweed (*Urena Lobata*)	38.85	13.83	17.65	3.00	12.58	52.94

4 Conclusion

Based on result study there were 7 types of grasses, 3 Type of legumes and 12 several types of leaves that eaten by Kuantan cattle in Benai district during grazing period. Crude protein content and nitrogen free extract forage eaten by Kuantan cattle during grazing period range of 4,33-22,76% and 37,63-59,07%, respectively.

Acknowledgements are extended to the ministry of religion for funding this research through the 2023 Kajian Strategi Nasional Research Cluster of Universitas Islam Negeri Sultan Syarif Kasim Riau.

References

1. R. Misrianti,S.H. Wijaya, C. Sumantri, J. Jakaria, Trop Anim Sci J. **45**(2), 134–140 (2022).
2. R.K. Putra , H.P. Nastiti, Y.H. Manggol, Jurnal Nukleus Peternakan, **5**(1), 42 – 48 (2018).
3. S. Schaub, R. Finger, F. Leiber, **11**, 768 (2020).
4. A.H. Prayitno, D. Pantaya, B. Prasetyo, (2020).
5. A. A. A. S. Trisnadewi, I.G.L.O. Cakra, I.M. Mudita, I.W. Wirawan, E. Puspani, I.K.M Budiasa, Udayana Mengabdi, **12**(1), 35-37 (2013).
6. A. Manu, Pastura, **3**, 25-29 (2013)
7. B.N. Utomo, E. Widjaja. J. Litbang. **31**(4), 153-161 (2012).
8. M. A. Al-Arif, A.T.S. Estoepangestie M. Lamid, The Veterinary Medicine International Conference, **2017**, 338-343 (2017).
9. D. Sawen,M. Junaidi, Seminar Nasional Teknologi Peternakan Dan Veteriner. 835-840. (2011)
10. O. Yoku, A. Supriyantono, T. Widayati and I. Sumpe, Pastura, **4**(2), 62-65 (2015).

11. D.N. Nawangsari, E. N. Hendrarti. Analysis of Proximate Native Grass as Ruminant Cattle Feed in Magelang Regency, Central Java. Jurnal Pengembangan Penyuluhan Pertanian, **18**(33), 25-31 (2019).
12. M. Rusdy, B. S. Garantjang, and I. Syarif, LRRD. **31**(6) (2019)
13. P.A. Aye, African Journal of Food Science and Technology, **6**(2), 68-74 (2015)

The feasibility of broiler farming in a partnership model

Puji Akhiroh[1*], *Jovan* Mochammad Dinata[2], and *Eko* Nugroho[1]

[1]Lecturer at the Faculty of Animal Science, Universitas Brawijaya, Malang, 65145, Indonesia
[2] Student at the Faculty of Animal Science, Universitas Brawijaya, Malang, 65145, Indonesia

Abstract. This research was conducted at Unggas Jaya Farm, located in Karangrejo Village, Wungu District, Madiun Regency. The purpose of this study was to determine the capital and production costs required to achieve profitability, and to analyze Break Even Point, production costs, revenue, income, Revenue Cost Ratio, and Benefit Cost Ratio. The respondents in this study were the employee and manager of Unggas Jaya Farm. The research was carried out in May 2023 by interviewing the owner of Unggas Jaya Farm by means of questionnaires. The results show that the total business capital was IDR 3,314,388,085; with the production costs was IDR 1,513,357,675. The total revenue was IDR 1,647,660,654; and the total income is IDR.124,901,770.58. The Break Even Point for broiler price is IDR 19,018/kg, with Break Even Point unit of 73,758.85 kg. The Revenue Cost Ratio (R/C) is of 1.08, indicating that the broiler farming business is considered feasible (R/C > 1). The Benefit Cost Ratio is 0.083, indicating a profit of IDR 0.083 for every unit of production cost. The study concluded that Unggas Jaya Farm's broiler farming business is feasible and profitable for further development.

1 Introduction

Indonesia is an agrarian country where agriculture plays a significant role in the national economy. Broiler farming is one of the most popular livestock sectors in Indonesia. In 2022, broiler meat production in Indonesia reached 3,765,573.09 tons/year or an increase of 18.2% compared to that in 2021. In East Java province, broiler population in 2022 was 493,647,833 heads with a total meat production of 586,703.35 tons so that East Java is the province with the largest population [1], [2], [3]

The impetus for farmers to join the partnership system is the availability of farmer production equipment, available experts, capital that is not too expensive.The company provides, the product market is guaranteed. The cooperation that must be formed must produce satisfactory profits so that the broiler farming business can be carried out in the long term and can generate profits for both parties. The weakness in the core plasma system is that farmers are still not very familiar with their rights and obligations. The production costs incurred by the plasma which change every production will determine the profit of the plasma. An increase in production costs will affect the profits earned by the plasma.

* Corresponding author: pujiakhiroh@ub.ac.id

Feasibility analysis in business is important. A livestock business can be called feasible if the benefits obtained are higher than the costs that have been incurred so that a livestock business can last a long time. Farmers must conduct a feasibility analysis to minimize risk, expand·the business, enlarge the business, and increase the profit to be obtained, in accordance with what is expected optimally

2 Research Methods

This research was conducted at Unggas Jaya Farm in Karangrejo Village, Wungu Subdistrict, Madiun Regency in May 2023 using the data for one period in March 2023. Based on these data, we calculated the farmer's income and analyzed the feasibility of the farm business using several formulas as follows:

Production costs are costs incurred to produce output.

$$TC = TFC + TVC$$

Description:
TC:Total Production Cost (IDR)
FC:Total Fixed Cost (IDR)
VC:Total Variabel Costs (IDR)

Broiler business revenue is the price (IDR) multiply with the total broiler weight (Kg).

$$TR = P.Q$$

Description:
TR : Total Revenue (IDR)
Q : Total Harvest Weight (Kg/period)
P : Contract price (IDR/kg)

Income are obtained from the total revenue deducted by the total production costs.

$$\pi = TR - TC$$

Description:
π: Income (IDR)
TR : Total Revenue (IDR)
TC : Total Production Cost (IDR)

Feed Convertion Ratio (FCR) is the comparison between total feed consumption and the body weight gain.

$$\text{FCR} = \frac{Total\ Feed\ Consumption}{Body\ Weight\ Gain}$$

Broiler mortality is the comparison between the number of dead chickens and the total number of chickens.

$$\frac{Number\ of\ Dead\ Chickens}{Total\ Number\ of\ Chivkens}\ x\ 100\%$$

Performance index is one of the parameters used to determine the level of success in broiler maintenance. Formulator calculating Index Performance as follows:

$$IP = \frac{(100-D) x\ BB x\ 100}{FCR\ x\ (\frac{A}{U})}$$

Description:
IP: Performance Index
D : Percentage of Depletion/Mortality (%)
BB: Average Body Weight (Kg)
FCR: Feed Conversion Ratio
A/U: Average Age of Harvest (Days)

Break Even Point is the cost that comes out equal to the value of the sales price of a product. The formula for calculating BEP unit & BEP price is as follows:

$$BEP\ UNIT\ = \frac{Total\ Production\ Cost\ (IDR)}{Contract\ Pice\ (IDR)}$$

$$BEP\ PRICE = \frac{Total\ Production\ Cost\ (IDR)}{Total\ Production\ (Kg)}$$

Revenue Cost Ratio is the comparison between total revenue and costs. The formula for calculating R/C is as follows:

$$R/C\ Ratio\ = \frac{Total\ Revenue}{Total\ Cost}$$

Indicators of Business Feasibility:
- RC > 1, Feasible Business
- RC < 1, Business Not Viable
- RC = 1, Break-even

Benefit Cost Ratio is the profit divided by the production costs. The formula for calculating B/C is as follows:

$$B/C = \frac{Total\ Benefit\ (IDR)}{Total\ Cost\ (IDR)}$$

B/C Ratio calculation criteria
- B/C > 0 then the business/project is profitable and not detrimental.
- B/C < 0 then the business/project will experience a loss.

3 Result and discussion

Assets and money companies is named capital that require to fund their standard operations and generate revenue. To produce their products or services company need capital to allows businesses to cover expenses. Company can use products and services that provide profit to generate new capital and continue to increase revenue.

Table 1. Business Capital of Unggas Jaya Farm

Fixed Capital	**IDR/Farm**
Land Rent/Taxes	15,000,000
Cage and Warehouse	1,560,000,000
Feed Bin	43,680,000
Drinking Place	23,850,000
Electrical Installation	90,000,000
Water Installation	30,000,000
Employee Salary	52,500,000
Total Fixed Capital	1,815,030,000
Working Capital	
DOC Seedlings	359,100,000
Feed	1,073,361,778
OVK	3,186,590
Disinfectant	840,000
Electricity and Water Costs	22,731,707
Employee Consumption	1,000,000
Cage Maintenance	1,008,000
Employee Bonus	5,000,000
Gas Brooder	18,630,000
CSR	2,000,000
Chaff	10,500,000
Daily Labor	2,000,000
Total Working Capital	1,499,358,085
Total Initial Capital	3,314,388,085

3.1 Production

Production costs are costs that have been incurred and used to produce a product for sale in one period. Production costs are divided into two, namely fixed costs and variable costs or variable costs. According to [13] , production costs include direct materials, factory costs and direct employee costs.

3.2 Fixed Cost

Fixed costs are costs incurred by farmers are costs. fixed costs that cannot change in value and amount such as the cost of making cages and cage equipment. According to [11] fixed costs affect the value of production and will never change in value. Some components of fixed costs such as depreciation costs. Depreciation of cages, depreciation of feed bins, depreciation of drinking bins, interest on capital loans and tax costs. Table 2 shows that the costs used by Unggas Jaya Farm in that period.

Table 2. Fixed costs of Unggas Jaya Farm

Fixed Cost Type	Economic Period	Depreciation (IDR/Period)
Land Rent	70	214,285.71
Cage	175	5,485,714.29
Place Feed	175	221,028.57
Drinking Place	175	121,428.57
Electrical Installation	175	342,857.14
Water Installation	175	114,285.71
Salary	7	7,500,000
Total Fixed Cost		13,999,600

3.3 Variable Cost

Variabel costs are costs whose value will always change according to changes in the amount of production carried out such as DOC seeds, feed, medicines, electricity, employee salaries and others. [7] said that the amount of variable costs depends on the amount of production. The amount of production carried out is large or small. The costs of DOC, feed, medicine, electricity, chaff, and labor are variable costs incurred to produce broilers. Variabel costs incurred by Unggas Jaya Farm can be seen in Table 3.

Table 3. Variable costs of Unggas Jaya Farm

Types of Variable Costs	Total	Total (IDR)
DOC	42,000	359,000,000
Feed	119,400	1,073,361,778
OVK	1	3,186,590
Disinfectant	20	840,000
Electricity and Water	1	22,731,717
Consumption Employees	1	1,000,000
Care Cage	1	1,008,000
Bonus Achievements	5	5,000,000
Gas	90	18,630,000
CSR	1	2,000,000
Chaff	1,500	10,500,000
Labor	10	2,000,000
Non-Fixed Costs		1,499,358,085

The total production costs incurred by Unggas Jaya Farm in one period obtained a value of IDR.1,513,357,675 / period. This value is obtained from the sum of total fixed costs and total variabel costs. Total production costs will change in several other periods because they are influenced by market prices at that time.

3.4 Revenue

Revenue is a result that has been obtained from the production process during one period in the form of selling a product. Revenue is influenced by the amount of production obtained so that the more the number of products produced, the more income the breeder gets. Revenue can be calculated by means of the price of goods set multiplied by the number of products or goods produced in that period [4]. Table 4 shows that Jaya Farm Poultry revenue in one period.

Table 4. Revenue of Unggas Jaya Farm

Type	Total	Total Revenue (IDR)
Sales Chicken	79,573.5 (Kg)	1,632,660,654
Bonus Partnership	1 (Period)	15,000,000
Total revenue		1,647,660,654

Table 4 shows that the revenue of Unggas Jaya Farm in one period or for 35 days after harvesting. The total production costs incurred by Unggas Jaya Farm in one period obtained a value of IDR.1,513,357,675 / period. This value is obtained from the sum of total fixed costs and total variabel costs. Total production costs will change in several other periods because they are influenced by market prices at that time. Amounting to 79,573.5 Kg/period. These results were multiplied by the contract price of IDR.20,518, resulting in an amount of IDR.1,632,660,654 per period. This agrees with [5] which states that the largest proportion of revenue on broiler farms is from the sale of chickens themselves which depends on the price of chickens at that time. Acceptance of one period with a certain period will change so that the revenue earned from chicken sales will change as well.

3.5 Income

Opinion or profit is the result obtained by the farmer in a certain period. Income is obtained from revenue in a farm minus the total costs incurred by the farmer. Table 5 shows the calculation of gross and net income at Unggas Jaya Farm.

Table 5. Unggas Jaya Farm Income

Calculations	Total (IDR)
Total Revenue	1,632,660,654
Total Production Cost	1,513,357,675
Total Revenue Gross 1 Period	134,302,979
Total Revenue Gross 1 Year	940,120,853.81
1% tax in 1 Year	9,401,208.79
Total Revenue Clean	124,901,770.58

Table 5 shows that the March period of income from Unggas Jaya Farm was IDR.1,632,660,654 / period so that minus the production costs in that period of IDR.1,513,357,675 per period, the gross income from Unggas Jaya Farm was IDR.134,302,979 per period so that the income of Unggas Jaya Farm in 1 year was IDR.940,120,853.81. Unggas Jaya Farm's income is less than IDR.4.8 billion so that the tax to be paid is 1%. According to [15] that the amount of tax for chicken farmers with income turnover in one year is less than IDR.4.8 billion, the tax obtained is 1% in accordance with PP 46 of 2013. If the turnover obtained is more than IDR.4.8 billion per year, the tax to be paid will be different in accordance with Law number 36 of 2008 article 17 of the Income Tax Law. Unggas Jaya Farm's net income after tax deduction is 1% or IDR.9,401,208.54 per year so that the amount after the reduction is IDR.124,901,770.58 per period.

3.6 Feed Corvertion Ratio

Feed Conversion Convertion Ratio (FCR) is a factor that will affect the success of animal husbandry. The function of FCR is to calculate the amount of feed that has been consumed to form and obtain one kilogram of meat. [10] said that FCR is used to see the productivity of livestock, if the higher the FCR value, the higher the consumption of feed required. The economic value of livestock business can be influenced by the FCR value. The FCR value at Unggas Jaya Farm can be seen in Diagram 1.

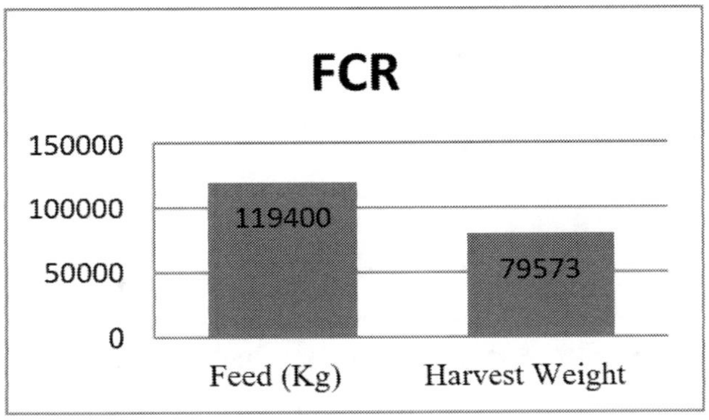

Fig 1. Feed Corversion Ratio

The FCR value is obtained from the amount of feed consumption in that period and then devided by the harvest weight. The FCR value of Unggas Jaya farm in that periode was 1.5. the FCR value multiplied by the feed price of IDR 8,990 will produce a value of IDR.13,485 per period so that it can be interpreted that to produce 1 kg of meat, farmers will need costs of IDR.13,485 per period. The selling price of broilers at Unggas Jaya Farm is IDR.20,158, so the difference between the selling price of broilers and the cost of feed needed is IDR.6,673 per period.

3.7 Mortality

According to [8], mortality is a measure of the number of deaths in a population. Mortality can be influenced by disease and genetics. Mortality factors are caused by climate/weather, disease, livestock behavior and farmer factors that reduce the use of vaccines and drugs to reduce existing production costs. Figure 2 is the initial population and population harvest at Unggas Jaya Farm.

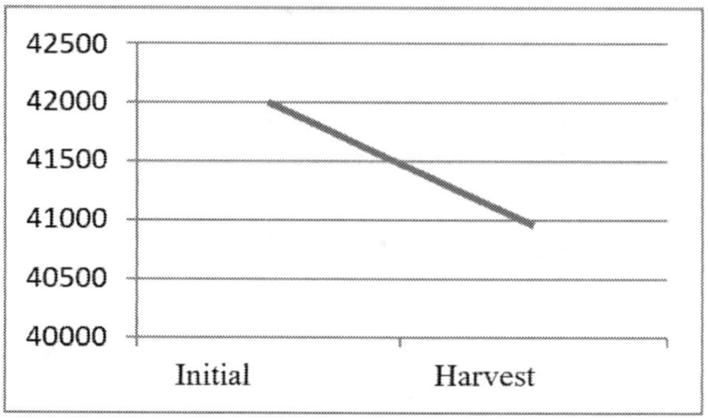

Fig 2. Broiler Population of Unggas Jaya Farm

The mortality value is calculated from the total initial population of 42,000 minus the harvested population of 40,956, so the number of chickens that died was 1,044 in that period. The dead population was divided by the initial population and then multiplied by one percent to get the mortality value for that period. The mortality value of Unggas Jaya Farm in that period was 2.49 percent.

3.8 Performance Index

The Performance Index (IP) is an indicator of success in the broiler farming business. [9] added that the Performance Index is one of the parameters used to determine the level of success in broiler maintenance. Table 6 shows the IP value at Unggas Jaya Farm.

Table 6. IP value of Unggas Jaya Farm

Live Chicken (%)	Average Weight (Kg)	Harvest Age (Days)	FCR
97.51	1.94	32.72	1.5
Performance Index (IP)			398

[12] said that the IP value is grouped into four <300 is said to be poor, the value of 300-350 is said to be quite good, the value of 350-350 is said to be quite good. 400 is said to be good and the value> 400 is very good. Unggas Jaya Farm received a score of 398, this value is higher than 350. Unggas Jaya Farm in that period has managed broiler farms well, so that the benefits that will be obtained by Unggas Jaya Farm are classified as profitable and efficient in managing the livestock business.

3.9 Break Even Point

Break Even Point (BEP) is one of the business feasibility factors. BEP is the break-even point of costs or expenses and income at a balanced point so that at that point the value does not experience a loss or profit. Business feasibility analysis using the Break Even Point (BEP) factor has two calculation formulas, namely price BEP and unit BEP. Diagrams 2 and 3 are the calculation of Break Event Point (BEP) and the comparison of production results and contract prices at Jaya Farm Poultry farm.

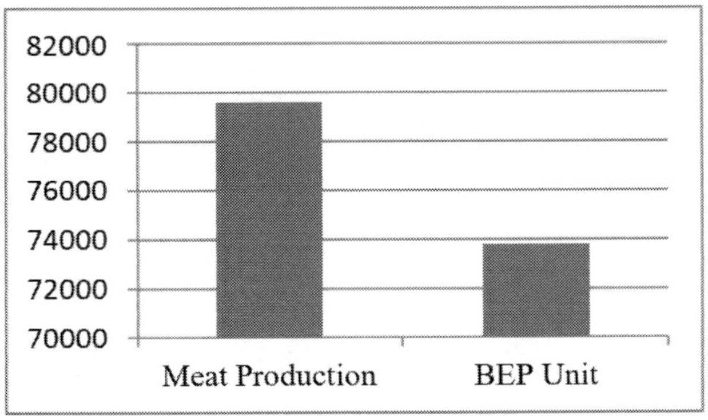

Fig 3. Comparison of BEP Unit and Meat Production Unggas Jaya Farm

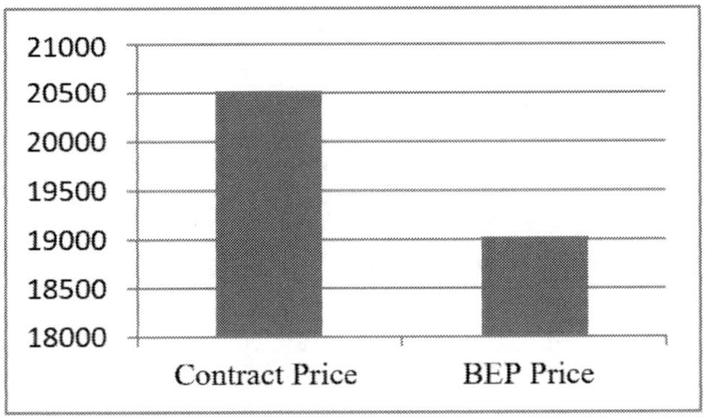

Fig 4. Comparison of BEP Price and Contract Price of Unggas Jaya Farm

During this period, Unggas Jaya Farm was able to produce more meat, namely 79,573.5 Kg per period compared to the BEP value of the unit of 73,758.85 Kg per period. The price BEP value is IDR.19,018 while the contract price agreed by Unggas Jaya Farm and the core company is IDR.20,518, so the contract price value is higher than the calculated price BEP value, so it is certain that Unggas Jaya Farm gets more profit.

3.10 R/C Ratio

Revenue Cost Ratio (R/C) is one of the business feasibility analysis factors. Revenue Cost Ratio is used to calculate and see the relative benefits of a business. According to Murti et al. (2020) R/C is the numerical value of the benefits and revenues obtained for every IDR.1 spent in the process of producing a product. The formula of R/C is the total revenue divided by the cost so that the R/C value is generated. The greater the R / C value, it is certain that the profit will also be greater. The R/C value of Unggas Jaya Farm can be seen in Table 7.

Table 7. R/C value of Unggas Jaya Farm

Total Revenue	IDR.1,632,660,654
Total Cost	IDR.1,513,357,675
Value (R/C)	1.08

The results of the calculation with the R / C formula, namely the total revenue for one period divided by the total production costs in one period, obtained a number worth 1.08. The results of these calculations mean that the R / C value of Unggas Jaya Farm has exceeded the value of 1. The R / C value can be interpreted that every IDR.1 of total production costs that have been incurred will generate a profit of IDR.1.08 per period, it can be seen from the criteria that the Unggas Jaya Farm business can be declared feasible and profitable.

3.11 B/C Ratio

Benefit Cost Ratio (B/C) is a business feasibility analysis that must be carried out in a livestock business. The B/C value is calculated to see how much profit is obtained in production costs. According to [6] if B/C gets a value > 0 the business can be considered profitable and feasible to implement, B/C gets a value < 0 the business can be considered unprofitable and not feasible to implement. Table 8 shows the value of the Benefit Cost Ratio (B/C) per period generated by Unggas Jaya Farm.

Table 8. B/C value of Unggas Jaya Farm

Net Profit	IDR 124,901,770.58
Total Cost	IDR.1,513,357,675
Value (B/C)	0.083

Unggas Jaya Farm gets a B / C value of 0.083, meaning that every production cost incurred by Unggas Jaya Farm of IDR.100, a profit of IDR.0.083 per period is obtained so that the Unggas Jaya Farm broiler farm can be considered profitable and feasible to continue.

4 Conclusions and Suggestions

4.1 Conclusions

Based on research on the fisibility of the Jaya Farm Poultry farm business in Karangrejo District, it can be concluded:

1. That the total business capital of Unggas Jaya Farm consisting of fixed capital and working capital amounted to IDR.3,314,388,085; (54.76%) fixed capital and (45.24%) working capital, total production costs of IDR.1,513,357,675, total revenue of IDR.1,647,660,654 and total income of IDR.124,901,770.58.
2. BEP price of Unggas Jaya Farm IDR.19,018/kg, BEP unit 73,758.85/kg. R/C of broiler livestock business Unggas Jaya Farm 1.08. So based on R/C> 1, it means that the broiler livestock business of Unggas Jaya Farm in Karangrejo District is declared feasible to be cultivated because the revenue obtained by farmers is sufficient to cover the total costs that have been incurred. Analysis of the feasibility of B / C broiler livestock business Unggas Jaya Farm worth 0.083 means that a profit of IDR.0.083 per period is obtained for each production cost incurred.

4.2 Suggesions

For further research, researchers can calculate in more detail and choose locations that have complete data from year to year. So that it can analyze with various business feasibility factors that have been determined.

References

1. Badan Pusat Statistik, Populasi Ayam Ras Pedaging menurut Provinsi(Ekor) 2018-2022 (2022)
2. Badan Pusat Statistik, Produksi Daging Ayam Ras Pedaging menurut Provinsi (Ton) 2018- 2022 (2022)
3. Badan Pusat Statistik, Rata-rata Konsumsi Perkapita Seminggu Menurut Kelompok Daging PerKabupaten/kota 2018-2020 (2022)
4. Fajarani, R. Jurnal Aves.**10** (1): 1-10 (2016)
5. N. Febrianto, J. A. Putritamara, B. Hartono. Agriekonomika. **7**(2):168-175 (2018)
6. Lailina, I Yurki., Sunarto, B Sudarmanto. Jurnal Agriekstensia. **19** (1) : 78 – 86 (2020)
7. Nurjana, I Nyoman.,R Anggraeni, I Kruniasih. Agros. **17** (2): 214- 221(2015)
8. N, Aisyah., M. A. Santi, N. Harahap, M. F. Harahap. Jurnal Ilmiah Peternakan Terpadu. **6**(3): 134-139 (2018)
9. T. Nuryati. Jurnal Peternakan Nusantara. **5** (2):77-86 (2019)
10. A. Qurniawan, I. I. Arie, R. Afnan. Jurnal Veterenier. **17** (4): 622-633 (2016)
11. Simanjuntak, M. Christiana. Jurnal Faperternak. **3**(1):60-81 (2018)
12. S, Darham dan N. I. Effendi. Stock Peternakan. **5**(1) : 1 -10 (2023)
13. A. Witjaksono, GrahaIlmu: Yogyakarta. (2006)
14. Zuhairoh., T. S. Aji, Ardianto. Jurnal Akademi Akuntansi. **2**(2) : 81 -92 (2019)

Effect of fermented laying hen manure and starter feed as growing media on black soldier fly larvae development

Shelfia Anggraini Wiyoso[1], *Hanief Eko* Sulistyo[2], *Faizal* Andri[2], and *Nurul* Isnaini[2*]

[1]Master Student, Faculty of Animal Science, Universitas Brawijaya, Malang 65145, Indonesia
[2]Faculty of Animal Science, Universitas Brawijaya, Malang 65145, Indonesia

Abstract. This study aimed to evaluate the development of Black Soldier Fly (BSF) larvae when reared on different feeding media, specifically using varying ratios of fermented laying hen manure and starter feed. The study employed a completely randomized design with five treatments and four replications. The treatments were as follows: T0: 100% fermented laying hen manure, T1: 75% fermented laying hen manure + 25% laying hen starter feed, T2: 50% fermented laying hen manure + 50% laying hen starter feed, T3: 25% fermented laying hen manure + 75% laying hen starter feed, and T4: 100% laying hen starter feed. Parameters measured in the study included larvae length and larvae width at 10 and 17 days. Data were analyzed using one-way analysis of variance followed by the Duncan post-hoc test. The study demonstrated that BSF larvae were successfully reared on growing media comprising 100% fermented laying hen manure (T0). However, the development of larvae in T0 remained relatively low. Notably, the use of laying hen starter feed particularly at 75% (T3) and 100% (T4) significantly improved ($P<0.05$) all larvae development parameters. It can be concluded that the Black Soldier Fly has significant potential as a bioconversion agent for laying hen manure. Moreover, the use of laying hen starter feed can be considered as a supplemental feeding media to enhance larvae development of Black Soldier Fly.

1 Introduction

The layer farms continue to face significant challenges in managing waste in the form of manure. Accumulated manure poses a considerable risk to animal health, which leads to the reduction of farm productivity [1]. Moreover, unproperly processed manure can also lead to environmental pollution [2]. The repercussions of this pollution have adverse effects on the environment, water sources, and can potentially cause health issues for communities [3]. This demands the development of sustainable solutions for the management of laying hen manure, striking a balance between farm productivity, environmental protection, and community welfare.

* Corresponding author: nurulisna@ub.ac.id

Black Soldier Fly (BSF, *Hermetia illucens* L.) farming is a key aspect of organic waste management. BSF can be reared on various types of organic substrates, such as food and household waste, abattoir waste, as well as livestock manure [4]. BSF farming results in BSF larvae, which can be used as an alternative feed for various livestock [5]. Additionally, the residual media from BSF farming can serve as organic fertilizer [6]. Implementing BSF farming with manure as a feeding media is expected to foster circular economies in the laying hen farming sector.

One crucial factor affecting BSF larva development is the nutrient content in the feeding media. On one hand, the use of laying hen manure can help mitigate environmental pollution. However, it is suspected that using laying hen manure may not optimally support larval growth due to suboptimal nutrient content [7]. An approach to address this issue is by supplementing the substrate with media containing more comprehensive nutrients. The laying hen starter feed is one such feed with well-balanced nutrients [8]. The use of laying hen starter feed as a supplement on the laying hen manure is expected to optimize larval development. The information on the use of laying hen manure and starter feed mixture as a feeding media for BSF is currently not available in the literature. Therefore, this study aimed to evaluate the development of BSF larvae when reared on different feeding media, specifically using varying ratios of fermented laying hen manure and starter feed.

2 Materials and methods

2.1 Experimental design

The study employed a completely randomized design with five treatments and four replicates. The treatments were different feeding media, specifically using varying ratios of fermented laying hen manure and starter feed. The treatments were as follows: T0: 100% fermented laying hen manure, T1: 75% fermented laying hen manure + 25% laying hen starter feed, T2: 50% fermented laying hen manure + 50% laying hen starter feed, T3: 25% fermented laying hen manure + 75% laying hen starter feed, and T4: 100% laying hen starter feed.

2.2 Fermentation of laying hen manure

Laying hen manure was obtained from a local laying hen farm. A batch of 6 kg of laying hen manure was mixed with 2 liters of a 2.5% sugar solution. It was stirred until evenly distributed and then tightly sealed in a plastic container. The fermentation process was carried out for 3 weeks with stirring done every 3 days. After 3 weeks, it is aerated and used as a feeding medium for BSF larvae.

2.3 Hatching process of BSF eggs

The hatching medium was prepared by mixing 200 grams of rice bran with 200 ml of milk waste. This hatching medium was then placed in a plastic container. A plastic cup with a 5 cm diameter was placed in the center of the hatching medium. A total of 0.5 grams of BSF eggs were placed on top of the plastic cup, lined with tissue. BSF eggs were then left to hatch for 3-4 days. Once the BSF larvae have hatched, they will descend into the hatching medium. The larvae were then kept in the hatching medium up to 7 days of age.

2.4 Growing process of BSF larvae

After 7 days of age, larvae were transferred to the separate plastic container containing different feeding medium in accordance with the experimental treatment. A total of 250 g feeding media was added once a week. Additionally, 50 ml of milk waste was also poured into feeding media every two days. At 10 and 17 days of age, a total of 25 larvae were carefully collected from each replicate for the evaluation of larvae development.

2.5 Evaluation of BSF larvae development

Parameters measured in the study included larvae length at 10 days, larvae width at 10 days, larvae length at 17 days, and larvae width at 17 days. The larvae development was evaluated using a ruler with centimeter units.

2.6 Data analysis

Data were tabulated in Microsoft Excel Program. After that, statistical analysis was conducted using IBM SPSS Statistics 22. Data were analyzed using one-way analysis of variance followed by the Duncan post-hoc test. $P<0.05$ was considered as a significant difference.

3 Results and Discussion

Effect of feeding media consisting of varying ratios of fermented laying hen manure and starter feed on BSF larvae length and larvae width at 10 and 17 days are presented on Table 1. The study demonstrated that BSF larvae were successfully reared on growing media comprising 100% fermented laying hen manure (T0). In line with this finding, other studies also acknowledged the potential for rearing BSF larvae on chicken manure [9, 10]. The use of chicken manure as feeding media of BSF larvae is logical since manure serves as the primary food source for many insects in their natural habitat [11]. However, it is crucial to note that despite this success, the developmental progress of larvae in T0 was comparatively lower than that observed in other treatments. Previous studies also noted that the growth response BSF larvae under chicken manure growing medium was relatively modest [7, 12]. The apparent limitations in BSF larvae development on this feeding media may be attributed to deficiencies in essential nutrients or an imbalance in the overall nutritional profile. These findings emphasize the importance of optimizing feeding media to enhance the BSF larvae development.

Table 1 showed a promising result for enhancing BSF larvae development through the use of laying hen starter feed as feeding media. Notably, as the proportion of laying hen starter feed increased, so did the positive impact on various parameters of larvae development. In particular, the inclusion of 25% laying hen starter feed (T1) resulted in a significant improvement ($P<0.05$) in larvae length at 17 days. Moving forward, at 50% inclusion (T2), a more substantial enhancement was noted ($P<0.05$) in larvae length at both 10 and 17 days, as well as larvae width at 17 days. The most significant improvements were witnessed with T3 (75% laying hen starter feed) and T4 (100% laying hen starter feed), where all larvae development parameters exhibited substantial enhancements ($P<0.05$).

Table 1. Effect of feeding media composition on Black Soldier Fly larvae development[1]

Treatments	LL10 (cm)[3]	LW10 (cm)[3]	LL17 (cm)[3]	LW17 (cm)[3]
T0[2]	0.34[a]	0.11[a]	0.93[a]	0.22[a]
T1[2]	0.42[ab]	0.11[a]	1.06[b]	0.24[ab]
T2[2]	0.49[bc]	0.13[a]	1.16[c]	0.25[b]
T3[2]	0.56[c]	0.17[b]	1.18[c]	0.29[c]
T4[2]	0.65[d]	0.24[c]	1.49[d]	0.39[d]
SEM	0.027	0.012	0.043	0.014
P value	0.001	0.001	0.001	0.001

[1]Data were presented as means of four replicates and each replicate consisted of 25 larva length and width measurement
[2]T0: 100% fermented laying hen manure, T1: 75% fermented laying hen manure + 25% laying hen starter feed, T2: 50% fermented laying hen manure + 50% laying hen starter feed, T3: 25% fermented laying hen manure + 75% laying hen starter feed, T4: 100% laying hen starter feed
[3]LL10: larvae length at 10 days, LW10: larvae width at 10 days, LL17: larvae length at 17 days, LW17: larvae width at 17 days
[a-d]Different superscripts in the same column showed significant differences ($P<0.05$)

The utilization of laying hen starter feed is postulated to offer additional sources of protein. Chicken manure was reported to have a total Kjeldhal nitrogen of 2.28% which was equivalent to 14.25% of crude protein [9]. Whereas, laying hen starter feed containing 20.15% of crude protein [8]. Improvement of protein in the feeding media could contributing to the observed improvements in larval development. These findings align with previous research by Logan et al. [7], who reported similar positive effects when incorporating chicken meat and egg wastes into chicken manure, thereby promoting BSF larvae development. Widyaswara et al. [13] also reported improvements in BSF larvae development by adding organic household waste into chicken manure. Moreover, Schreven et al. [14] also noticed that supplementation of camelina oilseed press cake into chicken manure could improve BSF larvae development. The consistent pattern of positive outcomes across different studies underscores the robustness of the observed effects.

4 Conclusion

It can be concluded that the Black Soldier Fly has significant potential as a bioconversion agent for laying hen manure. Moreover, the use of laying hen starter feed can be considered as a supplemental feeding media to enhance larvae development of Black Soldier Fly. Further investigations into the economic feasibility of combining laying hen manure and starter feed as feeding media of Black Soldier Fly are warranted for a comprehensive understanding of its practical applications in waste management and alternative feed production.

The authors acknowledge financial support from Faculty of Animal Science Universitas Brawijaya through Professor Grant with contract number 2044.11/UN10.F05/PN/2023.

References

1. R.B. Bist, S. Subedi, L. Chai, X. Yang, J. Environ. Manage. **328**, 116919 (2023)
2. A. Uwizeye, I.J. de Boer, C.I. Opio, R.P. Schulte, A. Falcucci, G. Tempio, F. Teillard, F. Casu, M. Rulli, J.N. Galloway, A. Leip, J.W. Erisman, T.P. Robinson, H. Steinfeld, P.J. Gerber, Nat. Food, **1**, 437-446 (2020)

3. J. Muhammad, S. Khan, M. Lei, M.A. Khan, J. Nawab, A. Rashid, S. Ullah, S.B. Khisro, Environ. Technol. Innov. **19**, 100909 (2020)
4. C. Lalander, S. Diener, C. Zurbrügg, B. Vinnerås, J. Clean. Prod. **208**, 211-219 (2019)
5. S.S. Raman, L.C. Stringer, N.C. Bruce, C.S. Chong, J. Clean. Prod. **373**, 133802 (2022)
6. S. Barbi, M. Montorsi, L. Maistrello, M. Caldironi, L. Barbieri, Sci. Rep. **12**, 20505 (2022)
7. L.A. Logan, T. Latty, T.H. Roberts, J. Appl. Entomol. **145**, 621-628 (2021)
8. M.K. Abou-Elnaga, S. Selim, J. Anim. Feed Sci. **27**, 268-275 (2018)
9. M.K. Awasthi, T. Liu, S.K. Awasthi, Y. Duan, A. Pandey, Z. Zhang, Sci. Total Environ. **737**, 139842 (2020)
10. X. Zhang, J. Zhang, L. Jiang, X. Yu, H. Zhu, J. Zhang, Z. Feng, X. Zhang, G. Chen, Z. Zhang, Curr. Microbiol. **78**, 303-315 (2021)
11. T. Liu, M.K. Awasthi, H. Chen, Y. Duan, S.K. Awasthi, Z. Zhang, J. Environ. Manage. **251**, 109593 (2019)
12. L. Mazza, X. Xiao, K. ur Rehman, M. Cai, D. Zhang, S. Fasulo, J.K. Tomberlin, L. Zheng, A.A. Soomro, Z. Yu, J. Zhang, Waste Manage. **102**, 312-318 (2020)
13. A. Widyaswara, L. Soetiarso, Y.D. Prasetyatama, U. Hapsari, Proc. 2nd Int. Conf. Smart Innov. Agric. pp.161-165 (2022)
14. S.J. Schreven, H. de Vries, G.D. Hermes, G. Zeni, H. Smidt, M. Dicke, J.J. van Loon, Appl. Environ. Microbiol. **88**, e00084-22 (2022)

Quality evaluation of waste cooking oil from restaurants in Malang as the potential for ruminants feed supplement

Ahlan Bangkit Asmoro[1] and *Hartutik* Hartutik[1*]

[1]Faculty Animal Science, Universitas Brawijaya, Malang 65145, Indonesia

Abstract. Derived from hypothesis of having more fat content in feed could inhibit microbial activity and will increase the energy for the ruminants, this research conducted to determine the contrasting quality of fresh and waste cooking oil. Analysis focused on organoleptic and iodometry to initiate a supplementation product. Material used were fresh and waste cooking oil from four chicken restaurant in Malang: A Restaurant (T_1), B Restaurant (T_2), C Restaurant (T_3), and D Restaurant (T_4). Data collected thrice a day regularly and apprised that different cooking oil brands used. Brand A used in A (T_1) & B Restaurant (T_2), brand B used in C Restaurant (T_3), and brand C used in D Restaurant (T_4). Analysis executed at Animal Nutrition and Feed Laboratory, Animal Science Faculty of Universitas Brawijaya with 15 panellists. Using randomized nested block design; four restaurants, two types of, and three replications (4x2x3); It can be perceived that the cooking oil waste on each restaurants has significant effects on the smell and texture. The panellists concurred that the best cooking oil waste as the potential for ruminants feed supplement, is the C Restaurant (T_3) with the greenish yellow colour, thick texture, slightly slack odor, and tasted a bit bitter. However, all the restaurants' fresh cooking oil, by iodine content analysis, higher than the standard of Indonesian National Standards. T4 hit the highest with 61.87 g Iod/100g compared to T1, T2, and T3 with 52.68, 48.44, and 53.90 g Iod/100g respectively. It can be concluded that the best cooking oil waste for the potential feed supplement is the C Restaurant (T3) with 53.90 ± 1.09 g Iod/100g iodine level.

1 Introduction

About 70% of the production cost in livestock farming is the cost of procuring feed. In addition, feed costs also play a role as a determining factor in the viability of the production system so that it is necessary to reduce costs on feed but on condition that nutrient supply is met according to the standard. Feed with good nutrition will produce good production and reproduction when given to livestock, one of the important nutrients is energy. Ruminants can digest forage as a source of fiber in the rumen into volatile fatty acids which can then become a source of energy. In addition, ruminants still need energy that is more quickly or easily degraded in the rumen such as starch or fat [1].

* Corresponding author: hartutik@ub.ac.id

Ruminant livestock is one type of livestock that is able to make a major contribution to human welfare by providing the most potential animal protein through meat and milk if the nutrients needed are not met, it can affect the product yield of the livestock. Ruminants use carbohydrates as their main source of energy, while reserve energy comes from fat [2]. Low production and reproductive performance of livestock is caused by insufficient supply of essential nutrients such as amino acids, fatty acids, minerals, and energy so that the addition of oil in feed is necessary. Adequacy of energy and essential fatty acids can be met by the addition of oil. The purpose of fulfilling essential nutrients is to increase ration energy and modify the composition of fatty acids in ruminant body tissues [3].

The addition of oil to the ration has several benefits including increasing ration energy, increasing the efficiency of energy use through inhibition of methanogenesis as a defaunation agent [4]. Oil added to the ration of ruminants can cause a decrease in microbial activity in the rumen. As it is known that the advantage of ruminants lies in their ability to utilize fiber, therefore it is necessary to effectively use feed with fat protection [5].

Meeting the need for oil in the ration can be met by utilizing waste cooking oil. Malang is one of the cities that is well known as a cuisine centre with many restaurants and eateries that produce a large quantities of waste cooking oil. The waste cooking oil is the oil residues that are left and discarded after deep frying and broiling food. It is usually disposed off and may cause environmental pollution. It has been shown that waste cooking oil can be used in feed as source of energy. In addition to reducing environmental pollution, the use of cooking oil waste can also reduce feed costs. However, the use of waste cooking oil in feed needs further research on its feeding value and quality. The objective of this study was to conduct a survey on the production of waste cooking oil from several restaurants in Malang district, and their colour, odor, texture, and flavour. The Iodometric tests on the oils were also carried out.

2 Materials and Methods

2.1 Experimental design

Oil sampling was carried out at four fried chicken restaurants. Organoleptic analysis was conducted at the Animal Nutrition and Feed Laboratory, Faculty of Animal Science, Universitas Brawijaya. Iodometric analysis was conducted at the Chemistry Laboratory, Faculty of Mathematics and Natural Sciences, Universitas Brawijaya.

2.2 Sampling of waste cooking oil

The waste cooking oil were obtained from four different restaurants, namely A, B, C, and D Restaurant. Iodometric Test were conducted on the oil samples following the method of [6].

2.3 Research method

This research was conducted using a survey method in four food restaurants, namely A, B, C, and D Restaurant which were restaurants with many customers. Data collected in this study include primary data obtained from direct interviews with respondents using questionnaires and direct observations, while secondary data are obtained from data on the amount and type and date of use of cooking oil by each restaurant. Organoleptic test data was obtained from the assessment of panellists with the criteria of panellists aged 18-25 years, male or female and have experience in cooking.

Two types of samples were taken, namely samples from unused oil (bottled) and waste oil samples (waste oil that was discarded). This used oil sample was taken from each restaurant as much as 500 mL. Sampling was done 3 times on different days as a replicate. Each sample taken is placed in a glass bottle labelled with the date of collection, the name of the restaurant, the type of oil (new or used), while waiting for the sample analysis schedule that is not studied is placed in the freezer so as not to stale. The collected samples were then analysed for physical quality and iodometric test.

2.4 Research variables

The tested variables include organoleptic and iodometric tests of waste oil from each restaurant and also compared with new oil unused in each restaurant.

2.5 Data analysis

The research data were analysed using Analysis of Variance (ANOVA) in Randomized Group Design (RAK) because the conditions of each restaurant were not homogeneous. The analysis aims to determine the effect of the number of puppies on the research variables. If the test concluded that the treatment or group had a real effect, it was continued with the Real Differential Test.

3 Results and Discussion

Sensory Characteristics of Waste Oil in Restaurants towards the Potential of Waste Oil as an Animal Feed Supplement. The following are the results of a survey of several restaurants in Malang.

Table 1. Survey results on the use of cooking oil in fried chicken restaurants in Malang City

Restaurant	Replication	Brand	Fresh cooking oil (liter/day)	Fried portion (pcs)	Usage ratio (ml/pcs)	Waste produced (liter/day)	Waste treatment
T_1 (A)	R_1	A	16.00	1800	7.2	3.00	Distributed to hoarder
	R_2		15.00	1700	7.1	3.00	
	R_3		18.00	2100	6.9	3.50	
T_2 (B)	R_1	A	22.00	2400	7.1	5.00	Disposed
	R_2		20.00	2100	7.6	4.00	
	R_3		19.00	2050	7.3	4.00	
T_3 (C)	R_1	B	20.00	2200	7.5	3.50	Disposed
	R_2		19.00	2000	8.0	3.00	
	R_3		23.00	2500	7.6	4.00	
T_4 (D)	R_1	B	15.00	1800	6.7	3.00	Disposed
	R_2		13.00	1600	7.2	1.50	
	R_3		14.00	1700	7.0	2.00	

According to the survey results, several restaurants produce waste cooking oil with an average of 15-20 liters of waste oil per day. Most of the restaurants disposed off the used oil onto the drains and sewage system.

3.1 Organoleptic Test

The number of panellists was 15 people with an age range of 18-25 years obtained from the provision of training to determine certain sensory properties before testing. The following are the results of the panellists' assessment of the new oil with waste cooking oil.

Table 2. Results of panellists' assessment of new cooking oil (M_1) with waste cooking oil (M_2)

Treatments		Colour Quality (1-5 scale)	Texture Quality (1-5 scale)	Odor Quality (1-5 scale)	Flavor Quality (1-5 scale)	Description (Color, Texture, Odor, Flavor)
M_1	T_1	4.800	5.000	5.000	5.000	Transparent yellow, thick, fresh
	T_2	4.800	5.000	5.000	5.000	Transparent yellow, thick, fresh
	T_3	4.800	5.000	5.000	5.000	Transparent yellow, thick, fresh
	T_4	4.800	5.000	5.000	5.000	Transparent yellow, thick, fresh
Average		5.000	4.800	5.000	5.000	Transparent yellow, thick, fresh
M_2	T_1	3.044	2.378	2.778	3.044	Greenish yellow, thick, rancid
	T_2	2.867	3.044	2.867	2.822	Brownish yellow, very thick, bit rancid
	T_3	3.067	3.156	3.044	3.178	Greenish yellow, thick, bit rancid
	T_4	2.711	3.156	2.778	2.756	Brownish yellow, very thick, bit rancid
Average		2.950	2.922	2.934	2.867	Brownish yellow, very thick, rancid

3.1.1 Colour

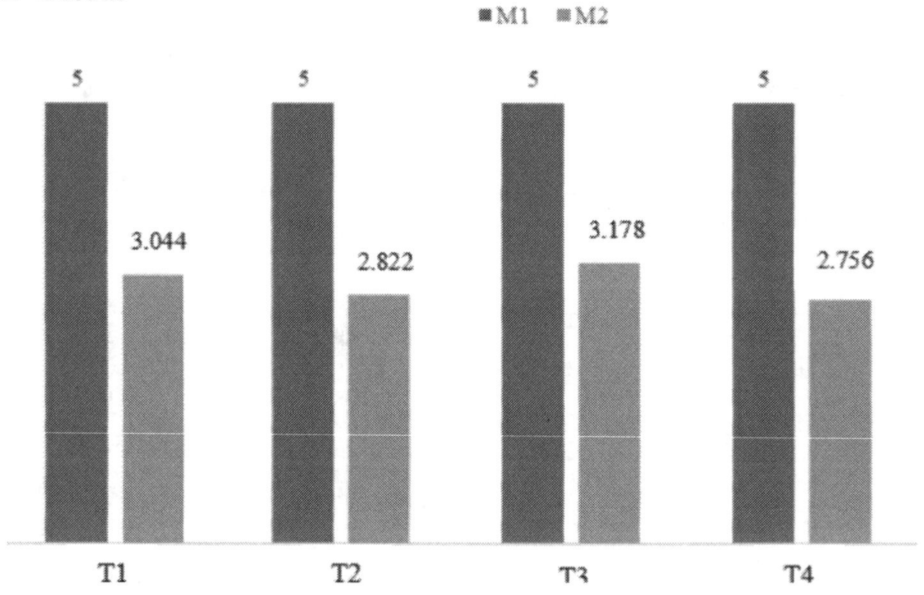

Fig. 1. Organoleptic of colour

 https://doi.org/10.1051/bioconf/20238100022

The colour organoleptic test was conducted based on the level of preference of 15 panellists. Based on the graph above, the colour value of the new cooking oil from T1 to T4 has the same average value of 5, which means that all new oils from T1 to T4 have a clear yellow colour. Meanwhile, in the graph, the panellists considered that the waste cooking oil experienced a change in colour, this can be seen from the results of the panellists' average value of 2.950, which means that the colour changed to a brownish yellow colour. The waste cooking oil that has the darkest colour among the others is waste oil C from restaurant D and the waste oil that has a fairly good colour is oil B from restaurant C.

Changes in used cooking oil are caused by oxidative damage [7]. Oxidation reactions occur between oxygen and double bonds of triglycerides / oil [8]. The presence of antioxidants (tocopherols) in oil is useful for diverting the oxidation process from oil to antioxidants so that oil bonds remain intact. In addition, the presence of molecular bonds of carbohydrates and proteins, or referred to as the Maillard Reaction, which is the reaction between the carbonyl group and the amine group of the protein, is also the cause of the brown colour of used cooking oil. The dark colour of the oil can also occur during the processing, storage and use of the oil. Heating is too high and repetitive, causing polymerization reactions and Maillard reactions that cause the oil to thicken and darken [8]. The greater the colour density of a solution, the more turbid the solution, and the worse the quality of the solution. It is necessary to purify the colour of waste cooking oil in order to improve the potential quality of waste oil as animal feed [9].

3.1.2 Texture

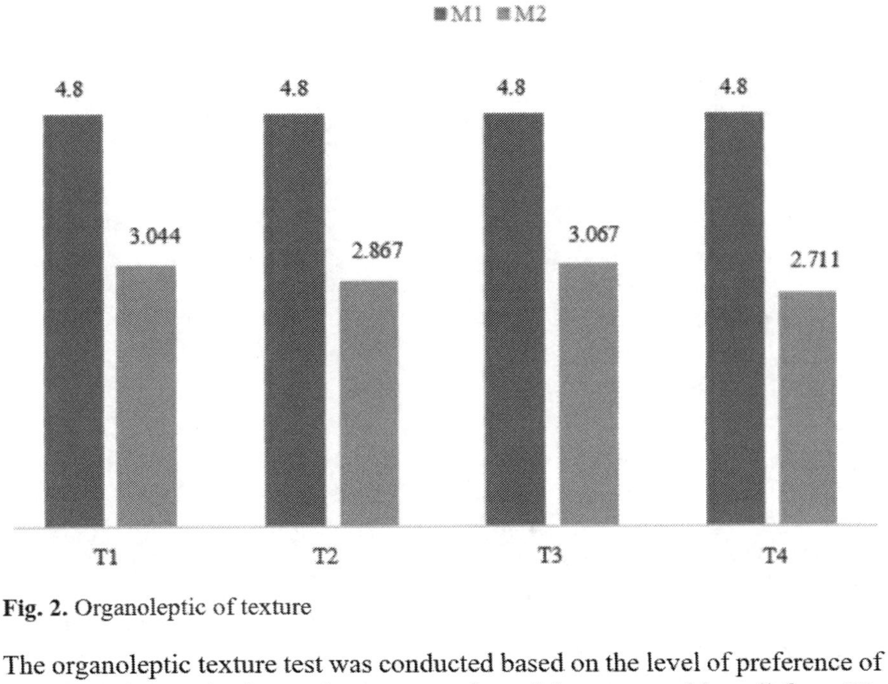

Fig. 2. Organoleptic of texture

The organoleptic texture test was conducted based on the level of preference of 15 panellists. Based on the graph above, the texture value of the new cooking oil from T1 to T4 has the same average value of 4.8, which meant that all new oils from T1 to T4 have a slightly thick texture. As for waste cooking oil, based on the graph above, the texture value with the lowest average was found in waste cooking oil D Restaurant (T4) which is 2.8 which means thick tends to be very thick while the highest average value is found in waste cooking oil C Restaurant (T3) which is 3.067 which means thick. T1 shows that waste oil brand A in A Restaurant is on average classified as a waste oil assessment with a thick texture with an average value of 3.044. Meanwhile, Waste oil A in B Restaurant (T2) still gets a value close to 3 so it is still very thick texture. Waste oil B from C Restaurant T3) is classified as a thick texture assessment because it has a value above 3. While waste oil C from D Restaurant (T4) has the lowest value with a value of 2.8 so it is still a very thick texture.

Oil becomes thicker, foamy, brown in colour, and leaves a slightly bad odor due to various kinds of damage due to oxidation, hydrolysis, polymerization, and metal reactions [10] Repeated heating of oil at high temperatures and a long enough time will produce polymer compounds that are solid in oil. Oil that has been damaged not only causes damage to nutritional value but also damages the texture of the flavour of the fried food [11]. The texture of cooking oil that has been used many times will have a thicker texture than fresh oil due to the formation of acidic polymers, dimers and glycerides during heating when used [12]. There is no difference in the odor of each treatment because it has an average value that is not much different (the same), which has a value close to 3, which means that the texture of the waste oil is rather thick.

3.1.3 Odor

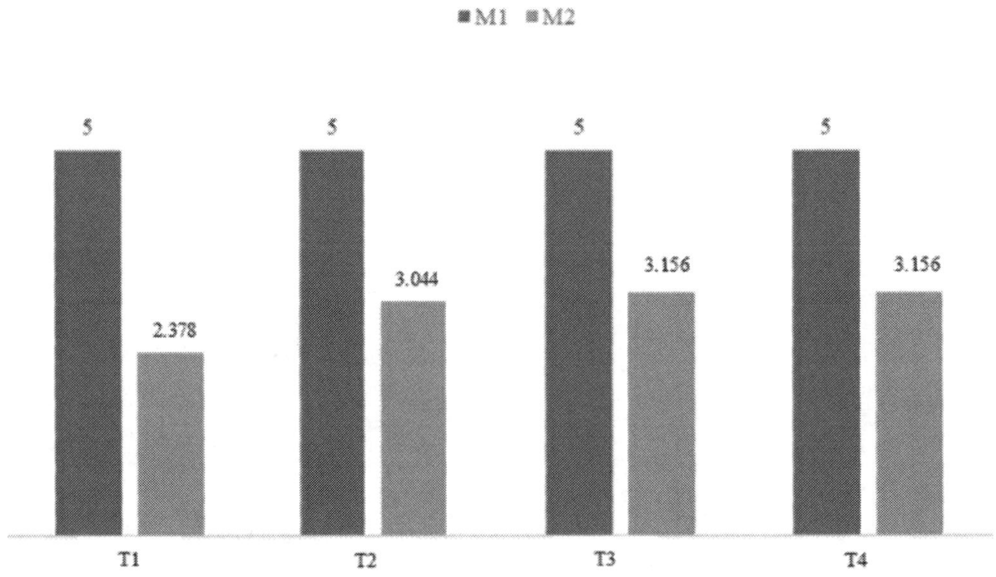

Fig. 3. Organoleptic of odor

The organoleptic odor test was conducted based on the level of preference of 15 panellists. Based on the graph above, the odor value of new cooking oil from T1 to T4 has the same average value of 5, which means that all new oils from T1 to T4 have the typical smell of new oil. While the waste cooking oil, based on the graph above has an average value of 3 so that from the assessment of the waste cooking oil panellists have a slightly bad odor. The restaurant that has the lowest score is restaurant A which has a score of 2.378 which means it has a worst odor than the other three restaurants.

This condition is possible because during the food frying process, the oil undergoes decomposition Fat hydrolysis reaction occurs when fat is hydrolysed into free fatty acids and glycerol. The cause of rancidity is due to the formation of free fatty acids. Rancidity in oil is felt when the oil contains free fatty acids, especially at low temperatures. Fat oxidation reactions occur in fats that contain unsaturated fatty acids in the carbon chain, have single or multiple double bonds. Changes starting from free radicals if oxygen is available will form peroxides which are also radical in nature, then form hydroperoxide compounds and are easily degraded into short-chain compounds from the aldehyde, ketone and alcohol groups which have an impact on the rancidity of the oil [3]. There is no difference in the odor of each treatment because it has an average value that is not much different, which is touching the value of 3, which means it is not slightly rancid. The deterioration of fat is the development of a rancid odor and taste, which is called the rancidity process. It is caused by factors that can accelerate the reaction, such as heat and light. Oxidation can also occur when there is contact between a certain amount of oxygen and oil or fat. The occurrence of this oxidation results in a rancid odor in oil or fat [13]. So, the oil is not good if consumed. Oil or fat refining is needed to eliminate bad taste and odor, unattractive color and extend the shelf life before consumption [14].

3.1.4 Flavour

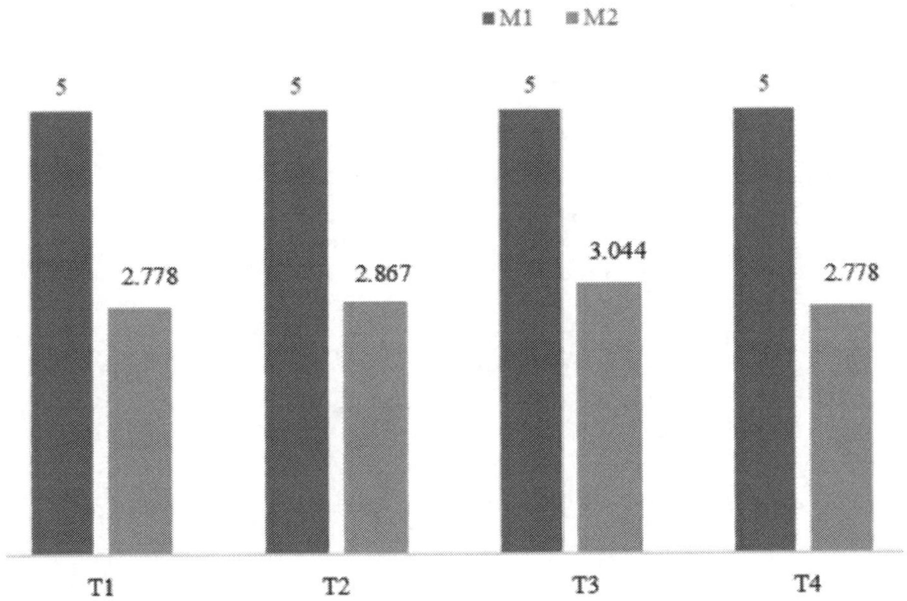

Fig. 4. Organoleptic of flavour

The organoleptic taste test was conducted based on the level of preference of 15 panellists. Based on the graph above, the taste value of the new cooking oil from T1 to T4 has the same average value of 5, which means that all new oils from T1 to T4 have a normal bland taste. While the average waste cooking oil has a value of about 2.867 which is classified as having a bitter taste. However, in T4, the C restaurant has the best value among other restaurants, which is 3.004, which is classified as having a slightly bitter taste.

3.2 Iodine Content

In determining the quality of oil there is a standard that becomes a reference. For domestic use, the standard used is SNI (Indonesian National Standard), but for cooking oil to be exported, the SPB (Special Prime Bleach) standard is used.

The average iodine content of new cooking oil is 55.183 g Iod/100g. According to SNI - Coconut Oil Technology - SNI - 3741 - 2013 and Special Prime Bleach (SPB), all new cooking oils from T1 to T4 have exceeded the minimum limit of iodine content in new cooking oil.

Table 3. Iodometric Test Results according to SNI-Coconut Oil Technology-SNI-3741-2013

Treatments		Iodin Content (g Iod/100g)	SNI - Coconut Oil Technology - SNI - 3741 - 2013	Special Prime Bleach (SPB)
M_1 (new cooking oil)	T_1	55.320	45 – 46 (g Iod/100g)	53 ± 1,5 (g Iod/100g)
	T_2	55.320		
	T_3	53.050		
	T_4	57.040		
Average		55.183		
M_2 (waste cooking oil)	T_1	52.680 ± 3.532^{ab}		
	T_2	48.443 ± 1.150^{a}		
	T_3	53.903 ± 1.098^{b}		
	T_4	61.870 ± 2.272^{c}		
Average		54.224 ±2.013		

A cooking oil used at A Restaurant (T1) experienced a decrease in iodine levels from new oil that became waste cooking oil with a decrease of 2.64 g Iod/100g. A cooking oil used at B Restaurant (T2) experienced a decrease in iodine levels from new oil which became waste cooking oil with a decrease of 6.877 g Iod/100g. Meanwhile, B cooking oil used in C Restaurant (T3) experienced an increase in iodine content from new oil which became waste cooking oil with an increase of 0.853 g Iod/100g. As happened in T3, C cooking oil used in D Restaurant (T4) experienced an increase of 4.83 g Iod/100g. Based on the results of this study, the average waste cooking oil had low iodine values. The decrease in iodine content is due to the heating of the oil which causes the breaking of unsaturated bonds to become saturated. this is also supported by the opinion of [15], At an increase in temperature the double bond is reduced because the double bond is split into single bonds during heating which results in the iodine number going down [16]. However, the iodine number produced is still above the SNI standard so it is still suitable for use. Meanwhile, the increasing iodine value in used cooking oil can be considered as the best quality oil and suitable for consumption.

4 Conclusion

Based on the results of this study, it can be concluded that waste cooking oil obtained from four restaurants showed changes in colour, texture, odor and taste. However, it had a high iodine content which indicate that the waste cooking oils from the restaurants selected were still suitable for consumption by livestock.

References

1. E. Wina, and I.W.R. Susana, *WARTAZOA. Indonesian Bulletin of Animal and Veterinary Sciences*, **23**, 4 (2013)
2. S. Sudibya, K.R. Akbar, W.P. Sabar and J. Riyanto, Sains Peternakan, **15**, 1 (2017)
3. A. Prarudiyanto, E. Basuki, A. Alamsyah, D. Handito, Jurnal Ilmiah Rekayasa Pertanian dan Biosistem. **3**, 41-48 (2015)
4. U.H. Tanuwiria, D.C. Budinuryanto, S. Darodjah and W.S. Putranto, *Jurnal Ilmu Ternak*, **10,** 32-38 (2011)
5. A. Abqoriyah, S.D. Widyawati, and L. Lutojo. Zoo Indonesia. **22**, 2 (2013)
6. Yeniza, and P.A. Anjar. Amina. **1**, 2 (2019)

7. I.O. Angelia, JTech, **4**, 32-36 (2016)
8. A.S. Suroso, Jurnal Kefarmasian Indonesia. **3**, 2, 77-88 (2013)
9. F. A. Viantini, Konversi. **4**, 2, 53-62 (2015)
10. R. Hartono and E. Suhendi, JIP. **9**, 1 (2020)
11. S. Chodijah and A.P. Harmansyah, Kinetika, **1**, 1 (2011)
12. M. Milawarni, and S. Saifuddin, Jurnal Teknologi Kimia Unimal, **6,** 2, 52-62 (2018)
13. P.V. Patty, Biopendix. **1**, 2, 146-152 (2015)
14. R.E.P. Gusti and Zulnely. Jurnal Penelitian Hasil Hutan **33**, 1, 61-68 (2015)
15. F. Estrada, R. Gusmao, N. Indraswati, Widya Teknik. **6,** 2, 121-130 (2017)
16. F. A. Viantini, 2015. Konversi. **4**, 2, 53-62 (2015)

Psychological resilience of dairy farmers, life satisfaction, and the moderating role of age; Is there a link?

Jaisy Aghniarahim Putritamara[1*], *Budi* Hartono[1], *Intan* Kamilia Hapsari[2], *Awang* Tri Satria[1], and *Tina* Sri Purwanti[1]

[1] Department of Livestock Socio-Economics, Faculty of Animal Science, Universitas Brawijaya, Malang 65145, Indonesia
[2] Department of Animal Husbandry, State Polytechnic of Lampung, Lampung 35144, Indonesia

Abstract. Psychological resilience, life satisfaction, and the moderating role of age in their relationship have been largely overlooked in the literature that discusses business resilience, particularly in small-scale farming in developing countries like Indonesia. This research has the purpose to fill a research gap by exploring the influence of farmers' psychological resilience on life satisfaction. It employs cross-sectional data, observing 210 dairy farmers in Indonesia who have been in their business for a minimum of 10 years. The data is subsequently tested for validity and reliability and analyzed using a quantitative multivariate approach, Structural Equation Modeling (SEM), with smartPLS software. The research demonstrates that psychological resilience has a positive influence on farmers' life satisfaction. Additionally, age has also been proven to moderate psychological resilience and life satisfaction. Based on these outcomes, we recommend that all stakeholders pay attention to formulating policies that support the enhancement of farmers' psychological resilience in facing uncertainties, such as endemic disasters like Foot and Mouth Disease (FMD) and Lumpy Skin Disease (LSD). The implications of these result can provide insights for stakeholders to enhance business resilience from intangible aspects, such as the psychology of farmers, which has been shown to impact life satisfaction, and the moderating role of age in this relationship. Thus, it is expected that entrepreneurs can establish a sustainable dairy cattle business amidst an increasingly uncertain business landscape in the future.
Keyword: Psychological resilience, life satisfaction, FMD (Foot and Mouth Disease), dairy cattle; Indonesia.

1 Introduction

In recent years, the business world has been struck by various forms of uncertainty, such as the fragility of supply chains due to restrictive policies during the pandemic, climate change causing disruptions in production, and threats of diseases.[1–10]. It can lead to entrepreneurs being unable to maintain their competitive advantages[11], as several international studies have

* Corresponding author: jaisyap@ub.ac.id

shown that when they cannot sustain, there is a decline in profit and business performance. For example, according to the empirical outcomes revealed by [12] conducted in the agricultural sector in the Netherlands, there is a connection between psychological issues and financial problems for entrepreneurs. Farmers experiencing psychological pressure often do so because they are facing financial difficulties. However, according to [13] attitudes, behaviors, environment, motivation, psychology, emotions, and business activities influence attitudes that indicate sustainability in entrepreneurship. This is also corroborated by the outcomes of research [14] which examined psychological resilience in the context of the sustainability opportunities of small businesses contributing to the circular economy. The research's results indicate changes in business sustainability in response to the financial vulnerability experienced by entrepreneurs during the pandemic. This is due to the impact of stress levels on entrepreneurs, making them unable to support business resilience.[15]. Therefore, studies related to psychological resilience have attracted the attention of researchers on the international stage for further investigation, particularly in relation to its connection to life satisfaction, enabling entrepreneurs to focus on the sustainability of their businesses amidst uncertainty threats. This is because empirical studies have proven that entrepreneurs who attain life satisfaction will have an impact on the resilience of their businesses[16]. Furthermore, according to the outcomes of research [17] business resilience is a crucial aspect in achieving subjective well-being. Research [16] provides evidence that even when entrepreneurs face uncertainty and business threats, they tend to be resilient in overcoming negative impacts and strive to restore their businesses to their previous state.

International studies have confirmed the association between psychological resilience and entrepreneurs' life satisfaction. Medically, the outcomes of research [18] have demonstrated a link between psychological resilience and the life satisfaction of mentally and physically healthy adults. The results of empirical research in the tourism sector in India, as shown in research [16] have also demonstrated a positive influence of psychological capital on life satisfaction. Empirical research conducted by [19] with a sample of university students in Turkey has similarly confirmed that psychological resilience has a positive impact on students' life satisfaction.

Unfortunately, to date, there has been no research confirming the role of psychological resilience in life satisfaction, especially in the livestock sector for dairy farming businesses in developing countries like Indonesia. However, along with dairy cooperatives, the dairy cattle industry in Indonesia has the potential to make a significant contribution to the country's economy [20]. On the other hand, the threats of climate change and the emergence of diseases, such as Foot and Mouth Disease (FMD) and Lumpy Skin Disease (LSD), which are significant epidemics, pose a serious threat to the resilience of ruminant businesses in Indonesia, causing social and economic harm to farmers due to a decrease in the quality and quantity of milk.[21,22]. Furthermore, to date, there has been no research considering age as a moderating variable that is suspected to either strengthen or weaken the relationship between psychological resilience and life satisfaction. However, according to some studies [23–25] it has been demonstrated that different age groups can influence an individual's level of life satisfaction, with life satisfaction being indicated to decrease in individuals aged over 65 years.

This research contributes to several important aspects. Firstly, it contributes to the existing literature related to the impact of psychological resilience on life satisfaction. Secondly, it contributes to assisting dairy farmers in achieving life satisfaction amidst business uncertainty. Thirdly, it contributes to the literature related to the middle theory of imperative strategies that aid entrepreneurs in sustaining their businesses by their ability to attain life satisfaction moderated by age.

2 Research Method

2.1 Measurement of research variables

The key variables in this research include psychological resilience and life satisfaction. Items related to the resilience variable are based on the outcomes of research [26] which encompass the entrepreneur's ability to bounce back from tough times, the challenges entrepreneurs face when dealing with cases of Foot and Mouth Disease (PMK) and Lumpy Skin Disease (LSD) that cause them stress, the difficulty entrepreneurs encounter in stepping back from their livestock farming business when adverse events occur, and the extended time required by entrepreneurs to overcome failures in life due to the serious impact of PMK and LSD on their business. As for the dependent variable in this research, which is life satisfaction, it is referred to the results of a research [26] comprising five statement items, including: entrepreneurs feel that in many aspects of what has occurred, their lives are almost approaching the ideal; entrepreneurs feel that the current condition of their life is very good; entrepreneurs feel satisfied with their lives; so far, entrepreneurs have achieved the important things they desired in their lives; and if entrepreneurs could rewind time, they would not change anything in their lives. This research also involves a moderating variable suspected to strengthen the relationship between psychological resilience and life satisfaction. The instrument used for measurement employs a questionnaire with a Likert scale scoring, where 1 corresponds to "strongly disagree" and 5 corresponds to "strongly agree."

2.2 Sampling

The aim of this research is to explore the role of psychological resilience in life satisfaction moderated by the age of entrepreneurs. Therefore, an instrument is needed to measure respondent's answers. Consequently, the construct is formed in accordance with the desired research model, where the dependent variable is influenced by independent variables, and there is a moderating variable to strengthen or weaken that relationship. This research was conducted over a period of two months, from July to August 2023. The research was carried out in the area with the highest population of dairy cattle in East Java, namely in Pasuruan regency, with a population of 97,101 head in 2021, increasing to 100,080 head in 2022. https://jatim.bps.go.id/statictable/2023/03/21/2590/-populasi-ternak-sapi-perah-dan-sapi-potong-menurut-kabupaten-kota-dan-jenis-ternak-di-provinsi-jawa-timur-ekor-2021-dan-2022.html. This research employs a multivariate approach with Structural Equation Modeling (SEM) analysis.

3 Empirical Result

3.1 Analysis of Validity and Reliability

Based on the outer model testing, it was found that the items forming the construct have valid and reliable contributions. In accordance with the minimum standards recommended by multivariate SEM quantitative experts [27] the minimum AVE value is 0.6, the minimum CA and CR values are 0.7, and the minimum factor loading value is 0.6.

Table 1. Outer Model Testing

Construct	Item	Loading	CA	CR	AVE
Psychologhy Resilience[26]	X1		0.765	0.826	0.640
I tend to bounce back from the tough times of Foot and Mouth Disease (PMK) and Lumpy Skin Disease (LSD).	x1.1	0.727			
I struggle to get through the incidents of PMK and LSD that cause me stress.	x1.2	0.780			
It is difficult for me to step back from the livestock farming business when bad things happen to my venture.	x1.3	0.724			
It takes me a long time to overcome failures in life due to the serious impact of Foot and Mouth Disease (PMK) and Lumpy Skin Disease (LSD) on my business.	x1.4	0.787			
Life Satisfaction[26]	Y		0.741	0.812	0.659
In many aspects of what has occurred, my life is almost approaching the ideal.	Y1	0.710			
My current life circumstances are very good.	Y2	0.767			
I am satisfied with my life.	Y3	0.716			
So far, I have achieved the important things I desired in my life.	Y4	0.725			
If I could rewind time, I would not change anything in my life.	Y5	0.731			

3.2 Structural Measurement Model

In the inner model testing, the results showed that hypothesis 1 was accepted with a t-value of 8.934, and similarly, hypothesis 2 was also accepted with a t-value of 3.666. This means that from the structural model testing, psychological resilience has a positive impact on the life satisfaction of entrepreneurs, and age plays a moderating role in the relationship between psychological resilience and life satisfaction. The following are the results of hypothesis testing.

Table 2. Hypothesis testing

Hypothesis	Relationship	Std.Beta	Std.Error	t-value	p-values	Decision
			Direct effect			
H1	PR >> LS	0.479	0.051	8.934	0.000	Yes
			Indirect effect			
H2	PR_age>>LS	0.217	0.063	3.666	0.000	Yes

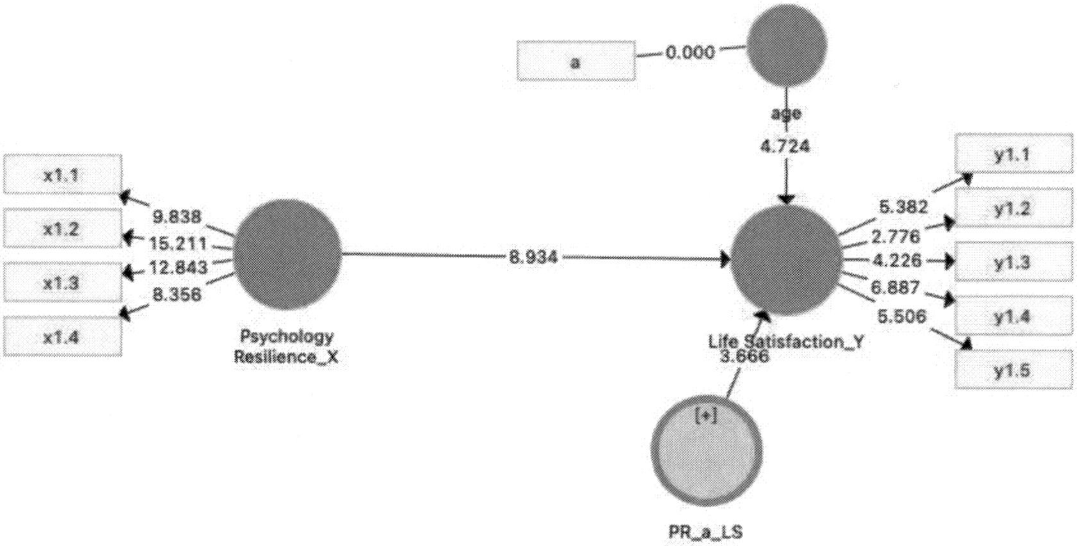

Fig. 2. Empirical Model of the Role of Psychological Resilience in Life Satisfaction Moderated by the Age of Dairy Farmers in East Java, Indonesia

4 Discussion

The result of this research indicate that psychological resilience has a positive impact on the life satisfaction of entrepreneurs. Psychological resilience refers to an individual's ability to cope with pressure, challenges, trauma, and their capacity to recover effectively from difficult experiences they have gone through.[28–30]. Based on the empirical research conducted by reference 31, it has been demonstrated that entrepreneurs who can endure tend to have the ability to manage their thinking and cope with stress, enabling them to handle difficulties with better actions and decisions. Entrepreneurs tend to engage in positive activities[32]. Therefore, these approaches can have a positive impact on individuals' levels of life satisfaction because entrepreneurs are considered better able to overcome obstacles and engage in more productive business activities with behaviors that make them happy.[33]. This is also triggered by several important factors such as social support, life experiences, and individual mindset. [34–36].

Furthermore, the second outcome in this research provides evidence of the significant moderating role of age between psychological resilience and life satisfaction. For instance, as suggested by the opinion of reference [37] if entrepreneurs have relatively stable financial conditions and a better economic situation, individuals are considered capable of achieving life satisfaction. According to several studies [38–40]it has been demonstrated that individuals experiencing financial problems tend to experience anxiety and lower life satisfaction. This is because an individual's achievements play a crucial role in motivating them to attain life satisfaction. Therefore, for individuals who can achieve their life goals, they will experience life satisfaction.[41]. In this context, age plays a crucial role in both bolstering and weakening the psychology of resilience concerning life satisfaction. [41–45]. As individuals age, their psychological resilience improves, allowing them to achieve a higher level of life satisfaction compared to younger entrepreneurs. This is because, throughout the course of life, an individual's priorities and goals tend to change with age[46,47]. If someone might be more focused on education and career at a young age, while at an older age, the focus can shift to social relationships, family, or seeking meaning in life. [48,49]. These changes can influence the level of life satisfaction. Similarly, an individual's ability to adapt to changes in life, those who tend to be adaptive to changes are actually more satisfied with their lives.[50–52].

4.1 Theoretical Contribution

This research contributes to several important aspects in the literature. First, the research identifies the role of psychological resilience in life satisfaction. Second, the research provides empirical evidence for the mediating role of age in the relationship between psychological resilience and life satisfaction, particularly in the livestock sector, focusing on dairy farming businesses currently experiencing the complexity of uncertainty factors such as climate change and diseases like Foot and Mouth Disease (PMK) and Lumpy Skin Disease (LSD) in a developing countries.

4.2 Limitation

This study is limited to the research of dairy cattle farming in the business development centers located in East Java. Furthermore, this research has not yet proven the measurement of each business based on the scale of operations, as demonstrated by previous research (reference 11), which indicates that the size of businesses in terms of assets can show differences in the resilience of entrepreneurs when facing threats.

5 Conclusion

This research concludes that psychological resilience has a positive influence on the life satisfaction of entrepreneurs. Similarly, age plays a moderating role in the relationship between psychological resilience and life satisfaction. The model in this research can be implemented in other sectors in both developing and developed countries to assess the psychological resilience of entrepreneurs seeking life satisfaction in the face of uncertainty.

The researchers express their gratitude for the financial support from the Research Grant of the Faculty of Animal Science, Brawijaya University, under contract number 1873.4/UN10.F05/PN/2023.

References

1. Babayani, N. D. & Thololwane, O. I. A qualitative risk assessment indicates moderate risk of foot-and-mouth disease outbreak in cattle in the lower Okavango

Delta because of interaction with buffaloes. *Transbound Emerg Dis* **69**, 2840–2855 (2022).

2. Slawinski, N., Pinkse, J., Busch, T. & Banerjee, S. B. The role of short-termism and uncertainty avoidance in organizational inaction on climate change: A multi-level framework. *Bus Soc* **56**, 253–282 (2017).
3. Ozdemir, O., Kizildag, M., Dogru, T. & Madanoglu, M. Measuring the effect of infectious disease-induced uncertainty on hotel room demand: a longitudinal analysis of US hotel industry. *Int J Hosp Manag* **103**, 103189 (2022).
4. Phillipson, J., Bennett, K., Lowe, P. & Raley, M. Adaptive responses and asset strategies: the experience of rural micro-firms and foot and mouth disease. *J Rural Stud* **20**, 227–243 (2004).
5. Lee, S. & Klassen, R. D. Firms' response to climate change: The interplay of business uncertainty and organizational capabilities. *Bus Strategy Environ* **25**, 577–592 (2016).
6. Irvine, W. & Anderson, A. R. The impacts of foot and mouth disease on a peripheral tourism area: The role and effect of crisis management. *Journal of Travel & Tourism Marketing* **19**, 47–60 (2006).
7. Blake, A., Sinclair, M. T. & Sugiyarto, G. Quantifying the impact of foot and mouth disease on tourism and the UK economy. *Tourism Economics* **9**, 449–465 (2003).
8. Bui, B. & De Villiers, C. Business strategies and management accounting in response to climate change risk exposure and regulatory uncertainty. *The British Accounting Review* **49**, 4–24 (2017).
9. Baker, S. R., Bloom, N., Davis, S. J. & Terry, S. J. *Covid-induced economic uncertainty*. (2020).
10. Leduc, S. & Liu, Z. Can pandemic-induced job uncertainty stimulate automation? in (Federal Reserve Bank of San Francisco, 2020).
11. Putritamara, J. A. *et al.* Do Dynamic Capabilities and Digital Transformation Improve Business Resilience during the COVID-19 Pandemic? Insights from Beekeeping MSMEs in Indonesia. *Sustainability* **15**, 1760 (2023).
12. Gorgievski, M. J., Bakker, A. B., Schaufeli, W. B., van der Veen, H. B. & Giesen, C. W. M. Financial problems and psychological distress: Investigating reciprocal effects among business owners. *J Occup Organ Psychol* **83**, 513–530 (2010).
13. Tur-Porcar, A., Roig-Tierno, N. & Llorca Mestre, A. Factors affecting entrepreneurship and business sustainability. *Sustainability* **10**, 452 (2018).
14. Chhatwani, M., Mishra, S. K., Varma, A. & Rai, H. Psychological resilience and business survival chances: A research of small firms in the USA during COVID-19. *J Bus Res* **142**, 277–286 (2022).
15. Chadwick, I. C. & Raver, J. L. Psychological resilience and its downstream effects for business survival in nascent entrepreneurship. *Entrepreneurship Theory and Practice* **44**, 233–255 (2020).
16. Pathak, D. & Joshi, G. Impact of psychological capital and life satisfaction on organizational resilience during COVID-19: Indian tourism insights. *Current Issues in Tourism* **24**, 2398–2415 (2021).
17. Bajaj, B. & Pande, N. Mediating role of resilience in the impact of mindfulness on life satisfaction and affect as indices of subjective well-being. *Pers Individ Dif* **93**, 63–67 (2016).
18. Kong, F., Wang, X., Hu, S. & Liu, J. Neural correlates of psychological resilience and their relation to life satisfaction in a sample of healthy young adults. *Neuroimage* **123**, 165–172 (2015).

19. Gundogan, S. The mediator role of the fear of COVID-19 in the relationship between psychological resilience and life satisfaction. doi:10.1007/s12144-021-01525-w/Published.
20. Wijers, G. D. M. Inequality regimes in Indonesian dairy cooperatives: understanding institutional barriers to gender equality. *Agric Human Values* **36**, 167–181 (2019).
21. Sutawi, A. *et al.* Re-emergence of foot and mouth disease outbreak in Indonesia: A review. *Adv. Anim. Vet. Sci.* **11**, 264–271 (2023).
22. Fadillah, A. *et al.* Smallholders' milk quality awareness in Indonesian dairy farms. *J Dairy Sci* (2023) doi:https://doi.org/10.3168/jds.2023-23267.
23. Spreitzer, E. & Snyder, E. E. Correlates of life satisfaction among the aged. *J Gerontol* **29**, 454–458 (1974).
24. Cotton Bronk, K., Hill, P. L., Lapsley, D. K., Talib, T. L. & Finch, H. Purpose, hope, and life satisfaction in three age groups. *J Posit Psychol* **4**, 500–510 (2009).
25. Chen, C. Aging and life satisfaction. *Soc Indic Res* **54**, 57–79 (2001).
26. Prayag, G., Spector, S., Orchiston, C. & Chowdhury, M. Psychological resilience, organizational resilience and life satisfaction in tourism firms: Insights from the Canterbury earthquakes. *Current Issues in Tourism* **23**, 1216–1233 (2020).
27. Chin, W. W. & Newsted, P. R. Structural equation modeling analysis with small samples using partial least squares. *Statistical strategies for small sample research* **1**, 307–341 (1999).
28. Harrison, D., Sarkar, M., Saward, C. & Sunderland, C. Exploration of psychological resilience during a 25-day endurance challenge in an extreme environment. *Int J Environ Res Public Health* **18**, 12707 (2021).
29. Shih, M., Wilton, L. S., Does, S., Goodale, B. M. & Sanchez, D. T. Multiple racial identities as sources of psychological resilience. *Soc Personal Psychol Compass* **13**, e12469 (2019).
30. Harvey, M. R. Towards an ecological understanding of resilience in trauma survivors: Implications for theory, research, and practice. *J Aggress Maltreat Trauma* **14**, 9–32 (2007).
31. Gibb, J. *et al.* Historical and social aspirations: influence of SME key decision makers' resilience, social skills and stress on attaining firm performance goals. *International Journal of Entrepreneurial Behavior & Research* **28**, 1414–1437 (2022).
32. Chhatwani, M., Mishra, S. K., Varma, A. & Rai, H. Psychological resilience and business survival chances: A research of small firms in the USA during COVID-19. *J Bus Res* **142**, 277–286 (2022).
33. Terzioğlu, Z. A., Çakır-Çelebi, S. G. & Yıldız, M. Effect of online mindfulness-based physical exercise program on psychological well-being and life satisfaction during the Covid-19 pandemic. *Current Psychology* 1–11 (2022).
34. Weckroth, M. & Kemppainen, T. (Un) Happiness, where are you? Evaluating the relationship between urbanity, life satisfaction and economic development in a regional context. *Reg Stud Reg Sci* **8**, 207–227 (2021).
35. Mathew, V. S. Happiness a Driver for Innovation at the Workplace. in *Understanding Happiness: An Explorative View* 335–344 (Springer, 2023).
36. Terzioğlu, Z. A., Çakır-Çelebi, S. G. & Yıldız, M. Effect of online mindfulness-based physical exercise program on psychological well-being and life satisfaction during the Covid-19 pandemic. *Current Psychology* 1–11 (2022).
37. Heo, W., Lee, J. M. & Park, N. Financial-related psychological factors affect life satisfaction of farmers. *J Rural Stud* **80**, 185–194 (2020).

38. Rojas, M., Watkins-Fassler, K. & Rodríguez-Ariza, L. The life satisfaction of owner-manager entrepreneurs when the business of business is not only business. *Appl Res Qual Life* **17**, 2251–2275 (2022).
39. Nickerson, C., Schwarz, N., Diener, E. & Organization, G. Financial aspirations, financial success, and overall life satisfaction: who? and how? *J Happiness Stud* **8**, 467–515 (2007).
40. Heo, W., Lee, J. M. & Park, N. Financial-related psychological factors affect life satisfaction of farmers. *J Rural Stud* **80**, 185–194 (2020).
41. Cotton Bronk, K., Hill, P. L., Lapsley, D. K., Talib, T. L. & Finch, H. Purpose, hope, and life satisfaction in three age groups. *J Posit Psychol* **4**, 500–510 (2009).
42. Peterson, C., Ruch, W., Beermann, U., Park, N. & Seligman, M. E. P. Strengths of character, orientations to happiness, and life satisfaction. *J Posit Psychol* **2**, 149–156 (2007).
43. Joshanloo, M. & Jovanović, V. Similarities and differences in predictors of life satisfaction across age groups: A 150-country research. *J Health Psychol* **26**, 401–411 (2021).
44. Berg, A. I., Hassing, L. B., McClearn, G. E. & Johansson, B. What matters for life satisfaction in the oldest-old? *Aging Ment Health* **10**, 257–264 (2006).
45. St. John, P. D., Mackenzie, C. & Menec, V. Does life satisfaction predict five-year mortality in community-living older adults? *Aging Ment Health* **19**, 363–370 (2015).
46. Stephan, Y. Openness to experience and active older adults' life satisfaction: A trait and facet-level analysis. *Pers Individ Dif* **47**, 637–641 (2009).
47. Hudomiet, P., Hurd, M. D. & Rohwedder, S. The age profile of life satisfaction after age 65 in the US. *J Econ Behav Organ* **189**, 431–442 (2021).
48. Swift, H. J. *et al.* Revisiting the paradox of well-being: The importance of national context. *Journals of Gerontology Series B: Psychological Sciences and Social Sciences* **69**, 920–929 (2014).
49. Drabe, D., Hauff, S. & Richter, N. F. Job satisfaction in aging workforces: an analysis of the USA, Japan and Germany. *The International Journal of Human Resource Management* **26**, 783–805 (2015).
50. Upadyaya, K., Vartiainen, M. & Salmela-Aro, K. From job demands and resources to work engagement, burnout, life satisfaction, depressive symptoms, and occupational health. *Burn Res* **3**, 101–108 (2016).
51. Santilli, S., Nota, L., Ginevra, M. C. & Soresi, S. Career adaptability, hope and life satisfaction in workers with intellectual disability. *J Vocat Behav* **85**, 67–74 (2014).
52. Prayag, G., Spector, S., Orchiston, C. & Chowdhury, M. Psychological resilience, organizational resilience and life satisfaction in tourism firms: Insights from the Canterbury earthquakes. *Current Issues in Tourism* **23**, 1216–1233 (2020).

Morphometric analysis of IPB-D1 chicken in free-range and intensive rearing systems in the tropics region

Jonathan Anugrah Lase[1,4], *Rudi* Afnan[2], *Zakiah* Wulandari[2], *Sri* Estuningsih[3], *Tike* Sartika[4], and *Cece* Sumantri[2*]

[1]Graduate School of Animal Production and Technology, Faculty of Animal Science, IPB University, Jalan Agatis, Kampus IPB Dramaga, Bogor 16680, West Java, Indonesia
[2]Department of Animal Production and Technology, Faculty of Animal Science, IPB University, Jalan Agatis, Kampus IPB Dramaga, Bogor 16680, West Java, Indonesia
[3]Division of Veterinary Pathology, School of Veterinary and Biomedicine, IPB University, Jalan Agatis Kampus IPB Dramaga, Bogor, 16680, West Java, Indonesia
[4]Research Center for Animal Husbandry, Research Organization for Agriculture and Food, National Research and Innovation Agency of The Republic of Indonesia (BRIN), Cibinong Sciences Center, Cibinong, Bogor, Indonesia

Abstract. IPB-D1 chicken is a local chicken breed developed for meat production and fast growth. This chicken is formed from crossbreeding broiler breeds (Cobb) with 3 other local chicken breeds, Kampung, Pelung, and Sentul Chicken. It has advantages in disease resistance and body weight. In this study, IPB-D1 chickens were studied for morphometric characteristics in two (2) different rearing systems, namely free-range and intensive systems. The material used was 120 IPB-D1 roosters aged 20 weeks, consisting of 90 heads reared in the free-range system and 30 heads intensively. The method used was observation. The parameters measured were morphometric variables of chickens in free-range and intensive systems. The rearing system treatment showed no difference in femur length, tibia length, tarsometatarsus circumference, third finger length, and wing length, but showed significant differences in tarsometatarsus length, maxilla length, and cockscomb height between intensive and free-range rearing systems in the tropics. Free-range rearing can increase tarsometatarsus length and maxilla length.

1 Introduction

Chicken is one of the most popular livestock among Indonesians, known as a meat producer and widely utilized as a source of protein. Chicken meat, especially local chicken, is highly demanded by the Indonesian people because it can be obtained easily in the market and is widely cultivated by the Indonesian people. One of the current superior local chicken breeds, IPB-D1 chicken, was developed as a meat producer [1]. This type of chicken is formed to

* Corresponding author: ceces@apps.ipb.ac.id

inherit genetic advantages from local chickens native to Indonesia, namely disease resistance, and from purebred chickens (Cobb) as meat producers. The superior performance of IPB-D1 chickens is expected to be able to adapt and produce high yields in tropical Indonesia.

The rearing system plays an important role in supporting chicken production. The maintenance system that is widely practiced is the intensive and semi-intensive system. The intensive system is characterized by controlled maintenance, feeding, and drinking, and the environment is arranged according to the comfort of the livestock. However, there are weaknesses in this intensive system, namely that livestock are less able to express their normal behavior, so they do not meet the principles of animal welfare [2]. Currently, the issue of animal welfare in poultry rearing systems is a major issue that must be accommodated. There are many reports that there is a positive correlation between semi-intensively reared animals and performance, as well as the meat quality of semi-intensively reared chickens with backyard access [3, 4].

Semi-intensive rearing is currently developing, known as free-range systems [5]. A free-range rearing system is one that is designed for chickens to thrive and express their natural instincts [6]. Free-range is one way of raising livestock that accommodates the principles of animal welfare formulated by The Royal Society for the Prevention of Cruelly to Animals (RSCPA). Livestock raised with welfare will be able to move more freely and express their natural behavior so as to reduce stress in livestock [7]. IPB-D1 chickens in this study were kept in an intensive system and a free-range system.

On the other hand, one form of effort to increase livestock productivity is through selection. However, characterization is necessary as a basis for selecting local chickens. Characterization is the first step for breeding livestock in order to identify important traits of economic value such as body weight and body weight gain as well as livestock characteristics. Characterization of local chickens including IPB-D1 chickens can be done by identifying morphometrics. Morphometrics are quantitative traits that can be used as selection criteria to increase the productivity of local chickens. Quantitative traits of local chickens based on morphometrics include body length, neck length, wing length, wingspan, breast circumference, breast width, head length, head width, beak length, cockscomb length, cockscomb height, tibia length, metatarsus length, metatarsus circumference, longest finger length, femur length, maxilla length, sternum length, and body weight [8-10].

These morphometric variables can characterize the size and body shape of local chickens that are useful for predicting production potential, opportunities to increase livestock productivity, and as a reference for standardizing local chicken traits such as IPB-D1 chickens. Morphometrics is a good indicator and has a close correlation with live weight [11]. Morphometrics can be used as an indicator at the chicken selection stage [12]. Through this research, the morphometrics of IPB-D1 chickens in free-range rearing will be studied and compared with the intensive system.

2 Materials and Methods

This research was conducted at the Animal Husbandry Education and Research Unit (UP3J) of IPB University, Jonggol District, Bogor Regency. The research was conducted from October 2022 to July 2023.

2.1 Materials

The materials used in this study were 120 IPB-D1 chickens aged 20 weeks as parent stock. The observed chickens consisted of 30 chickens reared in the intensive system, and 90 chickens reared in cages with free-range access. The main tools and materials used in this study consisted of 2 m × 6 m colony cages for males reared in portable cages with free-range access, and 4 m × 3 m colony cages for males reared in permanent (intensive) cages. The portable cages were equipped with a free-range area measuring 12 m × 12 m as many as 3 units of rotational grazing, and overgrown with various grass forages bordered by a net fence to avoid livestock from predators. Feed management was used in intensive system and free-range is same. Other supporting tools as observation instruments are digital vectors, rulers, and stationery.

2.2 Methods

In this study, measurements were made on linear body surface size variables, based on the method of Nishida et al. (1982) [13]. The variables used consisted of femur (FEM), tibia (TIB), tarsometatarsus (TAR), tarsometatarsus circumference (TC), third finger (F3), wing length (WL), maxilla (MAX), and cockscomb height (CH). An illustration of the measured variables is presented in Figure 1.

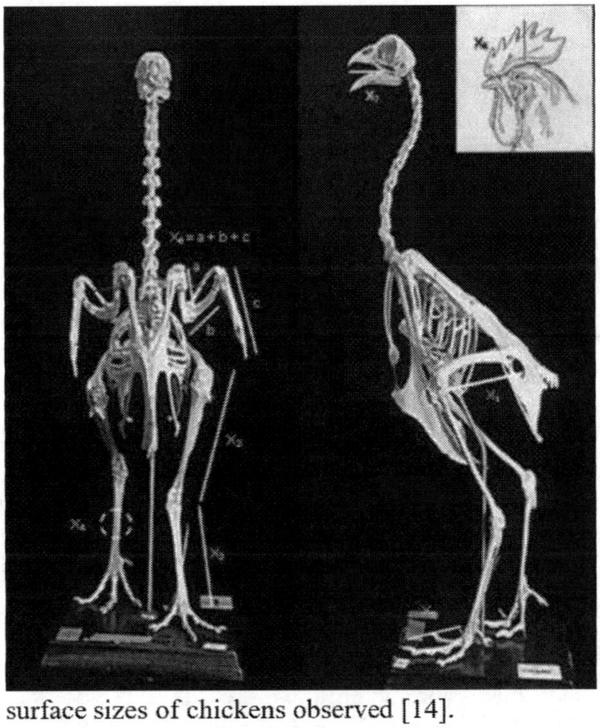

Fig. 1. Variable body surface sizes of chickens observed [14].

Description: X1= femur length; X2= tibia length; X3= tarsometatarsus length; X4= tarsometatarsus circumference; X5= third finger length; X6= wing length, X7= maxilla length; X8= cockscomb height.

2.3 Data Analysis

Descriptive analysis is used to determine the mean (x) and standard deviation (SD) [15]. Furthermore, the T-test is used to determine the average difference in the measurement data of roosters raised in an intensive system with roosters in a free-range system. The T-test formula is according to Steel and Torrie (1993) [16].

3 Results and Discussion

Chickens in this study were male, totaling 120 chickens, consisting of 30 intensively reared chickens and 90 free-range chickens. Morphometric observation data of chickens in two rearing systems are presented in Table 1.

Table 1. Average results of morphometeric measurements of IPB-D1 roosters

Variables	Morphometric	
	Intensive	*Free-range*
Femur Length	90.92±5.97 (6.57)	96.81±6.18 (6.38)
Tibia Length	150.88±9.81 (6.50)	149.69±9.93 (6.63)
Tarsometatarsus Length	108.02±7.13 (6.60)^{a}	110.51±5.19 (4.70)^{b}
Tarsometatarsus Circumference	39.67±3.60 (9.09)	41.78±2.77 (6.63)
Third Finger Length	65.25±4.00 (6.13)	61.85±3.98 (6.43)
Wing Length	247.33±13.70 (5.54)	256.44±14.04 (5.48)
Maxilla Length	36.48±2.47 (6.77)^{a}	36.48±1.64 (4.49)^{b}
Cockscomb Height	31.87±9.27 (29.09)^{b}	29.43±7.20 (24.45)^{a}

Description: Mean (mm) ± sd (coefficient of variation).
Different superscripts on the same line indicate significant differences (P<0.05)

3.1 Morphometric polymorphism

Morphometric observations of IPB-D1 chickens in the intensive cage system showed femur length (6.57). tibia length (6.50). tarsometatarsus length (6.60). tarsometatarsus circumference (9.09). third finger length (6.13). wing length (5.54). and maxilla length (6.77) were in the medium category. while cockscomb height (29.09) was in the high category. Furthermore. the free-range cage system showed that tarsometatarsus length (4.70) and maxilla length (4.49) were in the low category. while femur length (6.38). tibia length (6.63). tarsometatarsus circumference (6.63). third finger length (6.43). and wing length (5.48) were in the medium category. and comb height (24.45) was in the high category. This is in accordance with the opinion which states that the coefficient of diversity can be divided into three (3). namely low diversity (KK≤5). medium (5<KK<15). and high (KK≥15) [17]. This indicates that the maintenance of IPB-D1 chickens in free-range cages is expected to increase the morphometric uniformity of tarsometatarsus length and maxilla length. This is in accordance with the theory which states that genetic factors and environmental factors can cause diversity [18]. Genotype determines the potential of the character. while the environment determines the potential limit of the gene [19, 20].

Another research report states that native chickens have a wing length of 36.38 ± 7.81 (21) cm. femur length 88.88 ± 16.74 (19) mm. tibia length 118.67 ± 18.02 (15) mm. tarsometatarsus length 85.85 ± 21.35 (25) mm. third toe length 52.04 ± 10.81 (21) mm. tarsometatarsus circumference 4.74 ± 0.70 (15) cm [21]. Another opinion states that native chickens have a tarsometatarsus length 9.61 ± 0.89 (9.26) cm. a tarsometatarsus circumference 5.52 ± 0.65 (11.78) cm. and wing length 23.28 ± 2.18 (9.36) cm [18]. This shows that IPB-D1 chickens have a smaller level of diversity than native chickens as their parents. So. the performance of IPB-D1 chickens is good. but not as diverse as native chickens as its parents. This is the same as the results of other research. which repored that the new KUB chicken strains have low and medium diversity. such as the coefficient value. maxilla length 4.91. wing length 4.02. tarsometatarsus length 9.96. tarsometatarsus circumference 6.53. tibia length 6.76. and third toe length 8.62 [22].

3.2 Effect of intensive and free-range rearing systems

T-test results showed that femur length. tibia length. tarsometatarsus circumference. third finger length. and wing length of IPB-D1 chickens in intensive and free-range rearing systems were not significantly different ($p>0.05$). while tarsometatarsus length. maxilla length. and cockscomb height of IPB-D1 chickens in different rearing systems showed significant differences ($p<0.05$) with values of 108.02±7.13 and 110.51±5.19. 36.48±2.47 and 36.48±1.64. and 31.87±9.27 and 29.43±7.20. respectively. It is suspected that the free-range system causes chickens to move a lot. so that the legs of chickens are used more often and experience better growth than intensively reared chickens. In addition. chickens in free-range cages use their beaks more often to search for grass and other living things in the grazing area. thus requiring chickens to have longer beaks than animals housed in intensive systems. This is consistent with the statement from other studies. that chickens raised with free-range access have free movement space in the grazing area [5]. Chickens reared in a free-range system show greater body part activity and range of motion [23]. The phenotype of local ducks is influenced by their habitat and can be divided into several groups based on their place of life [24]. Habitat and climate differences also influence phenotypic and physiological effect in poultry [25].

Super native chickens. KUB chickens. and 8 (eight) week old native chickens have a maxilla length of 30.88 ± 2.5. 28.92 ± 1.42. 27.08 ± 1.65; wing length 170.08 ± 8.49. 166.07 ± 6.67. and 129.88 ± 5.71. respectively; tarsometatarsus length 64.96±7.37. 62.76±6.25. and 58.39±6.46; tarsometatarsus circumference 41.31±3.85. 36.44±2.38. and 34.10±3.19; tibia length 100.35±7.88. 97.76±6.61. and 90.45±6.7; third toe length 54.78±5.42. 52.11±4.49. and 48.26±5.13 [22]. The research of Putri et al. [22] has a smaller size than this study. This is because the chickens in the treatment have different ages. However. the new chicken strains (super chicken. KUB chicken. and IPB-D1 chicken) have morphometrics larger than the native chicken as the parent. Thus. the goal of making local meat-producing chickens with better performance has been achieved.

Another theory states that 20-week-old native chickens had a tarsometatarsus length of 74.59 ± 1.76. wing length of 55.90 ± 2.28. fibia length of 159.61 ± 0.99. third finger length of 76.55 ± 0.60. and tarsometatarsus circumference of 4.75 ± 0.49 [26]. Another report states that 24-week-old Sabung chickens have a maxilla length of 23.6 mm. wing length 115 mm. tibia length 92 mm. tarsometatarsus length 89 mm. tarsometatarsus circumference 69 mm. and third finger length 38 mm [27]. Research results from Nur et al. [26] and Nata et al. [27] showed smaller results compared to the results in this study. The Nunukan local chickens less than 24 weeks old had a cockscomb height of 54.21 mm. wing length of 252 mm. femur length of 109.7 mm. tibia length of 141.7 mm. tarsometatarsus length of 101.5 mm. tarsometatarsus circumference of 57.5 mm. and third finger length of 61.2 mm [28]. From the results of research. namely on the morphometrics of local Nunukan chickens. there are several variables. namely wing length. tibia length. tarsometatarsus length. and third finger length which have smaller values than the size of IPB-D1 in this study.

4 Conclusion

Body size of 20-week-old IPB-D1 chickens has a low coefficient of variation in tarsometatarsus length (free-range) and maxilla length (free-range); medium in femur length. tibia length. tarsometatarsus length (intensive). tarsometatarsus circumference. third finger length. wing length. and maxilla length (intensive); and high in cockscomb height. The rearing system treatments showed no differences in femur length. tibia length. tarsometatarsus circumference. third finger length. and wing length. but showed significant differences in tarsometatarsus length. maxilla length. and cockscomb height between intensive and free-range rearing systems in the tropics. Free-range rearing can increase tarsometatarsus length and maxilla length.

References

1. C. Sumantri. S. Darwati. *Perkembangan terkini riset ayam unggul ipb-d1.* in Prosiding Seminar Nasional Industri Peternakan I. 29-30 November. Bogor. Indonesia (2017)
2. L.M. Stadig. T.B. Rodenburg. B Reubens. J. Aerts. B. Duqueme. F.A. Tuyttens. Poult. Sci **9**. 12 (2016)
3. A.C Fanatico. P.B. Pillai. J.L. Emmert. C.M. Owens. Poult. Sci **86**. 10 (2007)
4. S.Z Jiang. Y. Jiang. G. Lin. F. Zhou. Chen. C. Zheng. Poult. Sci **52**. 3 (2011)
5. Z.H. Miao. P.C. Glatz. Y.J. Ru. As. Aust. J. Anim. Sci **18**. 1 (2005)
6. C. Castelini. C. Mugnai. A.D. Bosco. Meat. Sci **60**. 3 (2002)
7. S.H. Kiswanto. A. Fatikhunnada. M. Sholihudin. Peng. Sumb. Alam & Lingk **3**. 2 (2014)
8. M. Ashifudin. E. Kurnianto. Sutopo. J. Ilmu Ternak **17**. 1 (2017)
9. Hummairah. R. Hamdan. A.H. Daulay. J. Pet. Integ **3**. 3 (2016)
10. N.A. Rangkuti. Hamdan. A.H. Daulay. J. Pet. Integ **3**. 1 (2016)
11. Suparyanto. A.H. Martojo. P.S. Hardjosworo. L.H. Prasetyo. J. Ilmu Pet. & Vet **9**. 2 (2004).
12. E. Kurnianto. S. Sutopo. E. Purbowati. E.T. Setiatin. D. Samsudewa. T. Permatasari. J. App. Anim. Sci **3**. 2 (2013)
13. T. Nishida. Y. Hayashi. K. Nozawa. Thashiguchi. S.S. Mansjoer. The Research Group of Overseas Sci. Survey **73** (1982)
14. T. Koch. Anatomy of the chicken and domestic bird (Lowa State University. 1973)
15. A. Widodo. K. Andawaningtyas. Pengantar statistika (UB Pr. Malang. 2017)
16. R.G. Steel. J.H. Torrie. Prinsip dan prosedur statistika (Gramedia Pustaka Utama. Jakarta. 1993)
17. Kurnianto. Ilmu pemuliaan ternak (Universitas Diponegoro. Semarang. 2010)
18. Riyanti. K. Nova. T. Rafian. Sulastri. A. Suhadi. Mentari. Wah. Pet **7**. 1 (2023)
19. K. Nozawa. Trop. Agri. Res. Cent **4**. 3 (1980)
20. W. Yatim. Genetika Ed. IV (Tarsito. Bandung. 1991)
21. T. Rafian. J. Jakaria. N. Ulupi. J. Anim. Res. App. Sci **4**. 1 (2023)
22. A.B. Putri. Gushairiyanto. Depison. J. Ilmu & Tekno. Pet. Trop **7**. 3 (2020)
23. L.I. Chielo. T. Pike. J. Cooper. Anim **6**. 28 (2016)
24. D. Maharani. D.N. Hariyono. D.D Putra. J.H. Lee. J.H. Sidaodog. J. As. Pas. Bio **12**. 4 (2019)
25. P. Galeoti. D. Rubollini. P.O. Dunn. J. Evol. Biol **16**. 4 (2003)
26. A.M. Nur. H.D. Putranto. Murmeilisari. B. Brata. Suharyanto. Sutriyono. J. Pen. Pengelolaan Sumb. Alam & Lingk **11**. 2 (2022)
27. A.Y. Nata. C.D. Gaina. Y.T. Simarmata. J. Vet. Nus **6**. 17 (2023)
28. M. Alwi. C. Sumantri. S. Darwati. J. Vet **15**. 2 (2014)

Unlocking credit access for farmers (Case study of dairy farmers in East Java, Indonesia)

Tina Sri Purwanti[1], *Eko* Nugroho[1*], *Nanang* Febrianto[1], and *Suprih Bambang* Siswijono[1]

[1]Department of Socio-economy, Faculty of Animal Science, Universitas Brawijaya, Malang 65145, Indonesia

Abstract. Access to credit is crucial for the economic development and sustainability of smallholder farmers in rural areas. In the context of Malang, East Java, Indonesia, this study delves into the factors that influence smallholder farmers' access to credit. Smallholder farmers play a pivotal role in ensuring food security, agricultural, and livestock productivity in Indonesia, making it imperative to understand the determinants of their credit access. This research was conducted between July and August 2023, employing a Probit Model and descriptive statistics to analyze data collected through a survey questionnaire. A multistage random sampling technique was employed, encompassing 70 smallholder dairy farmers in Malang. Findings from the Probit Model unveiled that several factors significantly affect the decision of smallholder farmers to adopt credit access. These factors include education levels among farmers, cattle ownership, access to the internet, participation in livestock training programs, and the employment of family members. The positive influence of these factors emphasizes the importance of targeted interventions to enhance smallholder farmers' credit access. To promote the adoption of credit access among smallholder dairy farmers in Malang, collaborative efforts between stakeholders and government entities are essential. These efforts should focus on improving the availability of livestock-related training programs and enhancing internet accessibility, particularly to bolster the sustainability of the livestock sector, with a specific emphasis on dairy farming. Such measures can help uplift the livelihoods of smallholder farmers, foster agricultural growth, and contribute to overall rural development in the region.
Keywords: Credit access, dairy farm, smallholder farmers, probit model, Indonesia

1 Introduction

In the history of rural economic development, credit has become a key factor in creating employment opportunities for rural communities by helping expand production and improving their well-being [1]. So far, several international studies have emphasized the importance of access to credit services, which are considered capable of reducing poverty among low-income populations. Guirkinger [2] revealed that credit is a vital instrument in

*Corresponding author: eko_nug@ub.ac.id

poverty reduction, as it can assist rural communities, such as fishermen, in purchasing various production factors (fuel, supplies, and labor), thereby enhancing their productivity. However, participation in credit in rural areas, especially in coastal communities, remains low. Access to credit has been on the rise in various developing countries. In Asian countries, more than 515 million households have access to credit [3]. Furthermore, Beck et al. [4] explained that in African states, only 20 percent of households have access to credit. In coastal communities, low credit access has a negative impact on fisheries production [5]. Lack of access to credit is considered a significant reason why rural communities in various developing countries continue to live below the poverty line [6]. Livestock farmers are one of the groups facing obstacles in accessing credit. Therefore, livestock farming has become an interesting topic for researchers and policymakers, as it can help farmers break free from the cycle of poverty. Several previous studies have demonstrated the role of credit in rural communities. On the other hand, the allocation of credit to livestock farmers is seen as a challenge by financial institutions [2]. This is because the livestock sector is considered a high-risk investment due to its vulnerability to uncertainties like weather conditions and unpredictable yields [7]. Furthermore, Nordjo & Adjasi [8] pointed out that limited access to credit is due to the imperfections in the financial market caused by incomplete information.

Several previous studies regarding the impact of credit access on rural community productivity have been conducted, and the results obtained vary. Some of these studies indicate a positive and significant impact on rural community productivity [9, 10, 11]. In other words, with access to credit, fishermen are able to enhance their productivity by meeting input needs in their production activities. However, some researches have also found that rural communities face limitations in accessing credit, such as loan amounts not matching their requests, complicated requirements, and incomplete credit-related information [12, 13]. Therefore, this study tends to investigate the factors influencing credit access for dairy farmers in rural areas.

2 Materials and methods

2.1 Data collection

The study used cross-sectional data gathered from small-scale dairy farmers in East Java, Indonesia. The research sites were selected using a multistage random sampling method. Firstly, East Java province because it serves as the largest number of dairy cattle populations in the country. Then, the Malang District were selected based on the presence of farmers' associations with credit access from formal money lending institution. Finally, farmers in Wonokerto sub-district were listed and resulting in a total of 70 dairy farmers.

2.2 Data analysis

We applied the probit regression analysis to investigate the determinants of credit accessibility among dairy farmers. The study asserts that farmers' ability to access credit can be elucidated by their socio-demographic characteristics. Essentially, the model posits that:

$$ACC_i^* = \beta X_i + e_i, with\ ACC_i^* = 1\ if\ adoper, 0\ otherwise \quad (1)$$

where ACC is a livestock farmer's access to credit or not as described by a dummy variable (1 if farmer has access to credit and 0 otherwise); Xi is a socio-demographic variable, such as age, education, marital status, family members, and asset ownership; β is parameters to be estimated; and ei is an error.

3 Results and Discussion

3.1 Descriptive statistics

We observe several key variables characterizing farmers and their livestock context (Table 1). Credit access, represented as a dummy variable, is available to approximately 30% of the farmers. The average education level among farmers is 11.457 years, indicating a reasonable level of formal education. Farmers own on average land area of 0.104 hectares and possess on average of 18.093 years of farming experience, suggesting a mix of smallholders and experienced farmers. Regarding livestock ownership, farmers on average own 3.657 heifers, 3.014 calves, and 6.943 adult cattle. Internet access, as another dummy variable, is available to about 61.4% of the farmers. Training, rated on a scale from 1 (never) to 5 (very often), averages at 2.486, indicating some but not very frequent training. There is an average of 1.671 productive family members, highlighting family involvement, and access to market information, rated from 1 (very difficult) to 5 (very easy), averages at 3.043, indicating a moderate level of ease in accessing market data. These variables collectively provide valuable insights into the diverse circumstances and resources available to farmers, essential for understanding their livestock practices and livelihoods.

Table 1. Descriptive statistics

Variables	Measurement	Mean	Std. Dev
Credit Access	Dummy, 1 if the farmers had access to credit; 0 otherwise	0.300	0.462
Education	Farmers' education level in year	11.457	4.416
Total Area	Total cultivated area	0.104	0.196
Farming Experience	Total of farming experience in year	18.093	13.994
Calves	The number of calves owned by farmer (head)	3.014	4.066
Heifers	The number of heifers owned by farmers (head)	3.657	3.863
Adults	The number of adults cattle owned by farmers (head)	6.943	7.689
Internet	Dummy, 1 if the farmers had internet access; 0 otherwise	0.614	0.490
Training	Livestock training (1= never – 5=very often)	2.486	1.018
Employed Family Member	Total of employed family member (person)	1.671	0.737
Market Information	Access to market price, 1=very difficult-5=very easy	3.043	0.859

3.2 Descriptive statistics

Table 2 sheds light on several key factors that influence individuals' ability to secure credit. Firstly, education emerges as a critical determinant, with a coefficient of 0.197 and a highly significant p-value of 0.002. This indicates that higher levels of education positively impact access to credit, underscoring the importance of investing in education to enhance financial opportunities. Conversely, the total area of residence, as indicated by a coefficient of -2.237, does not significantly affect credit access, with a p-value of 0.159. Farming experience, with a negligible coefficient of 0.004 and a high p-value of 0.826, is shown to have no significant bearing on credit access. This implies that lenders may prioritize other factors over an individual's level of experience. Ownership of livestock, particularly adult cattle, emerges as an influential factor with a coefficient of 0.133 and a p-value of 0.046, indicating a statistically significant positive relationship between owning adult cattle and credit access. This suggests that livestock ownership may serve as collateral or a source of income, facilitating access to credit. Intriguingly, internet access, represented by a coefficient of -1.480 and a significant p-value of 0.005, is found to have a negative impact on credit access. This surprising result warrants further investigation into whether online financial behaviors or online sources of credit are associated with higher risks. Training, with a coefficient of -0.844 and a significant p-value of 0.006, negatively influences credit access. This suggests that some forms of training might not align with credit eligibility criteria or that training may not directly enhance financial prospects.

Table 2. Factor Influencing Farmers' ability to Access Credit: Probit Model

Credit Access	**Coef.**	**Std. Err**	**Z**	**P-value**
Education	0.197	0.065	3.030	0.002***
Total Area	-2.237	1.588	-1.410	0.159
Farming Experience	0.004	0.016	0.220	0.826
Calves	-0.039	0.089	-0.440	0.660
Heifer	-0.111	0.113	-0.980	0.327
Adults	0.133	0.067	1.990	0.046**
Internet	-1.480	0.524	-2.820	0.005***
Training	-0.844	0.309	-2.730	0.006***
Employed Family Member	1.134	0.315	3.600	0.000***
Market Information	0.106	0.283	0.370	0.709
_cons	-2.466	0.981	-2.510	0.012
Log likelihood	-30.118107			
LR chi2(10)	25.28			
Prob > chi2	0.0048			
Pseudo R2	0.2957			

Notes: **/*** indicate statistical significance at the 5 and 1% levels, respectively.

The presence of more productive family members is strongly associated with improved credit access, with a substantial coefficient of 1.134 and a highly significant p-value of 0.000. This underscores the significance of family networks and collaborative economic activities in securing credit. Access to markets, surprisingly, does not significantly affect credit access, as indicated by a coefficient of 0.106 and a high p-value of 0.709. This finding suggests that other factors, such as collateral or credit history, may play a more substantial role in credit decisions. This analysis reveals that education, ownership of adult cattle, internet access, training, and the presence of productive family members are notable determinants positively impacting credit access. Meanwhile, the total area, experience, ownership of heifer or calf cattle, and access to markets do not appear to be significant drivers of credit availability. These insights can inform policies and strategies aimed at improving financial inclusion and credit access for individuals and communities. Further research may be needed to explore the nuanced relationships uncovered, particularly the unexpected negative association between internet access and credit availability.

4 Conclusion

In this study, several key determinants of credit access among dairy farmers in East Java, Indonesia were identified. Higher education levels positively influenced credit access, while farmer experience had no notable effect on credit accessibility, and ownership of adult cattle was positively associated with credit access. Surprisingly, internet access was found to have a negative influence on credit accessibility, warranting further investigation. Training programs appeared to negatively affect credit access, calling for alignment with credit criteria. The presence of productive family members strongly correlated with improved credit access, underlining the importance of family networks. Access to markets did not significantly affect credit access, indicating the relevance of other factors in credit decisions. These findings have implications for policy and strategies aimed at enhancing financial inclusion and credit access in rural communities, and further research is needed to explore the unexpected negative impact of internet access on credit availability.

References

1. Luan, D. X., Thanh Vu, N., Thi Thu Huong, K., Hang, D. T. T., & Bauer, S. (2016). Is access to agribank credit an effective tool in improving household income? Evidence from the northern mountainous region of Vietnam. *Asian Social Science*, **12(8).**
2. Guirkinger, C. (2008). Understanding the coexistence of formal and informal credit markets in Piura, Peru. *World development*, *36*(8), 1436-1452.
3. Felsenthal, M., & Hahn, R. (2018). Financial inclusion on the rise, but gaps remain, global findex database shows. *Access at: http://www.worldbank.org/en/news/pressrelease/2018/04/19/financial-inclusion-on-the-rise-but-gaps-remain-global-findex-database-shows (8.9. 2018.).*
4. Beck, T., Demirgüç-Kunt, A., & Honohan, P. (2009). Access to Financial Services: Measurement, Impact, and Policies. *World Bank Research Observer*, ***24*(1).**
5. Guerrieri, V., Lorenzoni, G., & Prato, M. (2020). Schumpeter lecture 2019: Slow household deleveraging. *Journal of the European Economic Association*, ***18*(6),** 2755-2775.
6. Collins, A. (2018). Saying all the right things? Gendered discourse in climate-smart agriculture. *The Journal of Peasant Studies*, ***45*(1),** 175-191.
7. Rahman, A. (2017). Smallholder Agriculture and Inclusive Rural Transformation. *The Bangladesh Development Studies*, ***40*(1 & 2)**, 119-147.
8. Nordjo, R. E., & Adjasi, C. K. (2020). The impact of credit on productivity of smallholder farmers in Ghana. *Agricultural Finance Review*, ***80*(1),** 91-109.
9. Chisasa, J., & Makina, D. (2013). Bank credit and agricultural output in South Africa: A Cobb-Douglas empirical analysis. *International Business & Economics Research Journal (IBER)*, ***12*(4),** 387-398.
10. Gomina, A., Abdulsalam, Z., Maiyaki, D., & Saddiq, N. U. (2015). Impact of Savings and Credit Cooperative Societies on Poverty Status of Crop Farmers in Niger State, Nigeria. *Academic Research Journal of Agricultural Science and Research*, ***3*(6),** 142-150.

11. Wicaksono, E. (2014). The impact of agricultural credit on rice productivity. *Int J Adv Sci Eng Inf Technol*, **4(5)**, 20-23.
12. Dong, G., Lai, K. K., & Yen, J. (2010). Credit scorecard based on logistic regression with random coefficients. *Procedia Computer Science*, ***1*(1)**, 2463-2468
13. Reyes, A., Lensink, R., Kuyvenhoven, A., & Moll, H. (2012). *Impact of access to credit on farm productivity of fruit and vegetable growers in Chile.* e International Association of Agricultural Economists (IAAE) Triennial Conference, Foz do Iguaçu, Brazil, 18-24 August, 2012.

Production characteristics of red napier grass (*Pennisetum purpureum* cv. red) on different combination of soil and vermicompost as planting media

Rizka Muizu Aprilia[1], *Annisa Dian* Hayati[2], *Raihanah Ragillia* Annisah[2], *Hanief Eko* Sulistyo[3], *Hermanto* Hermanto[3], and *Ifar* Subagiyo[3*]

[1]Doctoral Degree Student of Animal Science Faculty, Universitas Brawijaya, Malang, East Java, Indonesia 65145
[2]Bachelor Degree of Animal Science Faculty, Universitas Brawijaya, Malang, East Java, Indonesia 65145
[3]Lecturer in the Department of Animal Nutrition and Feed, Faculty of Animal Science, Universitas Brawijaya, Malang, East Java, Indonesia 65145

Abstract. This study aimed at measuring production characteristics of Red Napier grass grown on planting media comprising different combinations of soil and vermicompost. The experiment was executed in greenhouse following completely randomized design with 5 treatments i.e. T0: 100% soil (control), T1: 75% soil + 25% vermicompost, T2: 50% soil + 50% vermicompost, T3: 25% soil + 75% vermicompost and T4: 100% vermicompost. Each treatment was replicated 5 times. The soil and vermicompost proportion were from 20 kg total planting media filled in polybags on which the grass stem cuttings were planted. Harvest was done 90 days after planting. Collected data were subjected to analysis of variance and least significant difference test. The results showed that treatments did not significantly affect ($P>0.05$) plant height, but significantly affect ($P<0.05$) leaf to stem ratio and very significantly affect ($P<0.01$) number of tillers, number of leaves, and nutrient production of stem, leaf and herbage. Highest production of dry matter was obtained at the treatment of T3.

1 Introduction

Forage is the main feed ingredient needed by livestock because it contains nutrients and minerals for growth, development and maintenance [1] Several types of forage that are widely cultivated are Bengal grass, King grass, Elephant grass and legumes. According to [2]. Elephant grass (*Pennisetum purpureum*) originating from Africa has great potential to be developed in Indonesia because it is well adapted to tropical climate and has the ability to grow quickly, produce high biomass and is palatable to ruminants. One of the cultivars that has not been widely studied in Indonesia is the Red Elephant grass (*Pennisetum purpureum*

* Corresponding author: ifars@ub.ac.id

cv. Red). Limited information on the grass production characteristics such as number of tillers, number of leaves, leaf-stem ratio, dry matter production and nutritional content hinder the socialisation of this grass to be cultivated by farmers communities even though many seeds (cuttings) have been offered through e-commerce.

Production of Elephant grass is largely determined by soil fertility which contains nutrients and water for plants. Lack of nutrients can interfere with plant growth and thus require additional fertilisers. Fertilisation aims to improve soil nutrient availability to meet the plant needs. Fertilisers made of organic ingredients can improve the chemical and physical fertility of the soil [3]. Vermicompost is one type of organic fertiliser available at most farming communities.

Vermicompost is an organic fertiliser produced from the degradation of organic materials in earthworm cultivation media and worm droppings which contain macro and micro nutrients that can fertilise soil. According to [4], the vermicompost contained complete macro and micro nutrients that were easily absorbed by plants. It also contains growth regulators such as gibberellin, cytokinin and auxin, as well as the nutrients N, P, K, Mg and Ca and Azotobacter sp which are non-symbiotic N-fixing bacteria that help to increase N content so that it can increase plant production [5]. It was reported by [6]. that applying vermicompost fertilisers can increase the growth of green mustard plants. However, information about the use of this fertiliser for the production of forage crops, especially Red Elephant grass in Indonesia is still difficult to obtain. Therefore, this research was conducted to determine the production characteristics of the Red Elephant Grass (*Pennisetum purpureum* cv. Red) cultivated on different combinations of soil and vermicompost as planting media.

2 Materials and Methods

2.1 Location

The experiment was executed from April to August 2022, in a greenhouse at the Sumbersekar field station of the Faculty of Animal Science, Universitas Brawijaya in Malang Regency of Indonesia. It lies in a tropical humid zone, 112o34'35'' East latitude and 7o55'07'' South longitude, at ± 400 m as altitude. Nutrient contents of the harvested herbage biomass were analysed at the Animal Nutrition Laboratory Faculty of Animal Science Universitas Brawijaya. Soil nutrients were analysed at the soil Laboratory Faculty of Agriculture Universitas Brawijaya..

2.2 Research Variables

The variables measured were plant height, number of tillers/clump, number of leaves/clump, leaves to-stem ratio and nutrient production of the aerial biomass.

2.3 Research Procedure

Stem cuttings of the Red Elephant grass were obtained from a farm at Gadingkulon Village, Dau District, Malang Regency. The cuttings, comprising three internodes, were taken from healthy plants aged 2 to 3 months. The research followed a completely randomised design with 5 treatments of planting media i.e T0 = 100% soil, T1 = 75% soil + 25% vermicompost, T2 = 50% soil + 50% vermicompost , T3 = 25% soil + 75% vermicompost and T4 = 100% vermicompost. Each treatment was replicated 5 times. The soil and vermicompost proportion were from 20 kg total planting media filled in a polybag on which the grass stem cutting was planted. Assuming that the weight of soil per ha is 2000 tons so the proportion of 25%

vermicompost fertiliser at T1 is equivalent to 500 tons of vermicompost/ha, the 50% vermicompost fertiliser (T2) is equivalent to 1000 tons of vermicompost /ha and the proportion of 75% vermicompost fertiliser (T3) is equivalent to 1500 tons of vermicompost/ha. The soil nutrient content under different treatments is presented in Table 1.

Table 1. Nutrient content of the soil under different treatments

Treatments	C-Organik (%)	N-Total (%)	Phosphor (%)	Kalium (cmol(+)Kg-1)
T0	1.52 (L)	0.2 (L)	62.0 (VG)	3.69 (VG)
T1	6.16 (VG)	0.57 (M)	46.77 (VG)	2.90 (VG)
T2	10.8 (VG)	0.94 (VG)	31.54 (VG)	2.11 (VG)
T3	15.44 (VG)	1.31 (VG)	16.31 (G)	1.31 (VG)
T4	20.08 (VG)	1.68 (VG)	1.081 (VL)	0.52 (M)

Note : VL = Very Low; L = Low; M = Moderate, G = Good; VG = very Good

The grass was harvested 90 days after planting. Plant height was measured just before the grass was harvested. After harvested, leaf blades and stems (including leaf sheaths) were separated and weighted. Composite samples of each plant part were taken and brought to the laboratory for Dry Matter (DM), Organic Matter (OM), Crude Protein (CP) and Crude Fibre (CF) analysis following the AOAC (2005) method. Leaf to stem ratio was measured as leaves DM production divided by stems DM production. Plant biomass production was measured by summing the leaves DM and stems DM production. Number of tillers was physically counted after harvest.

2.4 Statistical Analysis

The collected data were subjected to Analysis of Variance (ANOVA) using the formula

$$Yij = \mu + Ti + \varepsilon ij \qquad (1)$$

where Yij = Observation of experiment j that received the treatment I, μ = overall mean, Ti = effect of treatment i and Ɛij = error effect in experiment j in the treatment i. If the ANOVA indicated a significant effect from treatments then the Least Significant Difference (LSD) test was applied to find differences between means.

3 Results and discussion

3.1 Characteristics and production dry matter

Production characteristics of Red Napier (*Pennisetum purpureum* cv. Red) including plant height, number of tillers, number of leaves, leaf-stem ratio and DM production of leaves and stems are presented in Table 2.

Data on Table 2 shows that treatments did not significantly (P>0.05) affect plant height. The average plant height between treatments was 133.72 ± 13.7 cm which is lower than the average plant height of Elephant grass (124 cm) reported by [7]. [8]. mentioned differences in plant height can be caused by cutting height, type of growing media, water supply and species differences.

Table 2. Characteristics and production variables of Red Napier under different treatments

Variabel	Average at treatment:				
	T0	T1	T2	T3	T4
Plant height (cm)	117.8 ± 18.3	136.6 ±9.3	138.0 ± 13.5	140.8 ± 18.2	135.4 ± 9.1
Number of Tillers/clump	$3.0^{a} \pm 1.0$	$5.0^{a} \pm 0.7$	$6.0^{a} \pm 1.6$	$9.8^{b} \pm 3.0$	$6.8^{ab} \pm 2.1$
Number of Leaves /clump	$37.2^{a} \pm 10.1$	$61.6^{a} \pm 9.6$	$72.0^{ab} \pm 14.4$	$92.6^{b} \pm 31.3$	$83.2^{ab} \pm 7.6$
Leaf-Stem Ratio	$1.2^{b} \pm 0.2$	$0.87^{a} \pm 0.1$	$1.1^{ab} \pm 0.1$	$0.9^{a} \pm 0.2$	$0.9^{a} \pm 0.2$
Leaves DM production (g/clump)	$9.3^{a} \pm 1.6$	$23.89^{b} \pm 3.0$	$28.0^{b} \pm 2.8$	$31.6^{bc} \pm 3.2$	$23.3^{b} \pm 3.1$
Stem DM production (g/clump)	$8.0^{a} \pm 0.7$	$27.39^{b} \pm 2.1$	$26.4^{b} \pm 5.8$	$34.4^{b} \pm 7.5$	$27.1^{b} \pm 2.4$

Note : Different superscripts on same line indicate significant differences (P<0.05) [1]Data

Number of tillers/clumps as an indicator of ability for grass regrowth[9, 10] were significantly (P<0.05) different between treatments. It was highest (9.8 per clump) at T3 and lowest (3.0 per clump) was at T0. Similarly, the number of leaves/clumps was also significantly (P<0.05) affected by treatments and the highest (92.6 per clump) was found under the T3 treatment. The highest number of tillers/clump and leaves/clump occurred at the highest content of C-total and N-total with good levels of P and K in the T3 treatment planting media as shown in Table 1. Sufficient nitrogen and phosphor elements in the planting media will increase protein synthesis which is utilised by plants for meristematic cell division and enlargement forming of more tillers and leaves [11].

Table 2 shows that the leaves and stems DM production as well as leaves to stems ratio was significantly (P<0.05) affected by treatments. Highest leaves to stems ratio (1.2) was found at the T0 treatment. This was similar to the finding of [7] i.e. leaves-stem ratio of 1.2 for Elephant grass harvested at 90 days of age but lower than the leaves to stems ratio of dwarf Elephant grass reported by [12, 13]. However, data in Table 2 also shows that production of either leaves or stems DM at T0 treatments were significantly (P<0.05) lower than that obtained at the other treatments. Although statistically insignificant compared to those under T1, T2 or T4 treatments, the numerical value of average production of leaves or stems DM of T3 treatments were found to be the highest i.e. 31.6 ± 3.2 and 34.4 ± 7.5 g/clump those gave a leaves to stem ratio of 0.9 ± 0.2.[14] stated the high number of grass leaves indicate the potential of forage for animal feed. In this experiment, however, the highest leaves to stems ratio at T0 treatments coincided with lowest leaves and stems DM production. The T3 treatment, even though it gives lower leaves to stem ratio, produces higher leaves and stem DM biomass. This T3 treatment, as already mentioned above, also gives the highest number of tiller/clump and number of leaves/clump.

3.2 Nutrient Production

Analysis of variance found that treatments significantly ($P<0.05$) affect the DM, OM, CP and CF production. Data of plant parts production between treatments are presented in Table 3.

Table 3. Production of nutrients of the Red Napier plant parts between treatments

Nutrient	Plant Parts	Production (g/clump) at treatment:				
		T0	T1	T2	T3	T4
DM	Plant	$17.3^{a} \pm 2.2$	$51.3^{c} \pm 5.0$	$54.5^{c} \pm 8.4$	$65.9^{d} \pm 8.4$	$42.1^{b} \pm 6.3$
	Stem	$8.0^{a} \pm 0.7$	$27.4^{b} \pm 2.1$	$26.4^{b} \pm 5.8$	$34.4^{c} \pm 7.6$	$27.1^{b} \pm 2.4$
	Leaves	$9.3^{a} \pm 1.6$	$23.9^{c} \pm 2.9$	$28.1^{d} \pm 2.8$	$31.6^{d} \pm 3.2$	$14.9^{b} \pm 3.9$
OM	Plant	$15.4^{a} \pm 2.0$	$46.7^{b} \pm 4.5$	$50.2^{bc} \pm 7.8$	$61.9^{d} \pm 7.9$	$39.8^{b} \pm 5.9$
	Stem	$7.4^{a} \pm 0.7$	$24.8^{b} \pm 1.9$	$24.1^{b} \pm 5.3$	$32.1^{c} \pm 7.1$	$25.6^{b} \pm 2.3$
	Leaves	$8.1^{a} \pm 1.4$	$21.8^{c} \pm 2.7$	$26^{d} \pm 2.6$	$29.8^{e} \pm 2.9$	$14^{b} \pm 3.7$
CP	Plant	$0.7^{a} \pm 0.1$	$2.5^{b} \pm 0.2$	$3.3^{c} \pm 0.5$	$4.3^{d} \pm 0.6$	$3.1^{c} \pm 0.5$
	Stem	$0.2^{a} \pm 0.1$	$0.9^{b} \pm 0.1$	$1.1^{b} \pm 0.2$	$1.7^{c} \pm 0.4$	$1.7^{c} \pm 0.2$
	Leaves	$0.5^{a} \pm 0.1$	$1.6^{b} \pm 0.2$	$2.2^{c} \pm 0.2$	$2.6^{d} \pm 0.3$	$1.3^{b} \pm 0.3$
CF	Plant	$6.2^{a} \pm 0.8$	$16.7^{c} \pm 1.6$	$17.2^{c} \pm 2.7$	$20.4^{d} \pm 2.6$	$12.5^{c} \pm 1.9$
	Stem	$3.1^{a} \pm 0.3$	$9.1^{bd} \pm 0.7$	$8.5^{b} \pm 1.9$	$10.9^{d} \pm 2.4$	$8.3^{b} \pm 0.7$
	Leaves	$3.2^{a} \pm 0.6$	$7.6^{b} \pm 0.9$	$8.7^{bc} \pm 0.9$	$9.6^{c} \pm 1.0$	$4.3^{a} \pm 1.2$

Note : Different superscripts on the same line is significantly different ($P<0.01$)

Table 3 shows that the highest production of DM of each plant part is found at the T3 treatment. At this treatment, production of either leaves or stems DM is highest as already discussed in section 2.2. while the DM production of plant, as shown in Table 3, is also highest at T3 that reached 65.96 ± 8.37 g/clump. In line to this, highest production of OM, CP and CF also occurred under the T3 treatment. At this treatment, the planting media contained highest C-total and N-total with good levels of P and K and hence was able to provide nutrients as needed by the grass to optimise production. [15] mentioned that 80% of plant dry weight comes from carbon and nutrients supplied by organic fertilisers. The vermicompost application at T3 also provides relatively high phosphorus and nitrogenous nutrients to planting media. Availability of these two nutrients may play a role in the formation of Adenosine Triphosphate (ATP) which will function as an energy source and or nitrogenous compounds that contribute to cell division and growth that eventually increase DM and OM production of all components of the Red Napier grass. [16]. mentioned that nitrogen plays a role in forming chlorophyll needed for photosynthesis to produce assimilates such as protein and carbohydrate. In overall, the increased amount of DM produced under the T3 treatment was followed by higher production of other nutrients such as CP and CF.

4 Conclusion

On the basis of results and discussions above it could be concluded that the use of vermicompost up to 75% of total soil for growing Red Napier grass will lead to high number of tillers and leaves as well as highest amount of DM, OM, CP and CF production of the grass biomass.

The authors wish to express their sincere gratitude for the support via the scheme of Hibah Guru Besar granted by Universitas Brawijaya that financed this experiment.

References

1. Rochana, A., N. P. Indriani, B. Ayunningsih, I. Hernaman, T. Dhalika, D. Rahmat, and S. Suryanah. Animal Production. 18, 2 (2016)
2. Lasamadi, R.D., S.S. Malalantang, Rustandi, dan S.D. Anis. Zootek. 32, 5 (2013)
3. Roidah, I. S. Jur. Univ. Tul. BONOWORO. 1,1 (2013)
4. Mulat, T. Jakarta: Agromedia Pustaka (2003)
5. Oka, A. A. Jurnal Sains MIPA Universitas Lampung 13, 1 (2012).
6. Limbong, B., Putri, L. A. P., & Kardhinata, E. H. Agroekoteknologi, 2, 4. (2014).
7. Ansah, T., E.L.K. Osafo, and H.H. Hansen. Agric. Biol. J. N. Am. 1, 5 (2010)
8. Zewdu, T., R.M.T. Baars, and A. Yami. Journal of Tropical Science. 43, 1 (2003)
9. Santia, S.D. Anis, & C.L. Kaunang. Jurnal Zootek 37,1 (2017)
10. Lafarge, M & Loiseau, P. European Journal of Agronomy, 17 (2013)
11. Alfian, M.S and H. Purnamawati, Agrohorti 7, 1 (2019)
12. Zailan, M.Z., H. Yaakub, and S. Jusoh. Journal of Animal & Plant Sciences. 28, 1 (2018).
13. Sirait, J., Tarigan, A., & Simanihuruk, K. Wartazoa, 27 4 (2017)
14. Nyambati, E. M., F, Muyekho, C. M. Lusweti and G. Mureithi. Journal of Agricultural Extension and Rural Development 3, 1 (2011)
15. Lakkar, J. Y., Nastiti, H. P., & Suryani, N. N. Jurnal Peternakan Lahan Kering, 1, 1 (2019).
16. Riyawati. Skripsi. Program Studi Agroteknologi. Fakultaspertanian Dan Peternakan. Universitas Islam Negeri Sultansyarif Kasim, Riau (2012)

Effect of level of metabolizable energy on the breast and wing weight of local chickens: A meta-analysis

Anisa Ramadhani[1], *Veronica Margareta Ani* Nurgiartiningsih[1], and *Osfar* Sjofjan[1*]

[1]Faculty of Animal Science, Universitas Brawijaya, Malang, 65145, Indonesia

Abstract. The purposed of this study was to determine the effect of dietary energy and protein balance on the wing weight and breast weight of native chicken using meta-analysis technique. This study used meta-analysis technique to integrate 116 data from 8 articles that contain information about effect of energy and protein balance rations on wings weight and breast weight of local chicken. The meta-analysis technique begins with collecting data from several scientific publications that contain energy and protein levels. The data was tabulated in the form of database with the help of Microsoft excel and in this tabulation process the units of each data were equalized tabulated and analysed using SAS. The results of this study concluded that it took 17.04% of dietary crude protein to produce a minimum breast weight of 151.96grams and 2953.31 kcal/kg to produce 122.66grams of breast weight, as well as 17.43% of raw proteins to make a minimum wing weight of 88.25gram and 2970 kcal / kg to produce an 89.53gram of wing. In conclusion protein requirements and energy metabolism, each pile showed varying results. The existence of this diversity is influenced by genetic and environmental factors.

1 Introduction

The diversity of local chicken varieties in Indonesia is large. Indonesia has 31 local chicken groups, both native and introduced chickens that have undergone adaptation for decades or even hundreds of years. These local chickens are scattered all over the Nusantara region and serve the purpose of supplying village chicken meat, eggs, and companion animals [1]. Examples of local indigenous chickens are KUB chickens, Pelung chickens, Bekisar chickens, Kedu chickens, and Kate chickens. Some of these chickens are kept as pets and as favorite collections of enthusiasts because of the beauty of appearance, feathers, and their gentle voice.

The existence of diverse local chicken varieties encourages many farmers to raise and produce them for the production of meat and eggs which are highly in demand, as these products are known to be organically produced and possess certain flavours and highly in demand in niche market. Local chicken meat and eggs have unique characteristics and

* Corresponding author: osfar@ub.ac.id

flavours that are of great interest to the community. Besides, the abundance of Nusantara cuisines that use local chicken as their main ingredient, but the production of meat and eggs is limited and insufficient to meet market demands.

The development potential of this local chicken is hampered by its low productivity and poor growth rates. It's because of its longer growth than the commercial breeds. Improving local chicken productivity can be achieved by improving the management, supply, and quality of feed [2].

One of the important nutrients that local chickens need during growth is protein and metabolic energy. This is because the energy and protein of the feed are the main components that the poultry need to live and produce. Both components must be considered together in the preparation of the feed or, in other words, each energy and protein of the rotting have different levels of influence on growth and fatigue [3]. Research on feeding with protein and metabolic energy in various local chicken flocks has been extensive. The results obtained are also very varied and so far there is no data on the exact value of these two nutrients in particular that are needed by local chickens. That's why we need techniques to analyze and draw conclusions from existing data to address these differences. One method of answering the value of such nutrients is by using meta-analysis statistical techniques. This technique is a tool to integrate data from each choice made at different places, times, and researchers, but still with the same variable. Besides, the data used also has some consistency such as treatment, object, and so on [4].

The use of meta-analysis in poultry research can improve the power and precision of statistics as well as generate new hypotheses. Thus, through meta-analysis precise values of the amount of protein and metabolizable energy required by various local chicken varieties can be obtained based on several studies previously investigated. This study aims to determine the amount of protein and metabolizable energy needed by some native chickens in relation to the breast and wing weight of the carcass, through a meta-analysis approach.

2 Material and method

2.1 Materials

Data from several national and international journals that published research papers related to the influence of protein and metabolic energy levels in feed on breast and wing percentages in three local chicken varieties, namely Pelung x Native cross, Native chicken, and Jowo Persilangan chicken.

The database was compiled from several national and international articles published using searches such as Scopus, ScienceDirect, and Google Scholar, with keywords, native chicken, protein, metabolic energy, breast weight, and wing weight. The data collected from various subsequent articles were stored in a Microsoft Excel-based database. The tabulation was done by harmonizing the units and weighing data such as protein usage and metabolic energy as well as the variables studied as the weight of breast and wings. A total of relevant 116 data sets from eight articles were collected and used in the analyses.

The meta-analysis was conducted on data from published research articles that used proteins and metabolizable energy values that were predominantly from the local feed. The selected article lists the protein content and metabolizable energy derived from laboratory tests so that the data processed were valid. Protein levels and metabolizable energy in this study were used as continuous variables, while the percentage of breast and wing reported in the papers as response variables. The diagram flow of article selection in the meta-analysis using the SYRCLE method will result in the figure below.

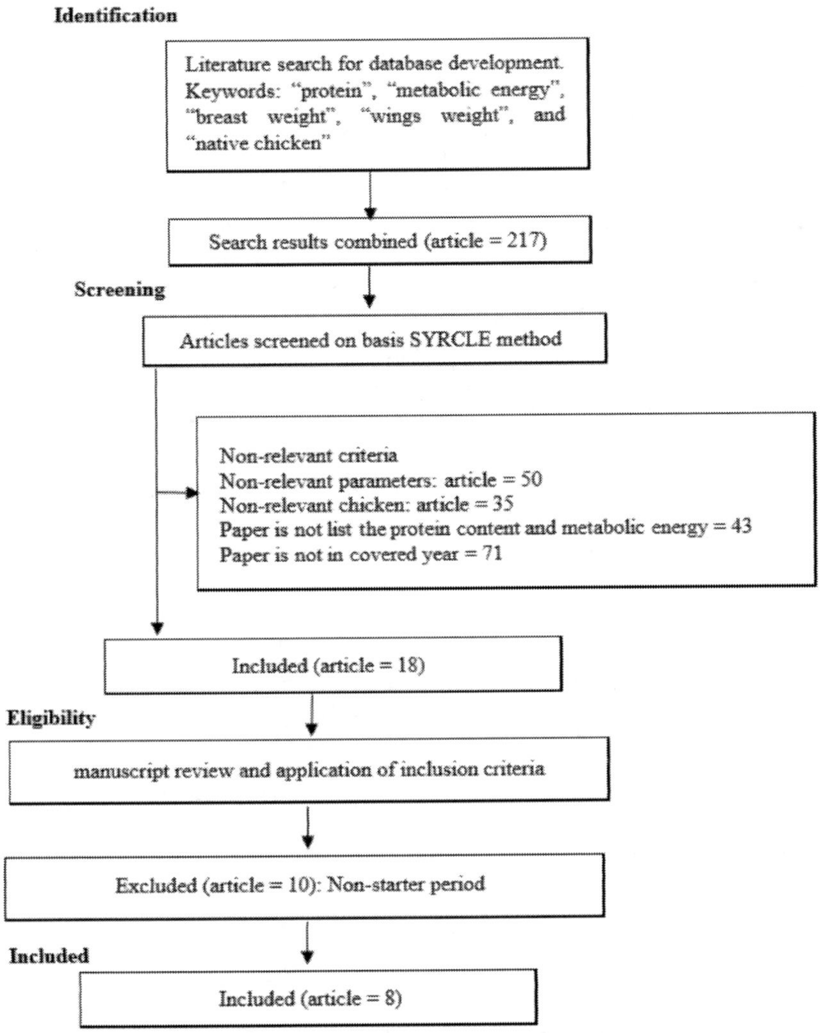

Fig. 1. Diagram flow of article selection in the meta-analysis using SYRCLE method (Sholikin et al., 2023)

Table 1. Studies included in the meta-analysis of the prediction of protein requirements and energy metabolic on breast weight and wings weight

No	References	year	Source of CP and ME	CP Level	ME Level	Phenotipe
1	Iskandar *et al.*	1998	Local feed + pollard	15.00-21.00	-	Pelung x Native
2	Ramdani *et al.*	2016	Coconut pulp meal	12.93-21.50	-	Native chicken
3	Sari *et al.*	2016	Soybean hall meal	13.66-20.05	-	Native chicken
4	Yuniza *et al.*	2011	Local feed with lisin	14.48	2903,9-2907,25	Native chicken
5	Iskandar	2006	Local feed	14.00-19.00	-	Pelung x Native
6	Ariawan *et al.*	2016	Papaya leaf powder	18.58	2985	Native chicken
7	Narayana *et al.*	2022	Bakery waste	18.38-18.63	2941-2995	Jowo Persilangan
8	Leke *et al.*	2015	Local feed + fish oil	14.66-20.46	2820,12-2988	Native chicken

CP: crude protein; ME: metabolizable energy

2.2 Data analysis

Statistical analysis designs used in the study of protein and energy balances for determining nutrition require a statistical model design [5], with a meta-analysis approach [6], as well as the SAS 2022 mixed model procedure. In this study model mixing procedures were used to combine fixed-effect methods and random effects. This combination was expected to reduce bias or too high an error. This excessive bias generally arises as a result of meta-analysis techniques in cattle using fixed-effect methods. This method only focussed on the probability value so the bias caused by the research variation was not taken into account, so the resulting bias is too high, on the other hand, there is a method that is able to analyze the variation present in the meta-analysis data, this method is called the random effect. Random effect methods not only concentrate on intra-study data but also take into account differences between studies. In fact, the difference between the resulting fixed effect and the random effect was not too high on homogeneous data, but when the data used was heterogenous then there was a high difference, therefore it is necessary to have a method capable of combining the two methods, this method is called a mixed model method. The statistical analysis model used from the mixed model in this study was as follows:

$$Y_{ij} = B_0 + B_1 X_{ij} + B_2 X_{2\,ij} + si + biX_{ij} + e_{ij} \quad (1)$$

Where *Yij*= the expected result or the bound variable observed at the level j of the variable X, the X free variable is continuously in accordance with the study *i*; B_0= the total intercept of all experiments; B_1= the linear regression coefficient Y on X (fixed effect); B_2= the quadratic coefficient of regression Y on the X (random effect); *Xij* = the synthetic date of the value j of X continuous variable in the research *i*; *si* = the random effect of study I; *bi* = the accidental effect of study i on the regression factor Y based on X; *eij* = overall research error.

3 Results and discussion

3.1 Descriptive formula model for protein and energy metabolic

The calculation of the model was done using XLSTAT. This calculation was conducted with the aim of knowing the formula so that the amount of protein and the metabolizable energy can be obtained. The calculation of a formula involving 8 articles with 116 data was done using a linear regression formula. The equation obtained using this linear regression analysis resulted in the model shown in the table below.

Table 2. Formula model for protein and energy metabolic to produce breast weight and wing weight

Variable	Breast weight formula model	Wings weight formula model
Crude Protein	15.7066+(0.0088 x Response means)	17.13363+(0.00332 x Response means)
Metabolic Energy	2923.8528+(0.24014x Response means)	2980.5996-(0.1184 x Response means)

In addition to calculating the formula model, we also conducted an analysis involving all data from the publications to find out the optimization values of breast weight and wing weight on various local chicken flocks. The diversity of data contained in the article makes the spread point and generates the size done using SAS 2022. The results of data analysis on protein and metabolic energy with variables of breast weight and wing weight can be seen in the following graph.

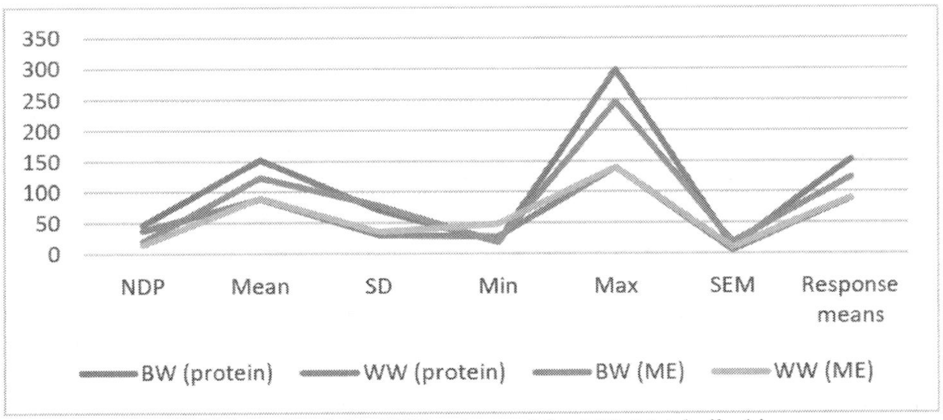

BW: breast weight; WW: wing weight; ME: Metabolizable energy

Fig. 2. Descriptive statistics of the effect different source of protein and metabolic energy

After obtaining the optimal point on each treatment on both breast weight and wing weight, the alignment was carried out using the model formula contained in Table 1. The calculation is done with the aim of acquiring protein and metabolic energy requirements in accordance with the optimal value of the results. The results of calculations using the formula model on both protein and metabolizable energy can be seen in Table 4 below.

Table 3. Prediction of protein and metabolizable energy requirement to produce breast weight and wing weight

Treatment	Protein need (%)	Metabolic energy needs (kcal/kg)
Breast Weight	17.04	2953
Wings Weight	17.43	2970

Based on the calculations that have been made, the protein and metabolizable energy requirements for breast weight and wing weight were 17.04% and 17.43% respectively as well as 2953.31 kcal/kg and 2970 kcal/kg. Through the amount of protein and the energy of the metabolism, it is expected to produce breast weight and wing weight according to the optimal response means or point produced. For example, the use of proteins with a percentage of 17.04% is expected to produce a minimum breast weight of 151.96 grams. The size of the protein and metabolic energy requirements is similar to the study before [7] which gave the protein percentages of 17% and produced a breast weight of 196 grams, as well as lower than the research before [8] which provided the protein of 17.66% and generated the breast mass of 298grams and the wing weight of 31grams. In terms of the volume of metabolic energy, the results of this study are almost similar to those of previous studies [9] which showed that metabolizable energy of 2959 kcal/kg and 2969 kcal /kg, resulted in 218.48 grams of breast weight and wing weight of 131.86 grams.

To date, the preparation of local chicken rations used by Indonesia is still based on recommendations from the research before [10] and breeding chicken standards [11]. The metabolic energy requirements of mild chickens aged 2-8 weeks are between 2600-3100 kcal/kg and protein between feeding 18%-24%, the metabolic energy and protein requirements are 2900 kcal / kg and 18%, respectively, but these standards are not protein and energy requirements for local chicks bred in tropical climates such as Indonesia [12]. So the amount of protein and the energy of metabolism on the results of this study can be considered as a prediction as the values were derived from data analysis based on many previous studies.

The need for protein to produce breast weight and wing weight must be observed. This is because protein is a nutrient that is needed for tissue development and composition as well as repair [13] ; [14]. The use of protein levels in feeding would affect the increase in chicken body weight. This is because feeding with low protein levels can not support maximum weight gain due to the limitation of the ability to meet protein needs in the body [15]. Weight gain in chickens can be achieved optimally if the supply of protein meets that required by the body, and hence satisfies the needs for muscle growth in the breast and wing [16]. The results showed that the requirement for protein for optimum breast weight and wing weight (Table 2).

As with protein, metabolic energy requirements as growth and production supports need to be considered. This is because the use of metabolic energy will affect the rate of feed consumption. The higher the energy content in the feed, the lower the consumption of the feed because the energy intake in chickens is heavily influenced by the energy contained in the feed. The accuracy in the calculation of the amount of metabolizable energy required is of great importance as excess energy may result in an increase in fat deposition [17]. Besides, metabolic energy is also correlated with proteins that can influence growth. In addition, the decrease in dietary energy and protein content leads to low protein digestion and decreased protein retention, thereby decreasing growth which ultimately increases the conversion value of the ration. As a result, the body growth will be affected and breast meat weight and wing weight will become low [12]. The percentage of breast weight would increase as body weight increases [18].

4 Conclusion

From the results of this study, it can be concluded the optimum requirement of dietary protein was 17.04% to produce a minimum breast weight of 151.96 grams. Also, the requirement of metabolizable energy was 2953.31 kcal/kg to produce 122.66 grams of breast weight. For wing weight, 17.43% of dietary proteins was required to achieve a minimum wing weight of 88.25 grams and 2970 kcal / kg to produce an 89.53 grams wing weight. The protein requirements and energy metabolism, each pile showed varying results. The existence of this diversity is influenced by genetic and environmental factors.

References

[1] Sjofjan, O., D. N. Adli, M. H. Natsir, Y. F. Nuningtyas, I. Bastomi, F. R. Amalia. JITAA, **46** (2021)

[2] Adli, D.N., O. Sjofjan, M.H. Natsir, Y.F. Nuningtyas, N. Sholikah, A. C. Marbun. Livest. Res. For Rur. Dev, **32** (7), (2020).

[3] Wiranata, G. A., I. G. A. M. K. Dewi, and R. R. Indrawati. Peternakan Tropika, **01** (02) : 87-100, (2013)

[4] St-Pierre, N. R. *J. Dairy. Sci*. **84** :741-755, (2001)

[5] D. N. Adli, S. Sadarman, A. Irawan, A. Jayanegara, T. M. Wardiny, T.R. Priambodo, A.P.A Yekti. Ital. J. of Anim. Sci, **22**, (2023).

[6] Sauvant D., P. Schmidely, J. J. Daudin, and N. R. St-Piere. Journal of Animal Consortium. pp. 1203- 1214, (2008).

[7] Iskandar, S., H. Resnawati, and D. Zainuddin. Jurnal Ilmu Ternak dan Veteriner, **04** (01) : 28-34, (1998).

[8] Sari, H., Hafid, and M. Tasse., JITRO. 3 (03): 67-80, (2016).
[9] Leke, J. R., Vony. Rawung, Jacqulein. Laihad, J. S. Mandey, and Linda., Sains peternakan, **13** (01): 52-56, (2015).
[10] Scott, M.L, Nesheim M.C., and Young R. J, Scott and associates ithaca, New York, (1982).
[11] NRC, Nutrient requirement of poultry.8th Ed, Nat, Acad of Sci, Washington D.C, (1994).
[12] Mahardika, I. G., Kristina. D, G. A. M, Sumadi. I. K, and Suasta, I. M, Majalah Ilmiah Peternakan, **16** (01), (2013).
[13] Herawati, M., Syaefullah, B. L., and Timur, N. P. V. T, Wahana Peternakan, **4** (2): 8-13, (2020).
[14] Handayani, I. M., Susanto. E, and Wardoyo. W, International journal of animal science, **3** (03) :76–85, (2020).
[15] Sidadolog, J. H. P and T. Yuwanta, Animal production, **11**(01) : 15-22, (2011).
[16] Faradila, S., L. Agustina, and M. Z. Mide, Jurnal agrisistem. **16** (02) : 55-60, (2020).
[17] Juliati, K., D. Sudrajat, and D. Kardaya, Jurnal peternakan nusantara, **1**(2): 159-164, (2016).
[18] Resnawati, H., Seminar nasional teknologi peternakan dan veteriner. balai peternakan ternak ciawi. Bogor. (2004).

The SNP (g.198655287 G>A) of AHSG gene polymorphism and its association with mineral composition in Indonesia lamb meat

Ganda Adi Septiyawan[1], *Kasita* Listyarini[2], *Ronny* Rachman Noor[2], *Katrin* Roosita[3] and *Asep* Gunawan[1*]

[1]Graduate School of Animal Production and Technology, IPB University, Bogor 16680, Indonesia
[2] Department of Animal Production and Technology, Faculty of Animal Science, IPB University, Bogor 16680, Indonesia
[3] Department of Community Nutrition, Faculty of Human Ecology, IPB University, Bogor 16680, Indonesia

Abstract. The AHSG (Alpha2-Heremans-Schmid Glycoprotein) gene is indicated to have an essential role in mineral composition. This study aimed to identify the polymorphism of the AHSG gene and its association with the mineral composition of Indonesian lamb meat. The samples were 85 rams Indonesian sheep, consisting of 70 Javanese thin-tailed sheep (JTTS) and 15 Jonggol sheep (JS). AHSG gene polymorphism was analyzed using the PCR-RFLP technique with the EagI restriction enzyme. The AHSG gene association was analyzed using the T-test. The Hardy-Weinberg equilibrium analysis showed that the AHSG was polymorphic with three genotypes, i.e., GG, GA, and AA. The AHSG gene with mineral composition was associated with significance ($P<0.05$) on the content of Fe (iron), Zn (Zinc), and Se (Selenium). The AA genotype is the recommended genotype due to its high mineral composition. The AHSG gene (SNP g.198655287 G>A) can be used as a genetic marker for the molecular-based selection of lamb with high mineral composition.

1 Introduction

Lamb meat is classified as a red meat that is preferred by consumers in each country, including Indonesia, due to its flavor, texture, and nutritional value [1]. Consumers have complex thoughts in assessing food products, especially meat, from several aspects, i.e., meat products have a tender, juicy texture, and quality flavor. Meat products also function as a source of protein and are nutritious, safe, and healthy food [2,3]. In addition to fatty acids and protein, meat provides an important dietary source for various other essential nutrients, including minerals needed by consumers. Genomic studies have explained that there is a relationship between mineral composition and muscle building, fat storage, and meat quality

* corresponding author: agunawan@apps.ipb.ac.id

[4]. High mineral content is now a positive trend as a food additive. Sensory properties, including flavor, tenderness, and cheesiness of meat, have been associated with and influenced by several mineral contents [5]. Interactions between Potassium (K) and Iron (Fe) with cholesterol, Polyunsaturated Fatty Acid (PUFA), and Monounsaturated Fatty Acid (MUFA) in black Angus cattle reported by Pannier et al. [6].

Improving lamb mineral composition using genomic studies is an effective and efficient way to identify genes that control mineral composition. The gene that is indicated to have an association is the Alpha2-Heremans- Schmid Glycoprotein gene referred to as Fetuin-A, which is a glycoprotein and is encoded by the AHSG gene, which is synthesized and secreted by the liver [7]. AHSG is a glycoprotein functioning in bone minerals; AHSG can bind Ca^{2+} (calcium ions) with a broad scope to prevent the formation of apatite minerals in circulation and plays an essential role in bone cell metabolism in the human body [8]. The AHSG gene is positively associated with limousin beef quality on tenderness and cheese properties [9]. The AHSG gene is implicated in the regulation of body fat and insulin sensitivity, which are thought to influence body fat content and metabolism [10]. In transcriptomic and proteomic studies conducted on Hu and Dorper sheep longissimus thoracis, the AHSG gene is associated and up-regulated in lipid synthesis [11]. The AHSG gene is a gene controlling some fatty acid composition in some Indonesian sheep breeds that have low saturated and high unsaturated fatty acids at SNP (g. 198655287 G>A) [12]. It has also been reported that the AHSG gene circulates Fetuin-A, which correlates strongly with plasma cholesterol levels in humans [13]. Identification of the AHSG gene has never been done on meat mineral composition, primarily Indonesian sheep. This study aims to characterize the AHSG gene at SNP (g. 198655287 G>A) on mineral composition through PCR-RFLP.

2 Materials and methods

2.1 Experimental Samples

A total of 85 DNA samples were extracted from 70 Javanese Thin-Tailed Sheep (JTTS), and 15 Jonggol sheep (JS). The DNA samples were obtained from the longissimus dorsi muscle of the rams. Animals were reared under an intensive farming system with forage (Napier grass) and concentrated feed and water ad libitum. The lamb samples of 85 animals were stored at -20^0C and analyzed for mineral composition. The animal ethics commission of IPB University has granted consent (approval number. 117-2018 IPB) for all procedures involving animals.

2.2 Mineral composition analysis

Meat was analyzed by the Atomic Absorption Spectrophotometry (AAS) method. The AAS method is an analytical method used to calculate the quantity of metal elements based on the absorption of radiation absorbance by free atoms in the gas phase. The mineral content of meat (longissimus dorsi) was analyzed in this study is potassium (K), iron (Fe), zinc (Zn), and selenium (Se). A total of 2.5 g of meat (longissimus dorsi) sample was put in an Erlenmeyer tube, and added 25 mL of concentrated HNO_3 and boiled for 30-35 minutes. The solution was cooled, and 10 mL of 70-72% $HClO_4$ was added. Then, the solution was slowly brought to a boil until the solution looked colorless. The solution was then cooled and added 50 mL of H_2O and boiled again until all NO_2 gas came out. The solution was then cooled and filtered into a 100 mL volumetric flask. The solution was then tested with an atomic absorption spectrophotometer (SSA) to obtain the mineral content of K, Fe, Zn, and Se in the meat.

2.3 DNA Extraction and Amplification

The longissimus dorsi samples were utilized for DNA extraction with the Geneaid gSYNC DNA extraction Kit protocol (Geneaid Ltd, Taiwan). The DNA extraction consisted of four stages, i.e., (1) Sample preparation, (2) Protein degradation, (3) Degradation of organic matter, and (4) DNA precipitation. After the extraction, the DNA was stored at -20 °C (Sambrook et al.1989). The SNP g.198655287 G>A and the design of the forward primer (5'-GGAGGAATCAGGGCATTTTC-3') and reverse primer (5'CCCATATCCTTACGCAATCC-3') used for amplification refers to the research Munyaneza et al. [12] procedure with a product length of 473 bp.

The DNA was then amplified using the PCR technique, which has two stages, i.e., (1) Sample preparation: 2 µL of extracted DNA was distributed as much as into a 0.2 mL tube, and prepared 14 µL of solution mixture of 6.1 µL NFW (Nuclease Free Water), 0.2 µL forward primer, 0.2 µL reverse primer, 7.5 µL My Taq HS Red Mix, then mixed using a mini centrifuge. (2) Samples incubation: The mixture is inserted in a thermocycler machine (ESCO Scientific, Singapore) for DNA amplification. The machine's operation commences with a 5-minute denaturation at 95 °C. The subsequent step involves 35 cycles of denaturation at 95 °C for 10 seconds, primary annealing at 58 °C for 20 seconds, and DNA extension at 72 °C for 30 seconds. Final extension is primer elongation at 72 °C for 5 minutes. The results of DNA amplification were visualized using electrophoresis (1.5% agarose gel).

2.4 Genotyping using PCR-RFLP technique

The genotyping procedure used the polymerase chain reaction-restriction fragment length polymorphism (PCR-RFLP) technique with the restriction enzyme EagI. The samples were incubated at a temperature of 37 °C for 4 hours (Thermo Fisher Scientific, EU, Lithuania). The DNA then were distributed as much as 5 µL into a 0.5 mL tube and mixed with 2 µL of a mixture consisting of 0.9 µL NFW (Nuclease Free Water), 0.7 µL buffer, and 0.4 µL EagI restriction enzyme. After incubation, the samples were visualized by electrophoresis technique (1.5% agarose gel). The genotypes marker of the AHSG gene were AA: 473 bp; GA: 473, 273, 200; and GG: 273, 200.

2.5 Data Analysis

The gene polymorphism was identified using Pop Gen 32 software to calculate the genotype frequency, allele frequency, and Hardy-Weinberg equilibrium. The AHSG gene association with the mineral composition was identified using the T-test statistical model of the SAS 9.2 software. The Allele frequency and genotype frequency were calculated based on the following procedure Hartl and Clark [14]. The Hardy-Weinberg equilibrium was calculated with the following approach Hartl and Clark [14]. The association of the AHSG gene genotypes with mineral content was described using the T-test [15]:

$$t = \frac{(X_1 - X_2)}{\delta^2 \frac{\sqrt{1}}{n_1} + \delta^2 \frac{\sqrt{1}}{n_2}} \delta^2$$

Note :

X1 and X2 = mean mineral composition of genotypes 1 and 2

n1 and n2 = number of individuals of genotypes 1 and 2

δ2 = combined variance

A normality test was conducted before a T-test using the Kolmogrov-Smirnov or Shapiro-Wilk test. If the data were abnormal, then the outlier data is removed. If all outliers data were already removed and the data was still abnormal, then the data were transformed before analyzing the data using the T-test [16].

3 Result and Discussion

3.1 AHSG gene polymorphism in the Indonesia sheep population

AHSG gene SNP (g.198655287 G>A) polymorphism was performed using PCR-RFLP technique with an EagI restriction enzyme and obtained three fragment products, i.e. GG, GA, and AA. Fragment product results were visualized using an electrophoresis technique, which can be seen in Figure 1. In this case, the restriction enzyme selection recognized and cut the target DNA sequence [17], resulting in different fragment products [18]. The mutated base of the AHSG|EagI gene is A instead of G. The fragment product produced by the GG genotype contains two fragments with lengths of 273 bp and 200 bp, while the fragment product produced by the GA genotype contains three fragments with lengths of 473 bp, 273 bp, and 200 bp, and the genotype has one fragment with a length of 473 bp.

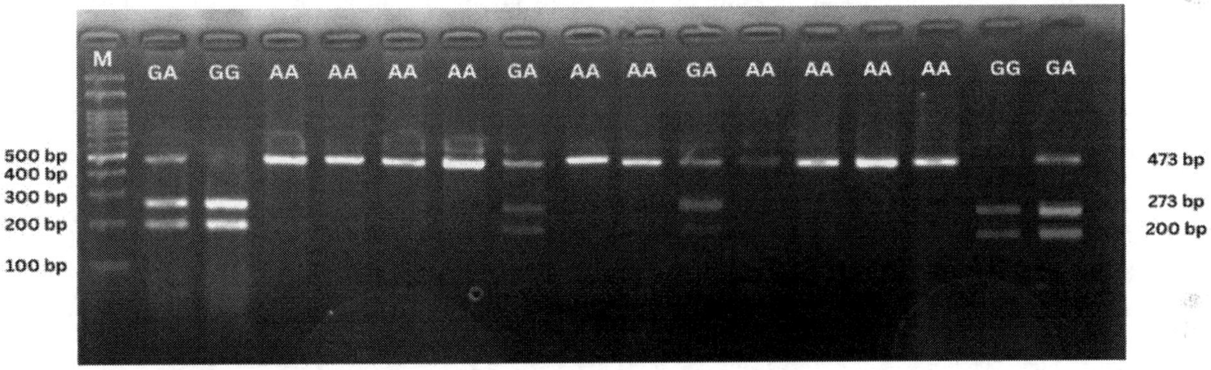

Fig. 1. Visualization results of AHSG gene (g.198655287 G>A) in Indonesian sheep with three genotypes: GG, GA, and AA using 2% agarose gel and 100 bp marker.

The AHSG|EagI gene polymorphism in Indonesian sheep produces GG, GA, and AA genotypes. The GA genotype was identified as the most dominant with a frequency of 0.46, followed by the AA genotype at 0.41 and the GG genotype at 0.13. This result indicates that the A allele is the most common compared to the G allele in the sheep population studied. The AHSG|EagI gene is categorized as a polymorphic because the overall genotype frequency of the sheep used less than 0.99, including the JTTS, JS, and total population.

The SNP in the gene are categorized as polymorphic if they have more than one allele with a frequency value of at least 1% [19,20]. The results of the chi-square analysis showed significant differences, indicating that the AHSG|EagI gene in the Indonesian sheep and Javanese thin-tailed sheep (JTT), Jonggol sheep (JS) and total population is the allele is not in equilibrium, which indicates the population may have experienced migration, mutation, selected mating, or selection [21].

Table 1. Genotype and allele frequencies and Hardy-Weinberg equilibrium of AHSG gene in Indonesian sheep population

No	Breed	n	Genotype frequency			Allele frequency		Chi-square (X^2)
			GG	GA	AA	G	A	
1	Javanese thin-tailed sheep (JTTS)	70	0.13 (4)	0.46 (42)	0.41 (24)	0.36	0.64	6.58
2	Jonggol Sheep (JS)	15	0.37 (4)	0.63 (11)	0.00 (0)	0.37	0.63	5.03
3	Total population	85	0.13 (4)	0.46 (53)	0.41 (28)	0.36	0.64	10.71

n = number of samples;
x^2 table= 3.84

3.2 AHSG gene with mineral composition

The AHSG gene SNP (g.198655287 G>A) has a significant association with mineral composition, including Ferrum (Fe), Zinc (Zn), and Selenium (S). The results of mineral composition in Indonesian sheep on the composition of Fe, Zn, and Se have values of 1.58-2.04 mg/100g, 2.42-3.05 mg/100g, and 0.81-0.89 mg/100g, respectively. The results of mineral composition in Indonesian sheep on the composition of Fe, Zn, and Se each have a value of 1.58-2.04 mg/100g, 2.42-3.05 mg/100g, and 0.81-0.89. mineral composition in Indonesian sheep has similarities to the minerals contained in three Croatian [22] and Marocco [23] indigenous sheep breeds, especially in Fe and Zn.

Table 2. Association of AHSG Gene with mineral composition

Mineral	GG ($\overline{x}$ ± sd) (n)	GA ($\overline{x}$ ± sd) (n)	AA ($\overline{x}$ ± sd) (n)	P value		
				GG vs GA	GG vs AA	GA vs AA
Fe	1.58 ± 0.3 (4)	1.71 ± 5.02 (50)	2.04 ± 0.44 (25)	ns	*	**
Zn	2.42 ± 0.48 (4)	2.51 ± 9.93 (53)	3.05 ± 1.01 (28)	ns	ns	**
K	268.85 ± 52.51 (4)	281.83 ± 92.68 (53)	256.45 ± 83.00 (28)	ns	ns	ns
Se	0.89 ± 0.06 (2)	0.73 ± 0.13 (46)	0.81 ± 0.06 (20)	ns	ns	*

$\overline{x}$: mean (mg/100 g)
Sd : standard deviation
P value : significancy between genotype
** : significant different at P<0.01
* : significant different at P<0.05
ns : not significant

Genotype AA on the composition of Fe, Zn, and Se is very significantly different compared to genotype GA (P <0.01), and especially on the composition of Fe also has a significant effect compared to genotype GG. Genotype AA has the highest value if with genotype GA and GG, especially on the composition of Fe and Zn. Fe and Zn composition in meat (leanness) is strongly influenced by muscle fiber type and muscle oxidative capacity [22]. Fe and Zn also have a positive effect on lipid metabolism [4]. Transcriptomic studies explain that the AHSG gene affects lipid synthesis (up-regulated) [11] and is indicated to be positively associated with Fe and Zn composition. The advantage of high Zn composition in lamb meat is that the body can absorb it well [24]. The influence on parameters with yield can be due to several factors, including breed, age, and muscle [23]. However, the composition of selenium is strongly influenced by feeding patterns and geographical location; case studies explain that sheep in pasture are rich in selenium composition [25] and Se composition in indonesian sheep are higher in value when compared to Croatian [22] and Marroco [23] indigenous breeds sheep.

4 Conclusion

The AHSG gene SNP (g.198655287 G>A) in Indonesian sheep, showed polymorphic with three genotypes, i.e., GG, GA, and AA, and the distribution alleles were not in the Hardy Weinberg equilibrium. In addition, the AHSG gene was significantly associated with Ferrum (Fe), Zinc (Zn), and Selenium (Se). The AA genotype could be a candidate marker for producing lamb with high mineral composition.

This research is funded by the Kedaireka Matching Fund program from the Indonesian Minister of Education and Culture year 2023 with grant number 18977/IT3.L1/HK.07.00/P/T/2023.

References

1. A. J. Mcafee, E. M. Mcsorley, G.J. Cuskelly, B.W. Moss, J. M. W Wallace, M. P. Bonham, A. M. Fearon. Meat Science **84** (**1**), 1–13 (2010)
2. D. W. Pethick, J. F. Hocquette, N.D. Scollan, F.R. Dunshea, Animal **15**, 100356 (2021)
3. D. W. Pethick, A. J. Ball, R. G. Banks, J. F. Hocquette, Animal Production Science **51**, 13–18 (2011)
4. W. J. D. Diniz, P. Banerjee, L. C. A Regitano, Physiol Genomics, **51** (**11**), 529-538 (2019)
5. C. Maltin, D. Balcerzak, R. Tilley, M. Delday, Proc Nutr Soc **62**, 337–347 (2003)
6. L. Pannier, D.W. Pethick, M. D. Boyc, A. J. Ball, R. H. Jacob, G. E. Gardner, Meat Sci **96**, 1111–1119 (2014)
7. E. Fisher, N. Stefan, K. Saar, D. Drogan, M. B. Schulze, A. Fritsche., H. G. Joost, H. U. Häring, N. Hubner, H. Boeing, C. Weikert, Circ Cardiovasc Genet. 607-613 (2019)
8. H. Jiang, S. F. Lei, S.M. Xiao, Y. Chen, S. Xiao, F. Yang, L. M. Li, S. Wu, H.W. Deng, Acta Pharmacol Sin **28**, 375–381 (2007)
9. Y. Zhu, M. Gagaoua, A. M. Mullen, D. Viala, D. L. Rai, A. L. Kelly, D. Sheehan, R, Meat Sci. **176**, 108488. (2021)
10. C. Lavebratt, E. Dungner, J. Hoffstedt, Journal of Lipid Research, **46**(**10**), 2278–2281 (2005).
11. L. Zhao, F. Li , X. Zhang, D. Zhang, X. Li, Y. Zhang, Y. Zhao, Q. Song, K. Huang, D. Xu, J. Cheng, J. Wang, W. Li, C. Lin, W. Wang, Meat Sci, **193**, 108930 (2022)
12. J. Munyaneza, A. Gunawan, R.R. Noor, IJSRST **2**, 351-360 (2019)
13. C. Marechal, G. Schlieper, P. Nguyen, T. Kru¨ger, E. Coche, A. Robert, J. Floege, E. Goffin, M. Jadoul, O. Devuyst, J Am Soc Nephrol **6**, 974–985 (2011)
14. D. L. Hartl, A. G. Clark, *Principle of Population Genetic*. (Sinauer Associates Sunderland, UK, 1997)
15. R. Walpole, *Pengantar Statistika* (PT Gramedia Pustaka Utama, Jakarta, ID, 1995)
16. P. Mishra, C. M. Pandey, U. Singh, A. Gupta, C. Sahu, A. Keshri, Ann Card Anaesth **22**, 67-72 2019.
17. A. Gunawan, C. Sumantri, R. Juniarti, *Gen dan Keragaman Genetik Ternak* (Bogor, IPB, ID, 2017)
18. W.S. Klug, M. R. Cummings, C. A. Spencer. 2006. *Concepts of Genetics 8th edition* (Pearson Prentice Hall, New Jersey, USA, 2006)

19. F. Rell, S. K. Widyastuti, I. N. Wandia.. Polimorfisme lokus mikrosatelit D10S1432 pada populasi monyet ekor panjang di Sangeh. Vet. Sci. and Med. J **1**,16-21 (2013)
20. A. Gunawan, Jakaria, K. Listyarini, A. Furqon, C. Sumantri, S. H. Akter, M. J. Uddin. 2018. Gene **676**:86-94 (2018)
21. Hadini, H. 2015. *Struktur genetika populasi jagung dalam keseimbangan hardy weinberg dan silang kerabatiyya* (Agro, UPY, ID, 2015)
22. A. Kasap, A. Kaić, I. Širić, Z. Antunović, B. Mioč, Italian Journal of Animal Science, **17** (**2**) 274–278 (2018)
23. K. Belhaj, F. Mansouri, B. A. Moumen, M. Sindic, M. L. Fauconnier, M. Boukharta, C. H. Serghini, A. Elamrani, Scientific World Journal **20**, 6633774 (2021)
24. M. C. Cabrera, A. Saadoun, Meat Science **98** (**3**), 435–444 (2014)
25. M. C. Cabrera, M. Pereiro, A. Saadoun, Annals of Nutrition and Metabolism, **63**, 1528 (2013)

Polymorphism of the SLC25A30 gene in Indonesian sheep and its association with carcass characteristics

Fajriani Mutmainnah Makmur[1], *Kasita* Listyarini[1], *Ronny Rachman* Noor[2], *Evy* Damayanthi[3], and *Asep* Gunawan[2*]

[1]Graduate School of Animal Production and Technology, Faculty of Animal Science, IPB University Jl. Agatis Dramaga, Bogor 16680, Indonesia
[2]Department of Animal Production and Technology, Faculty of Animal Science, IPB University Jl. Agatis Dramaga, Bogor 16680, Indonesia
[3]Department of Community Nutrition, Faculty of Human Ecology, IPB University Jl. Kamper Dramaga, Bogor 16680, Indonesia

Abstract. Carcass characteristics are one of the factors that significantly impact the quality of meat. Genetic improvement in carcass characteristics is required to encourage interest among consumers. The objective of this study was to discover the presence of the SLC25A30 gene polymorphism at SNP g.15912281 A>T and its potential association with carcass characteristics in Indonesian sheep. The study used 153 rams, consisting of 87 Javanese thin-tail sheep, 20 Javanese fat-tail sheep, 10 Garut composite sheep, 10 Compass Agrinak sheep, 10 Barbadoss cross sheep and 16 jonggol sheep. The PCR-RFLP method was used to investigate the SLC25A30|SspI polymorphism. The association between the SLC25A30 gene and carcass characteristics was analyzed using the T-Test method. The results showed that the SLC25A30 (g.15912281 A>T) gene polymorphism was polymorphic in all breeds. The genotypes identified in the polymorphism were AA, AT and TT. The SLC25A30 gene was significantly associated (P<0.05) with carcass characteristics (slaughter weight, hot carcass, carcass percentage, carcass length and cold carcass). In conclusion, the genotype that can be recommended for selecting good meat quality is the TT genotype respectively than the AT and AA genotype The SLC25A30 (g.15912281 A>T) gene indicates as a prospective gene for the purpose of identifying carcass characteristics in lamb meat.

1 Introduction

Lamb meat can be one of the alternative sources to meet the animal protein needs of the Indonesian population. Sheep meat is a good source of dietary protein (20 g/100 g of lean) that is rich in highly digestible essential amino acids in balance to human requirements;

*Corresponding author: agunawan@apps.ipb.ac.id

it is also a good source of nutritionally valuable micronutrients (iron, zinc, selenium, vitamins B3, B6 and especially B12) [1]. Badan Pusat Statistik (BPS) reported that 54,650.53 tonnes of lamb meat were produced in Indonesia in 2022. When compared to the prior year, when it was 50,702.06 tonnes, this amount climbed by 7.79%. Sheep carcass is a key criterion of a sheep's marketability and economic worth. The conformation, dressing (yield), fatness condition, meat colour, carcass linear measurements and fat cover quality (colour and consistency) are the major factors that determine it [2]. Multiple factors, including breed [3], slaughter age [4,5] and weight at slaughter [6,7] affect the quality of lambs' meat. As a result, selection objectives in sheep breeding now emphasise growth and carcass attributes more than in the past. In order to maintain consumer acceptability of lamb meat, producers must provide high-quality meat, which is determined by carcass measurements [8,9]. However, it happens frequently that consumers have expressed uncertainty regarding the higher quantities of cholesterol and saturated fatty acids contained in lamb meat. Both of these factors have been linked to the developing of cardiovascular disease and cancer [10]. Consumers also express displeasure with the taste or odor of lamb meat [11]. Thus, genetic enhancement is essential, and this can be achieved through different molecular techniques like selection and crossbreeding programs.

One candidate gene related to the nutritional composition of meat is the Solute Carrier Family 25 Member 30 (SLC25A30 gene). The SLC gene superfamily currently consists of 65 families with a total of 458 transport proteins found throughout the cell's organelles, with the most prominent family being SLC25, which has 53 members [12] SLC25A30 is a member of the SLC25 family and provides a function in the movement of various chemicals through the mitochondrial membrane [13]. Several key genes related to fatty acid production, adipogenesis, fat accumulation, and lipid metabolism were discovered, including LEPR, APOA5, GFPT1, CYP17A, FABP7, TGFBR2, GSTCD, and SLC25A30 [14].

According to the research results of Listyarini [15] there appears to be a correlation between the SLC25A30 gene and meat tenderness, as shown by RNA sequencing data. Nevertheless, the association between the SLC25A30 gene and sheep remains unexplored. Given this, exploring the association between carcass characteristics and the SLC25A30 gene is necessary. This research aims to examine the variations in the SLC25A30 gene and establish its correlation with various quality parameters in lambs.

2 Materials and methods

2.1 Ethical approval and Animal collected

All procedures involving animals in this study were ethically examined, and authorized IPB University's Animal Ethics Commission has given their clearance, and it is documented as 117-2018 IPB. In this study, 153 rams were examined to detect polymorphisms in the SLC25A30 gene. The sample consisted of 10 Barbados cross sheep (BCS), 20 Javanese fat-tailed sheep (GS), 87 Javanese thin-tailed sheep (JTTS), 10 Garut composite sheep (GCS), 16 Jonggol sheep (JS) and 10 Compass agrinac sheep (CAS). This research involved the slaughtering of sheep at the Slaughter House (PT Pramana Pangan Utama (PPU), a commercial abattoir. The slaughterhouse employee followed the halal guidelines and fulfilled animal welfare criteria guidelines in sluthering the animals. The sheep were raised on separate farms, provided unrestricted food access (ad libitum), and kept under comparable rearing conditions.

2.2 Carcass characteristics Analyses

Before to slaughter, the sheep must complete a 17-hour fasting period in order to settle the sheep and minimise the amount of undigested feed in their digestive system [15]. The sheep were slaughtered by cutting their jugular vein, oesophagus, and trachea in three conduits. The sheep blood is collected, quantified, and subsequently suspended to ensure optimal blood drainage. Slaughtered lamb is divided into non-carcass and carcass. All carcass and non-carcass were separated and weighed. In addition, carcass characteristics were measured. The warm carcasses were weighed and withered at 4°for 24 hours. The weight of the cold carcass was calculated and the spinal column was divided into two separate sections, one on the left and one on the right. The right carcass was divided into eight commercial cuts: shoulder, neck, rack, leg, shank, rib, flank and loin, while the left carcass is stored. The commercial pieces are separated between meat, bone, and fat (intermuscular fat and subcutaneous), then each part is weighed to determine the weight and percentage of the part. Carcass characteristics were observed by looking at the hot carcass, slaughter weight, cold carcass, carcass percentage, and carcass length.

2.3 DNA Extraction, PCR-RFLP Amplification and Genotyping

The Single Nucleotide Polymorphism (SNP) of the SLC25A30 gene was identified at position SNP g.15912281 A>T. The SNP was detected through RNA sequencing [16]. Muscle samples of the longissimus dorsi were processed with the Geneaid gSYNC DNA Extraction Kit (GS050/100/300) to isolate genomic DNA following the provided protocol. Amplification of SLC25A30 gene fragments was carried out using the GeneAmp ESCO PCR system. The DNA amplification procedure began with denaturation for one minute at 95 °C, then included 35 cycles consisting of denaturation at 95 °C for 15 seconds, annealing at 62 °C for 15 seconds, extension at 72 °C for 15 seconds, and final extension at 72 °C for 1 minute. The reaction was then cooled to 15 °C for 5 minutes. The PCR samples were verified by electrophoresis on an agarose gel containing 1.5% agarose, and the DNA bands were observed using a thermocycler and a UV transilluminator. The PCR products were subjected to PCR-RFLP (Restriction Fragment Length Polymorphism) using the SspI enzyme and incubated at 37 °C for 4 hours (Table 1). The resulting data were also electrophoresed on a 2% agarose gel to determine the genotype: TT (393 bp), AA (206, 187 bp), and AT (393, 206, and 187 bp).

Table 1. Primer sequence and size of PCR of gene SLC25A30

Gene	Accession number	Size of PCR	TA°C	Enzyme	Primer sequence
SLC25A30	NC_019467.2	393 bp	62	SspI	F: 5'-GCC CTG TGA GAG TAA CAG TA-3'
					R: 5"-TGG TTT AGT ACG AGG GAC GA-3'

3 Statistical Analysis

3.1 Genotype and allele frequencies

The genotype and allele frequencies were estimated using the calculation method of Nei and Kumar [17] and the Hardy-Weinberg equilibrium was identified through the methods of Hartl and Clark [18].

$$X_i = \frac{(2n_{ii} + \sum j \neq j \; n_{ij})}{2N} \qquad X_{ii} = \frac{n_{ii}}{N}$$

Description:
Xi = frequency of allele i;
Xii = frequency of genotype ii;
nij = number of individuals with genotype ij;
ni = number of individuals of genotype ii; and
N = total sample

3.2 Hardy-Weinberg equilibrium

Procedures were used to determine the Hardy-Weinberg equilibrium [18].

$$x^2 = \sum \frac{(O - E)^2}{E}$$

Description :
X^2 = chi-square
O = total the number of observations of the i-th genotype; and
E = total number of expected of the i-th genotype

3.3 Association of SLC25A30 gene with carcass characteristics

The effect of genotype (SPSS Statistics 26 software) on the phenotype was assessed by conducting T-Test procedures to test the differences value of carcass characteristics between breed on the same genotype. The data for carcass characteristics parameters in crossbred and local sheep were analyzed separately to determine if crossbreeding influenced the genotype associations obtained for the gene with carcass characteristics (slaughtered weight, dressing percentage, cold carcass, hot carcass and carcass length).

$$t = \frac{\bar{X}_1 - \bar{X}_2}{S\sqrt{\frac{1}{n_1} + \frac{1}{n_2}}}$$

$$S = \sqrt{\frac{\sum_{i=1}^{n}(\bar{X}_i - \bar{X}_1)^2 + \sum_{i=1}^{n}(\bar{X}_i - \bar{X}_2)^2}{n_1 + n_2 - 2}}$$

Description :
X1 and X2= the average traits for genotypes 1 and 2;
n1 and n2= individual numbers of genotypes 1 and 2; and
S = the combined standard deviation

Data analysis is completed by completing both the Kolmogorov-Smirnov and Shapiro-Wilk tests for normality test. To figure out how to interpet the results of the normality test, it is essential to determine the significance value. If the results of the normality tests indicate the data is not normal, a check for data outliers is done. Any data outliers found in each value are then taken out of the data. As a result of deleting outlier data, each parameter can have a different number of samples. Data transformation becomes necessary when a data check has been done and the distribution remains non-normal. After confirming that the data follows a normal distribution, a t-test is used to analyse it.

4 Results and Discussion

4.1 Polymorphism of the SLC25A30 gene

PCR-RFLP amplified a candidate marker for the SLC25A30 gene using the restriction enzyme SspI. PCR amplification with a length of 393 bp was obtained, as expected from the SLC25A30 gene primers. The SNP g. 15912281 A>T SLC25A30 of the SLC25A30 gene exhibits genotype polymorphism characterized by the occurrence of allele variations A and T, namely TT= 393 bp; AA= 206, 187 bp; AT= 393, 206, and 187 bp. The polymorphism of the SLC25A30 gene visualized in Figure 1 represents only a few randomly selected sheep samples.

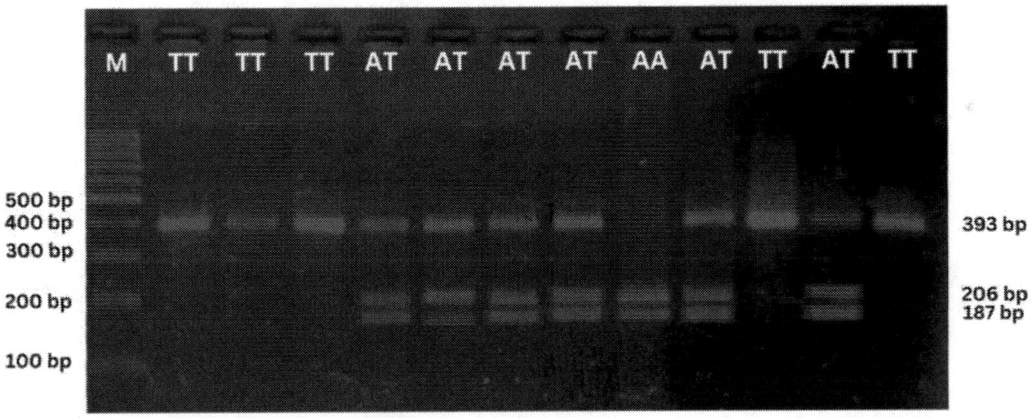

Fig. 1. PCR-RFLP result of SLC25A30 gene; M = marker 100 bp ladder size standard; TT (393 bp); AA (206, 187 bp); AT (393, 206, and 187 bp) genotype; bp = base pair

The SLC25A30 gene has three genotypes (AA, AT, and TT). All genotypes were found in every sheep breed except for Jonggol sheep, Barbados cross sheep, Compass Agrinac sheep, Javanese fat-tailed sheep, and Garut composite sheep, where the AT genotype was undetected (Table 2). In the Indonesian sheep population, at the SNP g. 15912281 A>T locus of the SLC25A30 gene, AA was the most common genotype in this population, while AT was just behind it. Crossbreeding may influence this, such as in Compass Agrinac sheep, where the AT genotype has been selectively bred out of the population. The SNP g. 15912281

A>T of the SLC25A30 gene in crossbred sheep, such as Garut Composite sheep, were found in Hardy Weinberg Equilibrium (HWE). However, in other local sheep groups, such as Garut Sheep, Jonggol Sheep in the SLC25A30 gene was not in Hardy Weinberg equilibrium. The non-equilibrium results from the unequal distribution of genotypes within these varieties. Multiple factors, including mutation, non-random reproduction, natural selection, gene flow, and genetic drift, can lead to deviations from HWE [19] or possibly as a result of these sheep's efforts to adapt to challenging ecological circumstances [20].

Table 2. Frequency of genotype and allele of the SLC25A30 gene

Breed of sheep	N	Genotype Frequency			Allele Frequency		X^2
		AA (98)	AT (8)	TT (47)	A	T	
Javanese thin-tailed sheep	87	0.57 (61)	0.37 (6)	0.06 (20)	0.75	0.25	55.81
Jonggol sheep (JS)	16	0.73 (12)	0.00 (0)	0.26 (4)	0.73	0.26	15.00
Barbados cross sheep (BSCS)	10	0.80 (8)	0.00 (0)	0.20 (2)	0.80	0.20	10.00
Compass agrinac sheep (CAS)	10	0.40 (8)	0.00 (0)	0.60 (6)	0.40	0.60	10.00
Javanese fat-tailed Sheep (GS)	20	0.40(8)	0.00 (0)	0.60 (12)	0.40	0.60	20.00
Garut composite sheep (GCS)	10	0.36 (5)	0.48 (2)	0.16 (3)	0.60	0.40	3.40
Total	153	0.64	0.05	0.30	0.67	0.33	115.83

N=number of samples
(..)=number of samples with genotypes AA, AT, TT
X^2 Table = 3.84

4.2 Association of SLC25A30 gene polymorphism with carcass characteristics

Considering the potential impact of crossbreeding and local sheep on the genotype associations with carcass characteristics (slaughtered weight, dressing percentage, cold carcass, hot carcass and carcass length). Data for carcass traits parameters in crossbred and local sheep were analyzed separately (Table 3).

The association between the SLC25A30 gene and carcass characteristics was analysed, showing significant differences ($P<0.05$) in relation to slaughtered weight, hot carcass, carcass percentage, carcass length and cold carcass (Table 3). Significant differences in dressing percentage carcass and carcass length are detected in each genotype in both local and crossbred breeds. The slaughter weight has an impact on the carcass proportion [21]. The growth rate directly impacts the rise in slaughter weight. Furthermore, the improvement in carcass percentage was additionally impacted by the increase in live weight and age. [22]. Slaughter weight and carcass percentage are directly correlated with the length of the carcass [23]. The body size of livestock will have an impact on the length of the carcass. The research data indicated that the carcass length of sheep with the TT genotype in crossbreed was longest (95,72 cm) ($P<0.05$) than local breed (63,22 cm; 59 cm; 62,62 cm). In the slaughter weight parameter, it is noticeable that the TT genotype has a higher slaughter weight (27,33 kg)

(P<0.05) in the local breed than other genotype, whereas the AT genotype has a greater amount of weight (27,70 kg) in the crossbreed.

Table 3. Association of the SLC25A30 gene with carcass characteristics

Parameter	Type of breed	Genotype of SLC25A30 ($\bar{x}$ ± Standard Deviation)		
		AA	AT	TT
Slaughter weight (kg) (n=153)	Local breed (n=123)	25.65±4.31 (82)	24.35±4.20 (6)	27.33±5.00 (35)
	Crossbreed (n=30)	20.51±5.85 (17)	27.70±2.12 (2)	21.69±5.71 (11)
	P value	0.00*	0.31	0.00*
Hot carcass (kg) (n=153)	Local breed (n=123)	10.69±2.55 (82)	10.10±1.97 (6)	11.88±2.84 (35)
	Crossbreed (n=30)	7.22±2.36 (17)	9.65±0.49 (2)	7.24±2.04 (11)
	P value	0.00*	0.81	0.00
Dressing percentage (%) (n=153)	Local breed (n=123)	43.25±3.86 (82)	41.97±2.01(6)	44.26±4.30 (35)
	Crossbreed (n=30)	34.91±3.02 (17)	34.87±0.88 (2)	33.29±1.47 (11)
	P value	0.00*	0.00*	0.00*
Carcass length (cm) (n=153)	Local breed (n=123)	63.22±6.81 (82)	59.00±4.19 (6)	62.62±7.43 (35)
	Crossbreed (n=30)	94.70±16.76 (17)	73.00±4.24 (2)	95.72±14.98 (11)
	P value	0.00*	0.00*	0.00*
Cold Carcass (kg) (n=146)	Local breed (n=116)	10.46±2.54 (80)	10.06±1.97 (6)	11.81±2.96 (30)
	Crossbreed (n=30)	6.70±2.31 (17)	9.25±0.35 (2)	6.80±2.03 (11)
	P value	0.00*	0.62	0.00*

N=number of samples
$\bar{x}$ = means of the meat quality values
*=significant

The SLC gene family regulates as transporters and has been relating to the ribeye area (REA) in Nelore cattle [24] the ribeye is a popular regarded cut of beef known for its tenderness, rich marbling and flavorful taste. Genes encoding solute carriers belonging to the SLC superfamily, including SLC11A1 and SLC12A4, were identified and explored as prospective candidate genes for determining meat colour in cattle [25]. Candidate genes from the solute carrier family (SLC), consists SLC37A4 and SLC3A2, have been identified as potential genes that regulate the drip loss characteristic in pigs [26]. The SLC gene's control over drip loss can be applied to lamb meat, as drip loss can impact the final cooked product. There is an expectation that the application of the SLC gene may have the potential to enhance the carcass quality in sheep.

5 Conclusion

The SLC25A30 gene SNP g. 15912281 A>T was discovered polymorphism among Indonesian sheep. The distribution of the SLC25A30 gene was not in Hardy-Weinberg equilibrium, increasing the percentage of the AA genotype. The carcass characteristics parameters (slaughter weight, hot carcass, dressing percentage, carcass length and cold carcass) were impacted by the SLC25A30 SNP g.15912281. In conclusion, the genotype that can be recommended for selecting good meat quality is the TT genotype respectively than the AT and AA genotype and variability in the SLC25A30 gene can affect the quality of carcass characteristics

References

1. Prache, S., Schreurs, N., Guillier, L. Animal. **16**. (2022)
2. Belhaj, Kamal. Farid Mansouri, Aziz Tikent, Yassin Taaifi, Mohamed Boukharta, Hana Caid Serghini, and Ahmed Elamrani. The Scientific World Journal (2021).
3. J.-B. Liu, J. Guo, F. Wang. Small Ruminant Research (2015).
4. Sañudo. Journal of the Science of Food and Agriculture (2019).
5. A. G. D'Alessandro, G. Maiorano, M. Ragni, D. Casamassima, G. Marsico, and G. Martemucci, Small Ruminant Research, (2021).
6. A. Sabbioni, V. Beretti, E. M. Zambini, and P. Superchi. Italian Journal of Animal Science (2016).
7. A. Camacho, A. Torres, J. Capote et al. Journal of Applied Animal Research, vol. **45**. 400–408, (2016).
8. Abubakar, A, CA Frima, V Silvia, CA Fitri, Z Hanum, and E Mariana. IOP Conference Series: Earth and Environmental Science (2023).
9. Ayeb, N, A Ghrab, A Barmat, and T Khorchani. Turkish Journal of Veterinary and Animal Sciences (2016).
10. Maharani, D, D-W Seo, N-R Choi, S Jin, M Cahyadi, C Jo, and J-H Lee. Korean Journal of Agricultural Science. (2013).
11. Gunawan, A, Jakaria, Listyarini K, Furqon A, Sumantri C, Akter SH, Uddin MJ.. *Gene*. **676** :86-94 (2018).
12. Pizzagalli MD, Bensimon A, Furga GS. *The Febs Journal*. 288 : 2784-2835 (2020)
13. Rochette L, Meloux A, Zeller M, Malka G, Cottin Y, Vergely C. 25:2417 (2020).
14. Gunawan A, Listyarini K, Harahap RS, Jakaria, Roosita K, Sumantri C, Inounu I, Akter SH, Islam, MA, Uddin MJ. PlosOne. 16(12) (2021).
15. Dagong MIA, Herman R, Sumantri C, Noor RR, Yamin M. IJAVS. 17:13-24 (2012).
16. Listyarini K, Sumantri C, Rahayu S, Islam MA, Akter SH, Uddin MJ, Gunawan A. Animals. **15** ;13(4) :674. (2023).
17. Nei, Kumar. New York (US): Oxford University Pr (2000).
18. Hartl DL, Clark AG. Sunderland (UK): Sinauer Associates (1997).
19. Graffelman, J, D Jain, and B Weir. Human Genetics. **136**(6):727–741 (2017).
20. Khasanah, H, A Gunawan, R Priyanto, MF Ulum, and Jakaria. Media Peternakan. **39**(2):95–103 (2016).
21. Lei X, Tao QJ, Lin R, Shuai CD, Xi T, Jun XS, Yan ZZ, Sheng HL. **22** :5 (2023).
22. Purslow P,P. Lawrie's meat science. 51-103 (2023)
23. Rajkumar V, Dass G, Verma AK, Das AK. Indian Journal of Animal Sciences. **84** (5) : 569-574 (2014).
24. Silva DBS, Fonseca L,FS, Pinheiro DG, Magalhaes AFB, Muniz MMM, Ferro JA, Baldi F, Chardulo L,AL, Schable RD, Taylor JF, Albuquerque LG. Sci Rep. **10** :14701 (2020).
25. Garzon NAM, Magalhaes AFB, Mota LFM, Fonseca LFS, Chardulo LAL, Alburquerque. Meat Science. **171** (2021).
26. Zhao X, Wang C, Wang Y, Lin H, Wang H, Hu H, Wang J. **20**:89 (2019).

Analysis system dynamic of chicken egg production system: Case in West Sumatra Indonesia

Dwi Yuzaria[1]*, *Fitrimawati* Fitrimawati[1], and *Muhammad* Ikhsan Rias[1]

[1]Department of Development and Animal Business, Faculty of Animal Science, Andalas University, Padang 25163, Indonesia

Abstract. A policy study is needed to guide the planning process to develop a sustainable agropolitan area for laying hens in Limapuluh Kota Regency. This research aims to design a policy for developing sustainable laying hen businesses through modeling the chicken egg production system. The dynamic systems approach was used to determine the production system for chicken egg production, which consists of two sub-systems: the layer chicken population sub-system and the business profit sub-system. Each sub-system interacts dynamically based on time. The research results reveal that the chicken egg production model in Limapuluh Kota Regency is accurate and can be used for decision-making activities. The model and its development can help determine the direction of chicken egg production policy to be developed into a sustainable agropolitan area. In the future, overall chicken egg production in Limapuluh Kota Regency is likely to increase, leading to an increase in profits for the egg-laying chicken farming business. However, an increase in feed prices without a corresponding increase in egg prices may hamper the rate of increase in egg.

1 Introduction

The purpose of producing agropolitans is for countries to encourage rural development, maximize the use of natural resources, raise farmer incomes, and boost regional economic growth. Growing laying hens is a popular and profitable business in many nations, including Indonesia. It is possible that they will help achieve the goals of sustainable development [1]. The Limapuluh Kota Regency (LKR) area has great potential in the subsector of laying hen farming [2]. Seeing this great potential, the development of agropolitan areas based on laying hens is one of the right choices as a balanced rural-urban development concept by adjusting the potential and characteristics of the area concerned. So that the *backwash effect*, both the drain of natural resources and rural human resources, can be avoided. According to[3], the development of agropolitan areas will encourage decentralization of development and authority in regions, villages, and cities that can strengthen each other, as well as the occurrence of economic development based on local resources through empowering local communities.

* Corresponding author: dyuzaria@ansci.unand.ac.id

Many complex problems are faced by the community in the development of agropolitan areas in the Limapuluh Kota Regency area, which are difficult to solve using only one specific method. One method that can be used to solve these complex problems is the system approach. [4], suggest that a systems approach can provide good problem-solving for complex multidisciplinary problems[5]. The systems approach is defined as a problem-solving methodology that begins with formulating goals and results as an operating system that can effectively be used to solve problems [6].

The development of agropolitan areas in the Limapuluh Kota Regency area faces several complex problems that cannot be solved with a single specific method. However, the system approach is one possible solution to these complex problems. According to [5], a systems approach can provide an effective problem-solving methodology for complex multidisciplinary problems. The systems approach involves formulating goals and results as an operating system that can be used to solve problems effectively [6]. To see the relationship between components in the development of agropolitan areas, it is necessary to build a model that is an implication of the system.

The agropolitan policy in Lima Puluh Kota Regency, West Sumatra, was initiated on June 6, 2005, when the Decree (SK) of the Regent of Fifty Cities Number 398 / BLK / 2005 was issued, designating Payakumbuh, Mungka, and Guguak Districts as agropolitan areas based on laying hen agribusiness. Initially, the laying hen industry showed promising economic development for laying hen farmers; however, in recent years, egg production has decreased and the number of farmers has also gradually decreased or even closed, which has a negative impact on the activities of residents who make a living from breeding laying hens. Sustainable development, as generally understood, is the ability to meet present needs without endangering the ability of future generations to meet their own, and agropolitan areas are typified by their capacity to maintain a balance between social, cultural, and political dynamics while guaranteeing the well-being of both present and future generations [7]

Overcoming these formidable obstacles, such as global climate change, resource depletion, and the need to meet the growing needs of the human population, requires concerted efforts, including the development of effective government policies and the adoption of more sustainable agricultural practice [8]. As the cumulative impacts of climate change become more evident across diverse sectors and regions, creative solutions are emerging and evolving to provide more comprehensive solutions [9] A lot of developing countries favor regional development that ignores environmental safety concerns in favor of excessive resource exploitation (Surya).

The sustainability of laying hen farming in LKR produces unsatisfactory results in a number of dimensions. A number of factors are linked to the decline in production, with a connection to the implementation of government policies [10] The agropolitan development of laying hen farms in LKR presents obstacles, including inadequate facilities, infrastructure, capital, and human resources [11].. The production levels are relatively low, insufficient to satisfy the demands of consumers from nearby regions like Riau, Jambi, Bengkulu, and Jakarta. Furthermore, the lack of infrastructure makes it difficult to market eggs and chickens because of the high cost of transportation to the surrounding areas [12].

A viable approach to improving the social and economic well-being of the community is the creation of agropolitan zones that include integrated laying hen farms. The goal of this research is to find a viable model for the growth of integrated laying hen farms in LKR in order to raise the community's standard of living. The purpose of this study is to develop a model for sustainable development of agropolitan areas based on integrated laying hen farms to improve the socio-economic living standards of the community. To achieve this goal, research activities in the second year consist of two stages, namely:

1. Analyze the regional system and build a model for the development of agropolitan laying birds in Lima Puluh Kota Regency, West Sumatra Province.

2. Formulate policy directions and strategy scenarios for the development of sustainable laying poultry agropolitan areas in Lima Puluh Kota Regency, West Sumatra Province

2 Research Method

2.1 Methods of Respondent Selection

Purposive sampling is a purposeful method of selecting respondents in order to obtain data and knowledge. It is customized to the surrounding conditions and the intended sample size, choosing respondents who are judged representative and have a thorough comprehension of the topics being studied. The respondents comprise experts and people from different agencies involved in the research topics who have a strong understanding of agropolitan problems.

2.2 Data Types and Sources

The data used for identifying agropolitan areas encompasses both primary and secondary sources. Primary data is gathered through questionnaires administered to selected respondents and experts, capturing information like total costs, farm revenues, and expert opinion scores. Secondary data is acquired from pertinent agencies in Fifty City Regency, including details on agricultural/livestock production, key commodities, population statistics, primary community activities in the agricultural/livestock sector, regional connectivity to neighboring areas, market proximity, adequacy of supporting facilities and infrastructure, the land potential for agropolitan development, GRDP, training, and extension education facilities, animal health services, security measures, and economic resources such as market availability and Village Unit Cooperatives (KUD). These secondary data sources are accessed through relevant agencies in Limapuluh Kota Districts such as Bappekab, and Central of Livestock and Agriculture

2.3 Analysis Methods

The employed approach involves utilizing a dynamic system with the assistance of Powersim Constructor software version 2.5. The process in a dynamic system encompasses requirements analysis, problem formulation, model simulation, and model validation.

The application of dynamic analysis in policy research facilitates a comprehensive understanding of policy evolution over time, intricate interactions among variables, dynamic system modeling, adaptation to environmental changes, trend identification, and long-term impact evaluation. This approach provides a more holistic foundation for designing and assessing effective and sustainable policies [13] This research aims to examine policy changes and model new policies to enhance the agropolitan area in LKR in the future;[14]. There are five steps in creating a dynamic system model [15], are as follows: (1) Issue formulation: in this phase, the problem to be solved is defined using a dynamic system model, and the associated variables are defined as well. (2) Dynamic Hypothesis: in this stage, a causal loop diagram is made to show the causal relationships between variables in the system. Once the causal loop diagram is made, it is changed into a simulation diagram or a flow diagram. Through simulations, dynamic hypotheses are produced by looking at the structural relationships between variables. (3) Formulation: in this phase, elements and variables in the system are converted into equations, levels, rates, and auxiliary variables. Estimates of initial values and parameter values are also made at this stage. Policy Formulation and Evaluation: The created model can be used as a reference for formulating or evaluating existing policies related to the modeled system. (4) Testing: This step is critical for comparing the output values generated by the simulation of the model with the output values from the real system. (5) Verification and validation of the model must be completed before performing this step. The next step is to create scenarios, simulate scenarios, and analyze scenarios. The scenario in creating a dynamic system model involves changing the values of specific variables to evaluate the results of the executed simulation. Scenario simulations are conducted to test several factors and variables that have been identified previously. Analysis and interpretation will identify which variables significantly influence productivity.

3 Result and discussion

3.1 Needs Analysis

Based on the literature review, several system actors who play a role in the egg production of layer chickens can be identified in Table 1.

Table 1. Identified System Actors and Their Needs

Actors	Their Needs
Breeder	The amount of production and business profits increased
Government	Eggs and chicken meat are available and meet the needs of the community
Livestock Industry	Production continuity and large profit margins
Community	The needs of eggs and chicken meat are met

Based on Table 1, it can be seen that the needs of each system actor. System actors and their needs have been adjusted to the limits of research.

3.2 System modeling with dynamic systems analysis

The problem of chicken egg production in Limapuluh Kota Regency is a fairly complex system problem involving various components and variables that interact with each other. Chicken egg production is seen as a matter of system dynamics that change over time and are influenced by factors that are also dynamic. The chicken egg production system in Limapuluh Kota Regency is depicted on a cause-and-effect diagram and can be seen in Figure 1.

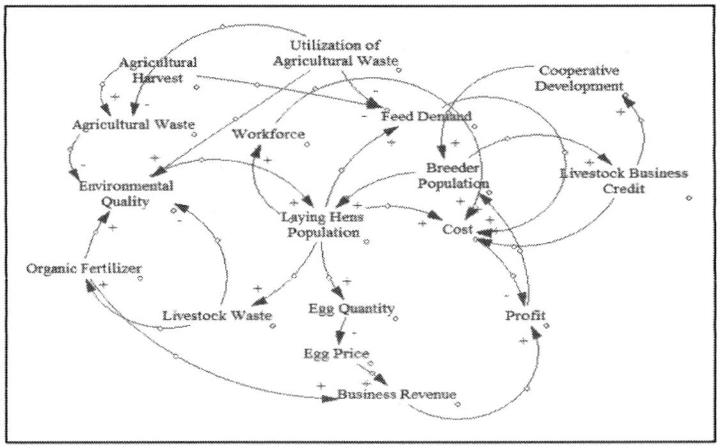

Fig. 1. Causal *loop* diagram of laying hen farming system in Limapuluh Kota Regency

Layer chicken population is a function of both the number of farmers and the environmental conditions in the farming area. The number of farmers is determined by the profitability of laying hen production and the business incentives offered by farmer-livestock cooperatives. Higher profits and more enticing business incentives will encourage more laying hen farmers to raise the population of layer chickens, which will increase egg production.

The profitability of the business depends on the ratio of revenues to costs; higher costs reduce profits. The majority of costs are incurred from the sale of animal feed, and the primary source of revenue is the sale of eggs. The amount of agricultural and livestock waste that is recycled into organic fertilizer and animal feed determines the environmental quality factors.

The conversion of refined waste into organic fertilizer offers a means of generating additional revenue for the business, whereas the conversion of agricultural waste into animal feed reduces the expenses linked to feeding animals. The dynamic system model that has been developed is limited to factors that are relevant to the production of chicken eggs, namely those that are impacted by variables pertaining to the number of layer chickens and business profits. The system for producing chicken eggs is divided into two main subsystems for modeling purposes: the subsystem of layer chicken population and the subsystem of business profits.

3.3 System Formulation

Model formulation is the formulation of a problem into a mathematical form that can represent a real system. Model formulation connects variables that have been identified in the conceptual model. Some of the assumptions used in modeling this study are:

3.3.1 The total population of layer-breed chickens

The total population in 2016 was 95,743 per week according to BPS Kab. Limapuluh Kota's 2016 report. The production rate, a variable flow, depends on the availability of labor and the productivity of the labor force in rearing day-old chick (DOC) livestock from the rearing phase until they are ready for production (16 weeks). The population growth dynamics assumptions take into account inputs in the form of production variables and outputs in the form of depletion variables. The variable rate of depletion is dependent on the number of layer chickens that die during the rearing process and the number of layer chickens that are considered to have completed their production cycle. The variable efficiency of workers is assumed to be 1,500 heads per person per week, meaning that each worker can tend to 1,500 day-old chick (DOC) livestock each week. The number of layer chickens killed in rearing is assumed to be 5% of the livestock population and the number of livestock estimated is assumed to be 37% of the population per week. The amount of value and assumption of the value used from each variable is sourced from statistical data from the Central Bureau of Statistics of Limapuluh Kota Regency in the 2016-2021 time frame.

3.3.2 Assumption of feed consumed by livestock

Since the feed mill supplies the concentrate feed, it is assumed that the feed consumed by livestock is a combination of concentrate feed (45% of the total feed requirement), rice bran (15% of the total feed requirement), and corn (40% of the total feed requirement). The feed for rice bran and corn is obtained from the agricultural yield of these crops in Limapuluh Kota Regency. The variable denoting the percentage of rice bran and corn harvest is assigned a value equal to the average value obtained from the rice and corn crops in Limapuluh Kota

Regency from 2016 to 2021. The livestock waste in this model is assumed to be used only as manure, with an assumed weekly value of 0.231 kg per head.

3.3.3 Variable demand for chicken culled

The market demand for chicken meat is represented by the demand for culled chickens and the value of this variable is influenced by the interaction between the variables "culled" and "time change demand." The value of the chicken demand variable affects the number of day-old chicks (DOC) needed, and the number of DOC that can be prepared for production as layer breed chickens are determined by the labor force availability and worker competency.

3.3.4 Business Revenue Variables

Assuming a productivity rate of 7.7 kg per head per week and an egg price of Rp 23,700 per kg, the egg acceptance variable in the model represents the sales value of eggs from purebred layer chickens; the values for laying hen productivity and egg prices are derived from the average values observed between 2016 and 2020; and the culled revenue variable represents the sales value of chicken meat from laying hens.

3.3.5 Variable total cost of business

The total cost variable of the business consists of the sum of the feed, other expenses, and operating costs. The variable feed costs are determined by the variable egg production, which is valued at Rp. 10,000 per kilogram of eggs, based on a Feed Conversion Ratio (FCR) of 2.30. The cost of animal feed for laying hens is expected to increase by 30% annually due to the yearly trend of the price of concentrate feed rising. The operational cost variables are those that pertain to the general running of the business. All of the costs that farmers incur during the production process are included in the operating cost variables. These costs include electricity, water, labor wages, maintenance, materials, social costs, health, security, fuel, and more. The value of the operational cost variables in the model is Rp. 1,000 per kilogram of eggs. The total cost of all of the expenses that farmers incur is included in the other cost variable. This includes sales, vaccines, medicines, miscellaneous costs, and depreciation on pullets. The value of the other variable costs in the model is Rp. 2,570 per kilogram of eggs.

3.3.6 Business profit variable

It is assumed that profits from livestock business operations will be allocated for credit installment payments and will also contribute to the overall profits of livestock cooperatives. The business profit variable is derived by deducting the total operating costs from the business revenue variables.

3.3.7 Labor variables

Labor variables are stock variables that are shaped by both inflows and outflows. Inflows are influenced by expected labor force variables and the amount of time needed for worker adjustments. The worker adjustment time variable is assumed to be 8 weeks, which represents the average amount of time workers must wait before beginning work in the cage. Outflows are influenced by the variable that represents the average amount of time workers stay in their positions, which is assumed to be 104 weeks, which indicates the average amount of time a worker stays in their position before quitting.

3.3.8 Labor productivity

Given that the weekly labor output is considered to be 1,500 heads per person, each worker can take care of 1,500 chickens in total.

3.3.9 Farmer variables

The term "farmer variables" refers to those who own laying hens in addition to workers; the number of farmers is assumed to be 25% of the total number of workers because farmers manage their livestock business in addition to acting as workers.

3.3.10 Livestock cooperatives

In the model, it is assumed that all laying hen farmers are members of Livestock cooperatives. Members of livestock cooperatives give 5% of business profits for the development of livestock cooperative businesses directly whenever farmers make profits.

3.3.11 All farmers

All farmers who raise laying hens are required to pay the financial institution on credit. It is anticipated that credit installment payments amount to twenty percent of business profits, provided that all farmers comply with timely payments in order to avoid any negative credit issues with financial institutions.

3.4 System modeling

Attempting to replicate the system is the goal of the modeling stage. It is a creative process that is not strictly linear but rather must follow scientific principles and logical-iterative thinking. The steps in the modeling procedure include redefining the problem to be addressed in accordance with the goals of the system study, formulating hypotheses, creating models, testing and analyzing those models, and so on. According to[16], successful modeling depends on establishing a relationship between the virtual world represented in the model and the real world, so as to guarantee that the model fulfills its function of simplifying the system. Figure 2 shows the model of the laying hen farming system in Limapuluh Kota Regency.

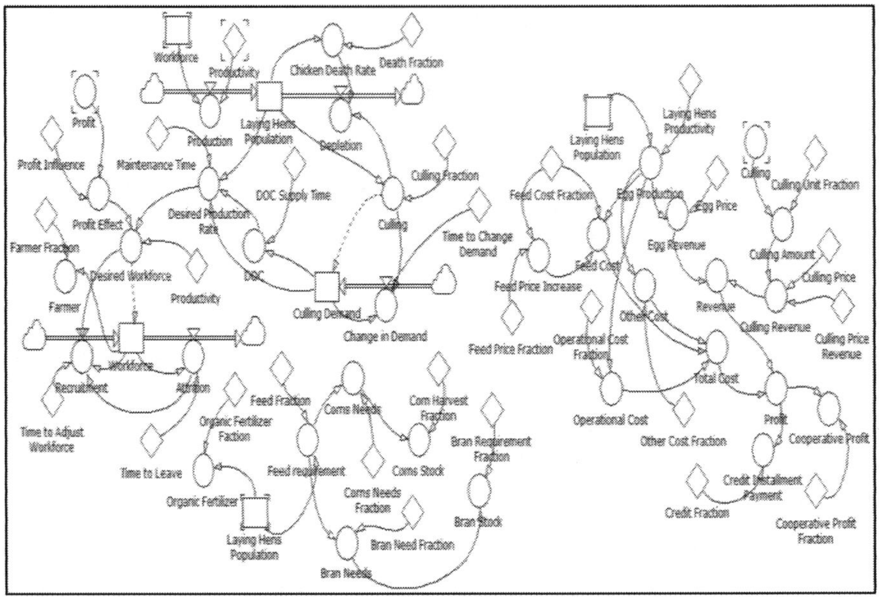

Fig. 2. Stock Flow Diagram of Chicken Egg Production Model in Limapuluh Kota Regency

3.5 Model Validation

According to [17] and [18], validation is the process of determining whether a conceptual model accurately reflects a real system. This involves determining whether the simulated conceptual model accurately represents the real system being represented.As a system simplification, the model necessitates that the model maker convince the model user that the built model is appropriate for solving the current problem, an essential step in enabling well-informed decision-making. The model's testing phase is divided into two components, according to[19] the structural validation test and the performance validation test. The purpose of the structural validation test is to determine the degree to which the model's structural similarity to the real structure matches. It is divided into two parts: construction validation and structure stability. Performance validation tests are carried out to provide assurance regarding the degree to which the model's performance matches the system's actual performance. This is necessary in order to determine whether the behavior of the model's output is consistent with the observed behavior in empirical data, thereby guaranteeing that the model conforms to factual representation. The degree of disagreement between the model's output and empirical data can be measured by evaluating its absolute mean deviation (AME = absolute mean error). It has been suggested by [20] that the acceptable deviation range is between 5 and 10%.

The formula for computing the value of AME is as follows:

$$AME = \frac{(Si - he)}{he} x\ 100\%$$

Where $Si = \frac{\Sigma Si}{N}$ and $He = \frac{\Sigma He}{N}$

The explanation is as follows: S stands for the simulation value, A for the real value, and N for the observation time.

AME analysis results show that the model for the laying hen farming system in Limapuluh Kota Regency performs admirably, displaying precision and meeting scientific standards. The AME test was used in this study's performance validation, using actual data pertaining to the population of laying hens (head/week) and egg production (kg/week) in Limapuluh Kota Regency.

Table 2. Comparison of layer laying hen population data and actual egg production with simulated model results of laying hen farms in Limapuluh Kota Regency

Year	Actual		Simulation	
	Population of *Layer* (tail/week)	Egg production (kg/week)	Population of *Layer* (tail/week)	Egg production (kg/week)
2016	95.742	735.072	95.743	737.221
2017	122.104	939.049	113.849	876.642
2018	143.278	1.102.300	130.513	1.004.957
2019	143.740	1.104.201	149.617	1.152.055
2020	149.011	1.148.878	171.517	1.320.683
Average	130.775	1.005.900	132.248	1.018.312
AME	**1,13 %**	**1,23%**		

Table 2 shows the results of the model output and performance validation. The AME value is below 5 percent, ranging between 1.13 and 1.23 percent. The results of the AME analysis show that the model of the laying hen farming system in Limapuluh Kota Regency has good performance, is relatively precise, and is scientifically acceptable.

3.6 Scenario of Sustainable Development of Laying Hens Farm System in Limapuluh Kota Regency

After taking sustainability and model performance into account, the analysis of system performance shows that the current system is positioned as reasonably sustainable. Time dynamics will introduce changes in system performance in the future. The application of scenario formulation for system development can help maximize the benefits of the laying hen farming system and improve system performance for sustainability. Seven significant factors influencing the interconnected system have been identified after an analysis of their interactions, and these factors are further explained and described as possible future changes or evolutions. Table 3 shows the prospective key factors for the development of laying hen farming systems in Limapuluh Kota District.

Table 3. Prospective key factors for the development of sustainable laying hen farming systems in Limapuluh Kota District

No	Factors	(*state*)		
		1A	1B	1C
1	Utilization of livestock waste	Unutilized	Increased	
2	Environmental carrying capacity	2A	2B	2C
3	Egg Price	3A	3B	3C
4	Farmer Revenue	4A	4B	4C
5	Workforce	5A	5B	5C
6	Cooperative advantages	6A	6B	6C
7	Credit disbursement	7A	7B	7C

Based on Table 3, several scenarios are formulated by pairing the anticipated changes and analyzing their implications on the system, based on the condition (state) of each factor. Three development scenarios are formulated for the development of a sustainable commercial layer chicken farming system in the Limapuluh Kota Regency: (1) conservative-pessimistic scenario, (2) moderate-optimistic scenario, and (3) progressive-optimistic scenario. There may be more development scenarios, but these three are the most likely to occur in the future based on the current state of each key factor. The profitability of layer farming and the commercial incentives provided by farmer-livestock cooperatives have an impact on the number of farmers [21]

The following variables in the model represent these key factors: (1) the manure fertilizer variable, which represents the key factor of livestock waste utilization; (2) the purebred chicken population, feed requirements, corn requirements, corn stock, requirements for rice bran, and rice bran stock, which represents the key factor of feed carrying capacity; (3) the egg production, egg receipts, and egg prices, which represents the key factor of egg price improvement; and so on. The key factors that represent farmer income are represented by the following variables: (4) business revenue, total business costs, and business profits; (5) labor; the key factor that represents labor absorption; (6) cooperative profit; the key factor that represents the availability of agribusiness infrastructure; and (7) business profit and credit installment payment variables.

This paper's research was financially supported by the Institute of Research and Community Service of Universitas Andalas under Contract No: T/8/UN16.19/PANGAN-PTU-KRP2GB-UNAND/2023. Gratitude is extended to the Chairman of LPPM at Universitas Andalas for providing the financial support

References

1. B. Bakrie, E.S. Rohaeni, Y. Yusriani, S. Tirajoh. The development of a newly formed superior local chicken in Indonesia Journal of Hunan University (Natural Science),**48**:9 (2014)
2. D. Bosco, Alessandro, Mattioli, Simona, C. Mancinelli, Alice, Cotozzolo, E. Castellini, Cesare. Extensive rearing systems in poultry production: The right chicken for the right farming system. A review of twenty years of scientific research in Perugia University, Italy. Animals, **12**: 5 (2021)
3. E. Rustiadi, S. Hadi, and and W. A. Muttaqien, Agropolitan area concept of rural-urban development is balanced. Bogor: IPB Press, 2006
4. B.K. Bala, F.M. Arshad, K. M. Noh. System Dynamics Modelling and Simulation. Book. Springer Texts in Business and Economics. (2014)
5. Manetch TJ, Park GL. (1977). System Analysis and Simulation with Application to Economic and Social System Part I. Third Edition, Departement of Electrical Engineering and System Science. Michigan: Michigan State University, East Lansing.
6. Eriyatno. 1998. Ilmu Sistem; Meningkatkan Mutu dan Efektifitas Manajemen. Jilid I Edisi Kedua. IPB Press.
7. Budiharjo, Eko dan Djoko Sujarto. Kota Berkelanjutan. Bandung: PT Alumni. 2009
8. Ricart, S. Gandolfi, C.Castelletti, Andrea. Climate change awareness, perceived impacts, and adaptation from farmers' experience and behavior: a triple-loop review. Regional Environmental Change **23**, 2023
9. H. Medina, D.Nunn, D. Patrick, B. Harriot. Challenges and opportunities for food systems in a changing climate: A systematic review of climate policy integration. Environmental Science and Policy **124**, 2021.

10. S. Bhanja, P. Bhadauria. Behaviour and welfare concepts in laying hens and their association with housing systems. Behaviour and welfare concepts in laying hens and their association with housing systems. Indian Journal of Poultry Science (2018) **53**(1) 1
11. L. Edwards, P. Hemsworth. The impact of management, husbandry and stockperson decisions on the welfare of laying hens in Australia. *Animal Production Science* **61**(10), 2021
12. M. T. Kidd, K. E. Anderson, Laying hens in the U.S. market: An appraisal of trends from the beginning of the 20th century to present Jpplied Poultry Rournal of Aesearch **28**;4, 2019
13. F. R. Jeffers, J. Jacobson, E. M. Searcy. Energy Policy. **52**, 2013
14. Hwang, S. Park, L. Moonseo, S. Hyun. Mathematical and Computer Modelling **57**:2013.
15. E. Suryani, R. A. Vinarti, and R. Files. Simulasi Perubahan Peringkat Webometric Indonesia Yang Dipengaruhi Kondisi Keuangan Institusi Pendidikan (Studi Kasus : Its). no. **2008**, pp. 1–10, 2011.
16. Hartrisari. Sistem Dinamik: konsep sistem dan permodelan untuk industri dan lingkungan. Bogor. Seameo Biotrop. 2007
17. Antonelli, G.Sciacovelli, L. Aita, A. Padoan, A. Plebani, Mario. Validation model of a laboratory-developed method for the ISO15189 accreditation: The example of salivary cortisol determination. Clinica Chimica Acta **485**: 2018.
18. R. D. Haloho, S. I. Santoso, S. Marzuki, W. Roessali, W. A. Setiadi. Profit function analysis of dairy cattle farming in getasan and west ungaran districts, semarang regency. Journal of the Indonesian Tropical Animal Agriculture 38 : 2013.
19. Purnomo, B. Herry, I. Aqidatul. Model Sistem Dinamis Perencanaan Bahan Baku Pada Produk Veneer Di Pt. Xyz. Agrointek **14**: 2020.
20. Muhammadi, Aminullah E, Soesilo B. 2001. Analisis Sistem Dinamis: Lingkungan Hidup, Sosial, Ekonomi, dan Manajemen. Jakarta (ID): UMJ Press.
21 Pavlova, Ivelina. Egg productivity of xl chicken population. Bulgarian Journal of Agricultural Science. **26** : 2020

Effects of ozonation time on the internal quality of milk: A meta-analysis

Maulidya Zahrina Qolby[1*], *Arif* Hendra Utama[1], *Firmansyah* Budi Pratama[1], *Lilik* Eka Radiati[1], *Wantamas* Jantasin[2], and *Danung* Nur Adli[1]

[1]Faculty of Animal Science, Universitas Brawijaya, Malang 65145, Indonesia
[2]Faculty of Animal Science and Technology, Maejo University, Chiang Mai 50290, Thailand

Abstract. The aim of that study was to evaluate the effectiveness and suitability of its use according to local conditions and ensure compliance with applicable food safety standards. A raw database had been constructed based on peer-reviewed and published research articles that reported the use of ozonation time on the internal quality of milk. The results of the observations were data collected using Microsoft Excel, followed by using Microsoft Word for formula processing before processing the data using the SAS application. That research used meta-analysis techniques to integrate 29 pieces of data from six pieces of literature that observed and researched the effects of the length of time of the ozonation process on milk quality. The variables observed were the effect of the long ozonation process for dairy products on the number of bacteria, protein levels, aflatoxin levels, and pH. The results of observing the effect of long ozonation on milk to determine the number of bacteria had an insignificant effect ($p > 0.05$). The protein results had a significant effect ($p > 0.05$) on the duration of ozonation; the results of aflatoxin levels found in milk had a significant effect ($p > 0.05$) on the duration of ozonation in milk; and the pH of milk also had a significant effect ($p > 0.05$) on the duration of ozonation in milk. In summary, using ozonation time was safe for the internal quality of milk.

1 Introduction

Milk is a product that is easily damaged or perishable because of its high nutritional content and its nature, which is easily affected by external factors. One process of handling milk that can cause damage is heating it at high temperatures. Heating milk at high temperatures can cause denaturation or changes in the protein structure of milk [1]. Milk proteins, especially whey proteins such as whey protein and casein, have complex structures and are susceptible to high temperatures. When milk is heated to too high a temperature, protein bonds can be disrupted and the protein structure can change. This can result in a loss of biological and functional activity of the protein as well as changes in the texture and taste of the milk. Apart from that, bacterial activity can also cause damage to milk. Bacteria in milk can produce enzymes that damage the nutrients in milk.

* Corresponding author: maulidyazahrina@student.ub.ac.id

Good milk quality is very important to maintain the cleanliness, safety, and nutritional value of the milk produced. In an effort to preserve milk, the preservation method commonly used is sterilisation. Sterilisation aims to destroy or inactivate pathogenic and spoilage microorganisms that can cause damage to milk. Commonly used sterilisation methods include thermal (heating) and non-thermal methods [2]. One technology that can be used to improve milk quality is ozonation. Ozonation is the process of using ozone to clean, disinfect, and remove unwanted odours from water, air, or surfaces [3]. In the context of milk, ozonation can be used to address various problems that can affect milk quality, such as bacteria, viruses, pathogenic microorganisms, unpleasant odours, and other contaminants.
The application of ozonation in milk processing has several significant benefits. Several aspects relate to ozonation and milk quality: killing of microorganisms, elimination of unpleasant odours, increasing shelf life, and reducing the use of chemicals. The application of ozonation in milk processing is beneficial not only for milk producers but also for consumers who consume dairy products. By improving the quality of milk, ozonation can help maintain the cleanliness and safety of dairy products, as well as the taste and nutritional value presented to consumers. It should be noted that ozonation must be carried out carefully and in accordance with appropriate guidelines. Therefore, it is important to follow the recommended instructions and guidelines and exercise strict supervision in the application of ozonation in milk processing.

In the context of the dairy industry in Indonesia, ozonation can be a technology that has the potential to help improve the quality of milk and, consequently, the quality of the milk products produced. However, the implementation of ozonation in the Indonesian dairy industry requires further study and research to evaluate the effectiveness and suitability of its use according to local conditions and ensure compliance with applicable food safety standards.

2 Materials and Methods

2.1 Literature search and Dataset development

A raw database was constructed based on peer-reviewed and published research articles which reported use of ozonation time on the internal quality of milk. Articles were selected based on the [4] method. An algorithm literature was constructed with a search in Scopus, Web of Science, PubMed, and Medline using the MESH terms "bacteria", "ozonation", "milk", "aflatoxin", and "pH". A single search from Google Scholar was also undertaken to identify additional studies that may have been relevant to our objectives.

.

2.2 Data analysis and coding

Prior to statistical meta-analysis, data analysis and coding were performed using SAS ODA Academics. The modelling used by following [5].

$$Y_{ijk} = \mu + S_i + \tau_j + S\tau_{ij} + \beta_1 X_{ij} + b_i X_{ij} + \beta_2 X^2{}_{ij} + b_i X^2{}_{ij} + e_{ijk} \quad (1)$$

Where: Y_{ijk} = dependent variable, μ = overall mean value, S_i = random effect of the *i*th study, assumed to be ~ N_{iid} $(0,\ \sigma_S^2)$, τ_j = fixed effect of the *j*th of τ factor, $S\tau_{ij}$ = random interaction between the *i*th and *j*th level of τ factor, also assumed to be ~ N_{iid} $(0,\ \sigma_{S\tau}^2)$, β_1 = overall value of the linear regression coefficient of Y to X (a fixed effect), β_2 = overall coefficient value of the quadratic regression of Y to X (a fixed effect), X_{ij} dan $X^2{}_{ij}$=

continuous values of the predictor variable (in linear and quadratic form, respectively), b_i = random effect of the study on the regression coefficient of Y to X, assumed to be ~ N_{iid} (0, σ_b^2), and e_{ijk} = residual value from unpredictable error. $S\tau_{ij}$ dan S_i are taken to be independent variables that are chosen at random. A validation and significance test were conducted on the model. The significance of the values was determined using a one-way analysis of variance. It is significant if the P-value (P or P-val) < 0.05 and tends to be significant if the P-value between 0.05 and 0.1. As a result, P_l represents the P-value for the linear constant (β_1) and P_q represents the P-value of quadratic constant (β_2). Therefore, the validation test was conducted using the root mean square error (RMSE) and Nakagawa determination coefficient (R2) or $R_{GLMM}(c)^2$

$$RMSE = \sqrt{\frac{\sum \quad (O-P)^2}{NDP}} \quad (2)$$

$$R_{GLMM}(c)^2 = \frac{(\sigma^2{}_f + \sum \quad (\sigma^2{}_l))}{(\sigma^2{}_f + \sum \quad (\sigma^2{}_l) + \sigma^2{}_e + \sigma^2{}_d)} \quad (3)$$

Note: O = actual value, P = estimated value, NDP = number of data point, $\sigma^2{}_f$ is the variant of a fixed factor, $\sum \quad (\sigma^2{}_l)$ is the sum of all variants of the component, $\sigma^2{}_e$ is the variant due to the predictor dispersion and $\sigma^2{}_d$ is the specific distribution of the variant.

3 Results and discussion

The results obtained using the SAS application to determine the number of bacteria showed that the length of time of ozonation of milk had insignificant effect (P > 0.05). on the across whole parameters. Before the ozonation process was carried out, milk had a very high bacterial content, according to the treatment results from previous research. The results of observations of total bacteria before and after ozone treatment are presented in Table 1.

Table 1. Studies included in the effects of ozonation time on the internal quality of milk

No	References	Level	Countries	Kind of Milk	Form	Time
1	[6]	0-10	Turkey	Milk concentrate	Liquid	5-10
2	[7]	0.10	Iran	Whey concentrate	Liquid	5-10
3	[8]	3-12	Indonesia	Fresh milk	Liquid	5-10
4	[9]	3-12	Indonesia	Dairy milk	Liquid	3-12
5	[10]	0-9	Indonesia	Goat milk	Liquid	3-9
6	[11]	0-15	Russia	Fresh Milk	Liquid	15-60

Table 2. Descriptive table of the effects of ozonation time on the internal quality of milk

No	Parameters	Unit	NDP	Mean	SD	MAX	MIN	SEM
1	AFM	(μg/kg)	19	0.36	0.13	0.62	0.20	0.03
2	pH	No unit	19	6.64	0.05	0.013	6.75	6.54
3	Bacteria	Log/Cfu	19	51390	71149	202500	938	0.19
4	Protein	No unit	19	3.54	0.52	4.27	2.99	0.14

AFM – Aflatoxin; CFU – colony forming unit; Kg – kilogram; Log- logarithm; NDP – Number data point; Max – Maximum; Min – Minimum; pH- potential hydrogen; SD – standard deviation; SEM- standard error mean, μg – microgram

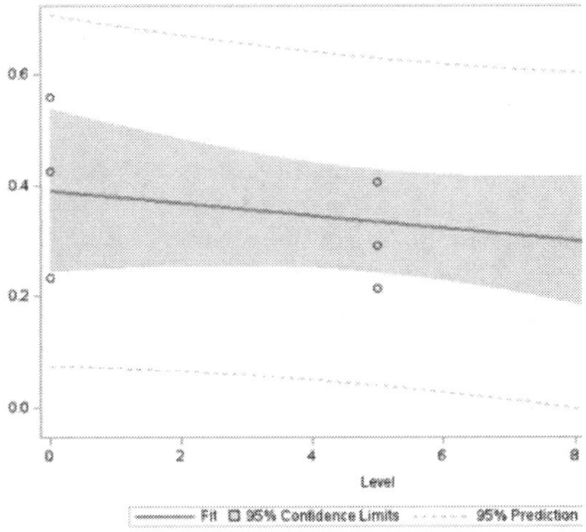

Fig. 1. the effects of ozonation time on the Aflatoxin

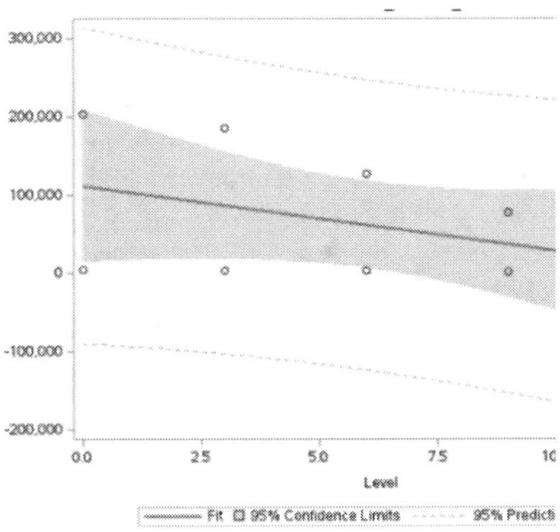

Fig. 2. the effects of ozonation time on the pH

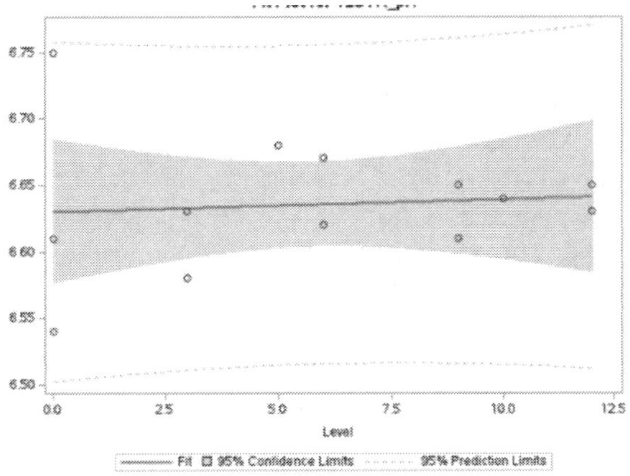

Fig. 3. the effects of ozonation time on the bacteria

The number of milk bacteria in the group that received ozonation treatment was much lower than the control group (without ozone treatment). From the literature results, it was found that before the ozonation process was carried out, the number of bacteria exceeded the number of SNI applied, based on SNI 3141:2011, with a maximum figure of 1 x 106 CFU/ml. The results after administering ozonation with an average time span of 3–12 minutes, reviewed from the three literatures, showed a process of reducing the number of bacteria to 0.098 x 10^6 CFU/ml. The amount of bacterial contamination decreased ($P > 0.05$), and increasing the time of the ozonation process would reduce the number of bacteria found in milk even though the thermal sterilisation process was not used. It is known that the longer the ozone exposure, the lower the number and activity of bacteria found in milk. The increasing exposure time to ozone in the process of sterilising consumption milk will influence the chemical and organoleptic properties as well as the activity of spoilage microbes found in pure milk, so it can be seen that ozone can indicate and provide significant effects [4].

The results of observations and research using the meta-analysis concept prove that the administration of ozone with the effect of a long exposure time can eliminate and sterilise the microbial content or number of bacteria in consumed milk. States that non-thermal sterilisation is an alternative process to replace thermal sterilisation, where the non-thermal sterilisation process will maintain product nutrition even though there has been a decrease in the number of bacteria found in milk [3].

Aflatoxin levels found in milk are the result of the process of breaking down the energy contained in milk nutrients by spoilage microbes. Aflatoxin itself is a type of mycotoxin, namely a group of natural poisons that can be produced by fungi such as Aspergillus sp. and have adverse effects on humans, animals, and plants that can result in disease and economic losses. The higher the level of aflatoxin contained in milk, the higher the level of bacteria contained in the milk processing process before consumption. In general, it is known that there are two types of fungi that produce aflatoxins found in milk, namely *Aspergilus flavus*, which produces aflatoxin B1 and *Aspergilus parasiticus*, which produces four types of aflatoxins. Some of the types of aflatoxin listed above are toxins that can damage agricultural products, especially livestock products. Aflatoxin itself is obtained from feed contaminated by the fungus *Aspergillus sp.*, which becomes a residue in the livestock's body, thus becoming AF contamination. The presence of AF contamination, especially AFB1, in animal feed, apart from endangering animal health, also causes residues of aflatoxin and its metabolites such as aflatoxin M1, aflatoxicol (Ro), aflatoxin Q1, and aflatoxin P1, which are deposited in meat, milk, and eggs [6]. The results of observations and analysis showed that for the aflatoxin content contained in the table above, there had been a decrease in aflatoxin residue during the ozonation process in milk.

According to the provisions of BSN 7385-2009 and Regulation of the Head of the Republic of Indonesia Food and Drug Supervisory Agency No. HK.00.06.1.52.4011 sets the maximum threshold for AFM content in milk and processed milk by-products at 500 ng/L in consumed feed and converted into residue left in milk to be 0.5 µg/L (ppb). [9-10]. Thus, the use of ozone can be ensured as a strong antimicrobial agent in meeting the needs of the food processing industry, which can maintain consumer health [12]. The most common way to stop the development of microorganisms in milk is to use a pasteurisation process or heat it to a temperature of 700–850°C, but it is found that the nutritional content will drop drastically or damage will occur. Therefore, it is necessary to have an effective process to reduce the development of milk-destroying microbes without reducing the nutritional content of the product too much. One of these methods is the ozonation method, where ozone technology is able to have a minimal impact on the nutritional quality of milk but can also kill bacteria found in milk [4].

Based on the results listed in table 4, it shows that the decrease in milk protein levels in terms of the length of time exposure to ozone had on milk protein levels had a significant influence, where after the ozonation process, the protein levels experienced a decrease that was not too sharp. The decrease in protein levels can be concluded that ozone given over a long period of time will prevent damage or loss of too much protein levels [4]. Based on table 4 above, it is known that the protein levels before being exposed to ozone correspond to a time range between 0 and 12 minutes, the protein levels in cow's milk have relatively the same amount. It is known that the ozonation process can result in protein denaturation but not too high, in contrast to the pasteurisation process, where during the production process it will have a very high protein denaturation effect due to the continuous thermal process given to the milk, but reviewed again the ozonation process described is appropriate. with different lengths of time, the level of protein denaturation is very low, so statistically the reduction in protein has no real effect on the nutritional content of milk products.

The results of other research strengthen the results of observations regarding the effect of ozonation as a substitute for developing non-thermal food processing technology in the field of dairy products, which stated that milk protein levels experienced a decrease that was not too significant by 0.10% for each length of time given and remaining above the established standard protein content, where the protein content is in accordance with SNI 3141.1:2011, processed cow's milk must contain a minimum of 2.8% protein from the total product [13]. This means that the process of using ozonation as a sterilisation for milk processing can minimise damage to protein levels in milk processing, being able to maintain it optimally and not experiencing a drastic decrease in the percentage of protein levels. The results of observations from the research cited showed that milk that received ozone exposure for a period of 3 to 12 minutes in accordance with the respective literature had relatively the same pH value, namely an approach to normal pH (8). These results show that the longer the treatment time, the milk tends to increase the pH value, although the increase is not very significant, and statistically, these results have no effect on milk nutrition. The cause of damage to milk is an increase in lactic acid production due to the breakdown of milk lactose, resulting in an increase in lactic acid production, which can cause the acidity of the milk to increase, and a decrease in the pH value, which causes the milk to become damaged and unfit for consumption. The acidity level of milk is an important indicator of the overall quality of milk [3].

4 Conclusion

To sum up, ozone treatment and exposure have a specific impact on the number of bacteria, aflatoxin levels, protein, and pH in milk and other dairy products. The use of ozone can be proposed as a powerful antimicrobial agent that can meet the needs of the food industry and achieve food regulatory agency approval and consumer acceptance. The long processing time of using ozonation as a non-thermal sterilisation technology can be used to improve the microbiological and nutritional quality of fresh milk and processed milk by-products.

References

1. Z. Wulandari, E. Taufik, M. Syarif, J. Ilmu Prod. dan Tek. H. Pet.. **5** (2017)
2. I. Y. Sigar, R. S. S. Santosa, Y. Subagyo. Prosiding Seminar Nasional Teknologi Agribisnis Peternakan (STAP). **9** (2022)
3. F. Listiana, R. S. S. Santosa, Y. Subagyo. Prosiding Seminar Nasional Teknologi Agribisnis Peternakan (STAP), **9** (2022)
4. D. N. Adli, S. Sadarman, A. Irawan, A. Jayanegara, T. M. Wardiny, T.R. Priambodo A.P.A Yekti. Ital. J. of Anim. Sci. **22** (2023).
5. Sholikin, M. M., A. Irawan, A. Sofyan, A. Jayanegara, B. Rumhayati, C. Hidayat, D. N. Adli et al. Poult. Sci. **102** (2023).
6. H. Mohammadi, S. M. Mazloomi., M. H. Eskandari., M. Aminlari, M. Niakousari. Ozone: Sci. and Eng. **39** (2017)
7. R. Widiastuti, WARATOZA. **24** (2014)
8. H. Mohammadi, Aflatoxins-Biochem Mol Biol. **397** (2011)
9. BPOM RI, Badan Pengawasan Obat dan Makanan Republik Indonesia 2012 (Jakarta, 2012)
10. BSN, Badan Standarisasi Nasional. 2009. (Jakarta Indonesia, 2009)
11. D. Sert, E. Mercan. *LWT.* **144** (2021)
12. Younis, F. I., Fayed, A. E., Elbatawy, O. I., Elsisi, A. J. of Agr.Sci.. **27** (2021)
13. BSN, Badan Standarisasi Nasional. 2011. (Jakarta Indonesia, 2011)

Measurement of some body parts of muscovy duck (*Carina moschata*) at age 3 weeks old fed by pellet feed under an intensive farming system

Firmansyah Budi Pratama[1], *Irfan H.* Djunaidi[2*], *and Eko* Widodo[2]

[1]Magister Degree Student of Animal Science Faculty Brawijaya University
[2]Lecturer in Animal Nutrition, Faculty of Animal Science, Brawijaya University

Abstract. This study was aimed to measure and determine the size of some body parts of the muscovy duck (Carina moschata) and their relationship with body weight. The research was be carried out for two months from August to September 2023 in local Intensive Muscovy Duck farm, Kademangan area, Blitar, East Java. The research material use were 60 muscovy ducks (unsexed), aged 3 weeks old, and reared in intensive farming system. This study used experiment quantitative methods (measurement). Muscovy duck sampling was carried out by random sampling. The variables observed were body weight, beak length, beak width, head length, head circumference, neck length, neck circumference, wing length, back length, chest length, chest width, shank length, shank circumference, and leg membrane width. Measurements were made on each individual using a measuring tape. The data obtained were tabulated and analyzed using normality test, correlation test and regression test. The results showed that there is a positive correlation between the morphological dimensions of livestock, including membrane width, shank length, beak width, wing length, head width, head length, neck length, shank circumference, beak length, breast length, breast width, neck circumference, back length, and body weight. And the length of the back has the largest contribution of 54% to the body weight of muscovy ducks (Carina moschata) aged 3 weeks old.

1 Introduction

Muscovy ducks are believed to have originated in Central and South America, were domesticated by Colombians and Peruvians [1], and are now found in most parts of the world, especially in the tropics [2]. In Indonesia, entok or mallard waterfowl are evenly distributed throughout the region, especially in agricultural areas ranging from plains to highlands. In English, the mallard waterfowl is called the Muscovy duck, named after the Moscow region, where it was first introduced before being introduced to Western Europe [3]. Entok ducks entered Indonesia through Manila, Philippines, so they were called Manila ducks, and they were then well developed in local farms in Indonesia [4].

* Corresponding author: irjuna@ub.ac.id

In general, the feather color of the Entok or Carina moschata, as reported by several researchers, is black, white, or a mixture of both with varying frequencies. Huang et al. [5] said that the feather color of domestic ducks is very diverse, namely black-white, blue, blue-white, brown, brown-white, white, black, black-white, purple, and calical (dull, dirty, and not shiny). Smaller variations were reported by Raji et al. [6] who found that feathers in northern Nigeria had 36.9 colored feathers, 30.6 white feathers, 6.4 black feathers, and 26.1 black feathers. Hassan and Mohammed [7] reported that only a small proportion of the observed feather colors were white. Chia and Momoh [8] found that the proportion of white ducks was lower at 30.33%; the rest were black ducks. Tamzil et al. [9] found that in female mallard ducks, white feather color was more dominant than white-spotted black feather color with a frequency of 62 and 38%, respectively, while in male ducks, white feather color and white-spotted black feather color had a balanced frequency (50:50).

Advantage of Muscovy duck is that its meat can be accepted by all religions and is known and accepted by the community as a meat producer and duck egg hatching service provider. Their wide and deep size allows them to incubate more eggs than native chickens. The body-size characteristics of Muscovy ducks are not widely known, so research is needed. Knowing about body size is critical to determine the bone development and body structure of Muscovy duck. Body size is related to body weight and can be used to estimate the body weight of livestock [10]. Muscovy ducks are one of the new species of birds that are raised for meat, so it is necessary to know their body size to have a method for estimating their actual body weight at the field level.

2 Materials and Methods

2.1 Location

The research was conducted from August to September 2023 in Kademangan, Blitar, East Java. The research material was 60 muscovy, age 3 weeks old, unsex. Sampling was random sampling, where the sample was taken intentionally to match the desired criteria. The criteria for livestock taken physically looked healthy, agile, and had good body shape proportions.

2.2 Research Variables

The observed variables were :

1.	Body weight (grams)	:	measured by weighing the ducks with a digital scale (g).
2.	Beak length (cm)	:	distance between the base of the maxilla to the tip of the maxilla, measured using a digital caliper (cm)
3.	Beak width (cm)	:	measured from the outer beak edge on the left and right, using a digital caliper (cm).
4.	Head length (cm)	:	measured from the base of the beak to the back of the head, using a digital caliper (cm).
5.	Head circumference (cm)	:	measured at the highest part of the head by wrapping a measuring tape (cm)
6.	Neck length (cm)	:	measured from the first cervical vetebrae to the last cervical vetebrae using a digital caliper (mm).

7.	Neck circumference (cm)	:	measured from the first cervical vetebrae to the last cervical vetebrae using a digital caliper (cm).
8.	Wing length (cm)	:	measured from the humerus bone to the tip of the phalanges using a measuring tape (cm)
9.	Back length (cm)	:	measured from the base of the neck to the base of the tail using a measuring tape (cm).
10.	Chest length (cm)	:	carried out from the tip of the chest in front to the tip of the back using a vernier term (cm)
11.	Chest width (cm)	:	obtained by measuring the distance from the left to the right (widest) sternal bone using a digital caliper (cm).
12.	Shank length (cm)	:	measured along the tarsometatarsus bone (shank) using a digital caliper (mm)
13.	Shank circumference (cm)	:	measured by wrapping the measuring tape around the center of the tarsometatarsus (shank) bone (cm)
14.	Foot membrane width (cm)	:	measured from tip to tip of the widest part of the foot membrane using a measuring tape (cm)

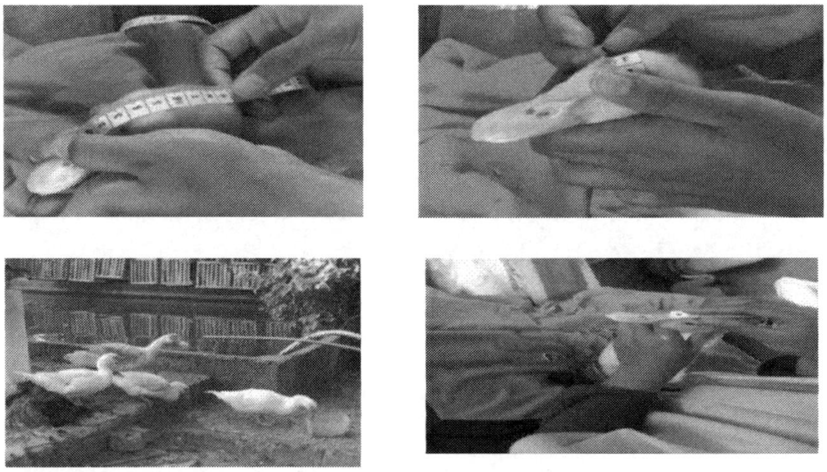

Figure 1. Measurement of Some body parts of Carina moschata

2.3 Research Procedure

This research procedure is to conduct preliminary research as a form of reference for pre-research reports, select 60 unsex Muscovy duck aged 3 weeks in Kademangan, Blitar Regency, measure the body dimensions of the livestock and weigh their body weight, conduct daily recording, and finally process the data.

2.4 Statistical Analysis

The data obtained were tabulated and analyzed by calculating averages to explain quantitative parameters and using normality, correlation and last use simple linear regression with the help of the IBM SPSS Statistic v26 software program to calculate the relationship between the measured variables.

3 Results and discussion

The Muscovy duck, categorized as a native avian species in Nigeria, stands out as the predominant waterfowl species in the country. Domestic Muscovy duck is valued throughout the world for its unique rich and flavourful taste, high yield of breast meat, and low calorie content. The male Muscovy duck has the potential to attain a substantial size, ranging from 4.5 to 5.5 kg, whereas the female typically weighs between 2.3 and 2.8 kg. A distinctive feature of this duck lies in its sexual dimorphism, where the male exhibits significantly greater size compared to the female. Morphometric measurements and associated indices play a pivotal role in characterizing, selecting, and genetically enhancing farm animals [11]. Body weight is an economically important trait for livestock and is highly valued by farmers [12]. These economic characteristics are the main determinants of livestock market prices, especially in developing countries where standardized marketing practices and regulations found in developed countries are not yet in place. The importance of these quantitative characteristics is due to their significant contribution to farmers' profit margins, especially on farms where market weight or meat for sale is the primary objective [12].

Table 1. Mean, Standard Deviation (SD), Carina mosschata age 3 weeks old

No	Variables	N	Mean	Std. Deviation
1.	Beak length	60	3.67	0.47
2.	Beak width	60	2.72	0.45
3.	Head Length	60	8.55	1.59
4.	Head width	60	2.65	0.48
5.	Neck Length	60	7.12	1.43
6.	Neck Circumference	60	7.67	1.96
7.	Wing Length	60	7.82	2.13
8.	Back Length	60	20.08	3.01
9.	Chest length	60	7.60	1.38
10	Chest Width	60	6.75	1.28
11.	Shank Length	60	3.62	0.69
12.	Shank Circle	60	3.88	0.58
13.	Membrane Width	60	4.8	0.93
14.	Body Weight	60	520.48	196.15

As can be seen in Table 1, the average morphological value of body size in Carina moschata aged 3 weeks each beak length 3.67 ± 0.47, beak width 2.72 ± 0.47, head length 8.55 ± 1.59, head width 2.65 ± 0.48, neck length 7.12±1.43, neck circumference 7.67±1.96, wing length 7.82± 2.13, dorsal length 20.08± 3.01, chest length 7.60±1.38, Chest width 6.75±1.28, calf length 3.62±0.69, calf circumference 3.88±0.58 and membrane width 4.8±0.93 and body weight 520.48 ± 196.15.

Table 2 shows that the correlation coefficient (r) between beak length and body weight has a value of 0.433 ($P < 0.001$), which means that there is a medium and positive relationship between half-life length and body weight. The correlation coefficient (r) between beak width and body weight has a value of 0.307 ($P < 0.017$), which means that there is a weak and positive relationship between beak width and body weight. The correlation coefficient (r) between head length and body weight has a value of 0.297 ($P < 0.021$), which means that there is a weak and negative relationship between head length and body weight. The correlation coefficient (r) between head width and body weight has a value of 0.174 ($P < 0.184$), which means that there is a weak and positive relationship between head width and body weight. The correlation coefficient (r) between head width and body weight has a value of 0.402 ($P < 0.001$), which means that there is a medium and positive relationship between head width and body weight. The correlation coefficient (r) between neck circumference and body weight has a value of 0.56 ($P < 0.000$), which means that there is a moderate and positive relationship between the neck of the head and the weight of the body. The correlation coefficient (r) between wing length and body weight has a value of 0.56 ($P < 0.000$), which means that there is a moderate and positive relationship between wing length and body weight. The correlation coefficient (r) between back length and body weight has a value of 0.74 ($P < 0.000$), which means that there is a strong and positive relationship between back length and body weight.

The correlation coefficient (r) between chest length and body weight has a value of 0.497 ($P < 0.000$), which means that there is a moderate and positive relationship between chest length and body weight. The correlation coefficient (r) between wing length and body weight has a value of 0.592 ($P < 0.000$), which means that there is a medium and positive relationship between wing length and body weight. The correlation coefficient (r) between shank length and body weight has a value of 0.193 ($P < 0.141$), which means that there is no positive relationship between shank length and body weight. The correlation coefficient (r) between shank circumference and body weight has a value of 0.574 ($P < 0.000$), which means that there is a moderate and positive relationship between body weight and entok body weight. The correlation coefficient (r) between membrane width and body weight has a value of 0.643 ($P < 0.000$), which means that there is a strong relationship. In essence, all the data above refers to the opinion of Luthfiana et al. [13] The positive correlation value means that if the X variable increases, the Y variable will increase. A positive sign (+) indicates the same direction or unidirectional genetic correlation, while a negative sign (-) indicates the presence of the opposite direction. The correlation value is negative; if variable X increases, then variable Y will decrease.

Based on Table 2, the value of the coefficient of determination in the regression model can be explained using the coefficient of determination, assuming that the greater the value, the better the model. The coefficients of determination of beak length, beak width, head length, head width, neck length, neck circumference, wing length, dorsal length, chest length, chest width, shank length, shank circumference, and membrane width are 0.181, 0.942, 0.088, 0.030, 0.161, 0.313, 0.313, 0.547, 0.247, 0.350, 0.037, 0.329, and 0.413, respectively. The coefficient of determination of the dorsal length value is the highest compared to the length of the beak, beak width, head length, head width, neck length, neck circumference, wing length, chest length, chest width, shank length, shank circumference, and membrane width. This data shows that the variable back length has a greater effect on body weight than other variables. The value of the coefficient of determination (R2) of 0.547 shows that there is an influence of back length on body weight of 54.7%. Likewise, the influence of beak length, beak width, head length, head width, neck length, neck circumference, wing length, chest length, chest width, shank length, shank circumference, and membrane width on body weight was 18.7%, 9.4%, 8.8%, 3%, 16.1%, 31.3%, 31.3%, 24.7%, 35%, 3.7%, 32.9%, and 41.3%, respectively.

Carina Moschata's body weight is: back length, beak length, beak width, head length, head width, neck length, neck circumference, wing length, thoracic length, thoracic width always increases as body size values increase, including lower limb length, circular shaft and membrane width. Ayeni et al [14] stated that egg size affects the quality of eggs and chicks. While the parameter that has the greatest influence on the hatching weight of chicks is evidenced by the coefficient of determination (R2) of 0.57 between hatching weight and hatching weight. This means that 57% of hatching weight is affected by egg weight, and the rest is influenced by other factors such as environment, habitat, and maintenance management. This is in line with the statement of Djego and Kihe [15] stating that the correlation between body weight and back length (0.32) and between body weight and shank length (0.31) in this study is included in the medium category. This is thought to be because along the spine, relatively many muscles or flesh are attached. This is in accordance with Budiarto's opinion [16] that, in addition to supporting the body, the spine also functions as a place for muscles to attach, so the longer the spine or back, the more muscles or flesh.

Table 2. Correlation and coefficient of determination between body size and body weight Carina moschata

Variables	Body Weight (g)			
	Correlation Coeffient (r)	**Determination Coefficint (R2)**	**P-Value**	**N**
Beak length	0.433**	0.1874	0.001	60
Beak width	0.307	0.0942	0.017	60
Head Length	0.297*	0.088	0.021	60
Head width	0.174	0.0302	0.184	60
Neck Length	0.402**	0.1616	0.001	60
Neck Circumference	0.560**	0.3136	0.000	60
Wing Length	0.560**	0.3136	0.000	60
Back Length	0.74**	0.5476	0.000	60
Chest length	0.497**	0.2470	0.000	60
Chest Width	0.592**	0.3504	0.000	60
Shank Length	0.193	0.0372	0.141	60
Shank Circle	0.574**	0.3294	0.000	60
Membrane Width	0.643**	0.4134	0.000	60

*. Correlation is significant at the 0.05 level (2-tailed).

**. Correlation is significant at the 0.01 level (2-tailed).

The analysis of factors affecting the body weight of ducks, or Carina moschata, in this study includes variables of membrane width, shank length, beak width, wing length, head width, head length, neck length, shank circumference, beak length, chest length, chest width, neck circumference, and back length. The analysis used is a multiple regression test, as can be seen in Table 3.

The regression model can be described by utilizing the coefficient of determination, where it is assumed that the higher the value, the more optimal the model performance will be. Referring to Table 3, it can be seen that the R square value reaches 0.716, indicating that the coefficient of determination reaches 71.6%. This means that variable X (membrane width, shank length, beak width, wing length, wingspan, head length, neck length, shank circumference, beak length, breast length, breast width, neck circumference, back length) has an impact and contributes to variable Y (body weight) of 71.6%. Therefore, the remaining 28.4% is influenced by other factors not included in variable X.

Table 3. Analysis of Factors Affecting Carina Moschata's Body Weight

Model Summary[b]

Model	R	R Square	Adjusted R Square	Std. Error of the Estimate
1	0.846[a]	0.716	0.636	118.354

a. Predictors: (constant), Width of membrane, length of shank, width of beak, length of wing, width of head, length of head, length of neck, Circumference of shank, length of beak, length of breast, width of breast, circumference of neck, length of back
b. Dependent Variable: Body Weight

4 Conclusion

According to the findings of this study, it can be concluded that there is a positive correlation between the morphological dimensions of livestock, including membrane width, shank length, beak width, wing length, head width, head length, neck length, shank circumference, beak length, breast length, breast width, neck circumference, back length, and body weight. And the length of the back has the largest contribution of 54% to the body weight of muscovy ducks (Carina moschata) aged 3 weeks old.

References

1. Cherry P, Morris TR. Domestic duck production: Sience and practice. Oxfordshire (UK): CABI.(2008).
2. Wojcik, E., and Smalec, E. Folia biologica (Kraków), 56 (3-4), 243-248. (2008).
3. Holderread D. Oregon (US): Dept. of Poultry Science, Oregon State University. (2001).
4. Ayuningtyas, G., Jakaria, J., Rukmiasih, R., & Budiman, C. (2016). Jurnal Ilmu Produksi dan Teknologi Hasil Peternakan, 4(2), 280-285.
5. Huang JF, Pingel H, Guy G, Łukaszewicz E, Baéza E, Wang SD. Worlds Poult Sci J. 68:551-563 (2012).
6. Raji, A. O., Igwebuike, J. U., and Usman, M. T. (2009). ARPN J. Agric. Biol. Sci, 4(3), 58-62.
7. Hassan WA, Mohammed MS. In: Proceedings of the 8th Annual Conference of Animal Science Association of Nigeria (ASAN). Minna, 16-18th September 2003. Niger State (Nigeria): Animal Science Association of Nigeria. p. 23-24 (2003).
8. Chia SS, Momoh O. Proceedings 37th Annual Conference of Nigerian Society for Animal Production. Lafia, Oktober 2013. Nasarawa State (Nigeria): NSAP. p. 20-22 (2012).
9. Tamzil, M. H. WARTAZOA,28 (3), 129-138 (2018).
10. Musa A.M., N.Z. Idam, and K.M. Elamin. 2012. Regression analysis of linier body measurements on live weight in Sudanese Shugor sheep. Online J.Anim. Feed Res. 2(1):27-29
11. Oguntunji, A. O., and Makram, A. Agricultural Science & Technology, 11 (4), 1313-8820 (2019).
12. Oguntunji AO. Genetika, 49, 743-753. (2017).

13. Luthfiana, N. A. Skripsi. Universitas Tidar. Magelang (2021).
14. Ayeni A. O., Agbede, J. O., Igbasan, F. A., Onibi, G. E., and Adegbenro, M.: International Journal of Environment, Agriculture and Biotechnology 3(3), 987–993 (2018).
15. Djego, Y., dan Kihe, J. N Jurnal Nukleus Peternakan, 7(1), 51-54 (2020).
16. Budiarto AB.. Skripsi. Institut Pertanian Bogor. Bogor. (2015).

Evaluation of nutrition of Black Soldier Flies Larvae (BSFL) using Van Soest and fatty acid profile reared at different media

Inaayah Novitri Cahyawati[1], *Osfar* Sjofjan[1*], *Heli* Tistiana[1], *and Jaiwei* Lee[2]

[1]Faculty of Animal Science, Universitas Brawijaya, Malang
[2] Department of Tropical Agriculture and International Cooperation, National Pingtung University of Science and Technology, Taiwan

Abstract. The current study aimed to evaluate nutrition of black soldier fly larvae (BSFL) using Van Soest and fatty acid profile reared at different media. A total of five different treatments were used in this study as follows: fruit waste, food waste, tofu by-product, 50% fruit waste + 50% food waste, and 50% fruit waste + 50% tofu by-product. The parameters observed were fatty acid profile (saturated and non-saturated fatty acid) and fiber composition (Neutral Detergent Fiber, Acid Detergent Fiber, Cellulose, hemicellulose, and lignin). Fatty acid profile analysis was done using gas chromatography and fiber composition using Van Soest method. The result presented the composition of fatty acid in BSFL dominated by saturated fatty acid, namely lauric acid (34.28 – 38.39%), palmitic acid (17.45 –19.25%), arachidic acid (8.8 – 10.41%), myristic acid (7.35 – 9.36%), heptadecanoic acid (2.03– 2.91%), and decanoic acid (1.23 – 1.42%). Non-saturated fatty acid; linolelaidic acid (13.51 – 16.86%), oleic acid (1.64 – 1.91%), nervonic acid (1.34- 1.46%), dan linolenic acid (0.65 – 1.03%) measured higher as compared to the rest. Fiber components measured were NDF (20.19 - 29.11%), ADF (10.43 – 17.55%), hemicellulose (8.86 – 11.56%), cellulose (2.51 –10.49%), and Lignin (0.14 – 0.75%). Respectively, in conclusion composition of growth media of BSFL influence the profile of fatty acid and fiber in BSFL. In addition, the media composed of 50% food wastes and 50% tofu by-product gave highest and most complete fatty acid profile as well as fiber composition.

1 Introduction

Indonesia is a developing country, populated by 270.20 million people [1] and still growing. This population growth is accompanied by the increasing production of waste [2]. The amount of organic waste produced by Indonesia reaches up to 60% of the nation's total waste. A large portion of organic waste comes from food waste. The amount of food waste of Indonesia is around 300 kg per person annually, placing the nation second in global food waste production in large category [3]. Aside from food waste, organic waste types that are

* Corresponding author: osfar@ub.ac.id

abundant and still minimally utilised are market waste, for instance, fruit waste, vegetable waste, and industrial waste such as tofu by-product [4]. Each of the mentioned waste are still containing large amount of nutrition. For example, tofu byproduct still contains around CP 21%, fat 3.79%, DM 48.37%, and ash 1.21% [5].

The use of insects has been one of the references to be a solution to utilisation of organic waste. It is also mentioned to be beneficial to the livestock industry. Insect is a natural organic bio converter that consumes organic wastes to produce beneficial products. [6] stated that bio-conversion by insects is more efficient, environment- friendly, as well as requiring minimum space. Out of some of the insects that have been the subject of research such as Black Soldier Fly Larvae (BSFL), Mealworm larvae (*Tenebrio Molitor L.*) and crickets (Orthoptera: Gryllidae). BSFL possesses a level of conversion and resilience that are superior to those of mealworms and crickets, which allow less difficult cultivation [7]. BSFL is also known to have the ability to consume higher diversity of organic matter as compared to other larvae [8] therefore, BSFL could be a solution for solving organic waste problem. In the adult phase, BSF is not considered a vector of pathogens to humans, animals, and plants [9]. According to [10] BSFL contains excellent nutrients, it possesses high protein, complete fatty acid, high feed conversion, short reproduction cycle, mineral, as well as high in vitamin. This implies that BSFL has the potential for a poultry feed alternative.

The composing nutrients of BSFL is influenced by the applied media on which it was reared. Good growth media results in BSFL with higher quality of nutrition. BSFL contains over 28% fat [8]. According to [11] the fat of BFSFL could be utilised for biodiesel, oleo chemical, soap, and consumption. A number of countries have started using BSFL fat as condiments for bread and margarine. According to a study of [12] the nutritional content in form of crude fat and crude fibre of BSFL reared on food waste respectively around 31.95% and 17.85%. Nutritional content of maggot reared in tofu by-product showed crude fibre 8.72% and crude fat 39.41% [2]. In comparison to other insects, BSFL contains high fat and fibre with the fat mainly in the form of saturated fatty acids (SFA) [13]. Since the imposed of antibiotic growth promoters (AGPs) in Europe 2006, it has continued with developing countries as well. There is an alternative to seeking replacement of AGPs [14].

The earlier finding reveals the possibility of modification to be done in the composition of maggot nutrients such as the fatty acid and fibre fragments, which is due to the absence of information of which nutrition the maggot itself produce, or to what extent the nutrition of the media influences the fatty acid and fibre composition. Understanding of this matter is highly important if the maggots are to be reared in different media and to be applied to animal feed or other industries. This study aimed to evaluate the effect of different organic wastes as media of maggot growth in the fatty acid and crude fibre composition profile.

2 Materials and Methods

2.1 Materials

2.1.1 Material

The maggot was reared in Malang. Fatty acid profile test was done in Laboratorium Penelitian dan Pengujian Terpadu (LPPT) Gajah Mada University, Yogyakarta. Fiber content analysis was conducted in Laboratorium Nutrisi dan Makanan Ternak Universitas Brawijaya, Malang.

Black Soldier Fly Larvae (BSFL) was used priory reared for 28 days on organic waste media (fruit, food waste, and tofu byproduct). Fruit waste was obtained from markets (from apple, pear, dragon fruit, and starfruit) that were no longer in any quality for sale, food waste was obtained from restaurants and households, Tofu byproduct was obtained from tofu factories, dried and ground shortly after, before being given to BSFL.

2.1.2 Media

Media of growth applied are fruit waste, food waste, and tofu by-product. Feed was provided at 5 kg per day. Nutrition content of every media is displayed on table 1.

Table 1. Nutritional content of BSFL growth media.

Media	DM (%)	Ash (%DM)	CP (%DM)	Fat (%DM)	CF (%DM)	NFE (%)
Fruit Waste	93.07	2.68	7.76	1.09	6.03	82.44
Food Waste	40.44	1.67	15.03	12.45	19.77	51.08
Tofu Byproduct	11.00	4.27	20.03	2.89	2.48	70.33

Source: Central Laboratory, Muhammadiyah University, Malang (2023)
DM – Dry Matter; CP – crude protein; CF – Crude Fibre; NFE – Nitrogen-Free Extract

2.2 Research method

Analysis of fatty acid profile by gas chromatography and fibre fragments analysis by van-soest method. Research method is descriptive with 5 treatments. The treatments tested include fruit waste (T0), food waste (T1), Tofu byproduct (T2), 50% fruit waste + 50% tofu byproduct (T3), and 50% fruit waste + 50% food waste (T4).

3 Results and discussion

3.1 Fatty Acid Profile

Table 2. shows the effect of different media of BSFL growth being determinant to fatty acid profile of BSFL. The dominant fatty acid in all treatments of BSFL media exhibited lauric fatty acid (34.28-38.39%), followed by palmitic acid (17.45-19.25%) and linoleic acid (13.52-16.86%). The larger fraction of BSFL contains a high saturated fatty acid (SFA) percentage, followed by poly-non-saturated fatty acid (PUFA; up to 18.57%), and mono-non- non-saturated fatty acid (MUFA: up to 4.2%). The high level of SFA in BSFL is a way to store energy for the next phase in the life cycle [16]. Storage of energy in the form of SFA is backgrounded by the lack of its vulnerability against oxidation as compared to unsaturated fatty acid (NSFA) [17]. Lauric Acid from SFA is the main element (up to 38.39%) in BSFL. In this study, the highest lauric fatty acid was obtained from T4 and followed by P3.

Table 2. Fatty acid content of BSFL

Fatty acid type		**T0**	**T1**	**T2**	**T3**	**T4**
SFA	Butyrate acid	0.35	7.84	<0.1	0.16	0.16
	Lauric acid	34.28	35.14	35.99	37.56	38.39
	Miristic acid	8.58	7.35	8.9	9.36	9.31
	Palmitic acid	19.25	17.74	18.34	17.45	18.43
	Heptadecanoic acid	2.57	2.03	2.84	2.82	2.91
	Decanoic acid	1.37	1.42	1.31	1.23	1.31
	Arachidate acid	10.41	8.98	10.11	10.22	8.8
	Heneicosanoic acid	0.44	0.24	0.51	0.49	<0.1
	Docosanoate acid	<0.1	0.51	<0.1	0.58	<0.1

	TOTAL	**77.25**	**81.25**	**78**	**79.87**	**79,31**
MUFA	Palmitoleate acid	0.42	0.41	0.4	0.35	0.37
	Oleate acid	1.84	1.64	1.69	1.91	1.72
	Myristoleic acid	0.22	0.18	<0.1	<0.1	<0.1
	Erucate acid	0.28	0.28	<0.1	0.31	0.41
	Methyl nervonate	1.44	1.44	1.34	1.44	1.46
	TOTAL	**4.2**	**3.95**	**3.43**	**4.01**	**3.96**
PUFA	Linoleic acid	16.86	13.51	16.68	14.42	15.26
	Linolenic acid	0.91	0.7	0.89	1.03	0.65
	DHA	0.76	0.59	1	0.65	0.83
	EPA	<0.1	<0.1	<0.1	<0.1	<0.1
	TOTAL	**18.53**	**14.8**	**18.57**	**16.1**	**16.74**

Lauric acid (C:12) is a type of fatty acid commonly found in BSFL that functions for antimicrobial, antivirus, and antifungal. That makes BSFL a potential feed component because in addition to the high protein content, the use for antibiotic is also applicable. The high lauric acid in BSFL is suspected to be because the mentioned fatty acid is synthesized by the BSFL itself through carbohydrate synthesis in a substrate and could increase as per decomposition of the organic substrate [17] The claim is supported by [18] that carbohydrate from substrate forms a number of fatty acids through glycolysis, oxidative decarboxylation, and later becomes fatty acid. The process exhibits the ability of BSFL to synthesize lauric acid coming from non-fat from its body. The NFE and lauric acid is very closely related due to NFE being group of fractions of digestible carbohydrates. However, in the current study, any evidences of high NFE being able to increase lauric acid content in BSFL are yet to be found. The provision of fruit waste mixed with food waste or tofu by- product (T3 and T4) could increase lauric acid in BSFL as compared to each as their own media (T0, T1, and T2). In comparison, the result of a study by [17] showed higher lauric acid in similar media of food waste (57.56%) and fruit waste (60.89%). the result is far higher in comparison to that of the current study. The content of lauric acid could vary depending on the nutrition of the media and environment of rearing. In a number of studies, lauric acid in food waste media showed 13.44%, 39.9%, and 7.4% [18,19,20] In the current study, the proceeding high SFA content is arachidate acid (8.8 – 10.41%), palmitic acid (17.45-19.25%) and myristic (7.35%-9.36%). This aligned with [23] that SFA content showed highest after lauric is myristic (5,71-8,89%) and palmitic (5.84-11.51%). According to [24] SFA composed of C12:0, C14:0, C16:0 could be synthesized by BSFL. Other SFA fatty acid such as butyrate (<0.1 – 7.84%), Heptadecanoic acid (2.03 – 2.91%), decanoic acid (1.23 – 1.42%), heneicosanoic acid (<0.1 – 0.51), and docosanoate acid (<0.1 – 0.58) was also detected in BSFL, however in very low quantity.

Unsaturated fatty acid (USFA) is split into Monounsaturated fatty acid (MUFA) and Polyunsaturated Fatty Acid (PUFA). The division is based on the bond, MUFA has a single bond, while PUFA has more than two. In the current study, the volume of MUFA is smaller as compared to PUFA. MUFA was dominated by oleic acid (1,64-1,91%) and nervonic acid (1,34-1,46%). Oleic acid could represent essential fatty acids that are important to consider in formulation of drugs and could potentially benefit wound recovery. Consumption of MUFA could lift production of high-density lipoprotein (HDL) in blood, suppressing low-density lipoprotein (LDL) in blood and supporting heart health. The highest amount of MUFA in the current study is obtained from T0, despite generally fruit does not produce MUFA. This implies that BSFL could synthesise MUFA on its own. This aligns with [24] that MUFA such as oleic acid (C18:1) could be produced by BSFL on its own without interference of nutritional content from substrate.

Table 2. displayed that PUFA content in all treatments are high (14.8 – 18.57%). The composing components of PUFA include essential fatty acids, namely omega-3 and omega-6. Linoleic acid is a type of omega-6 fatty acid. In this study, the amounts of linoleic acid obtained from each treatment are T0 16.86%, P1 13.51%, P2 16.68%, P3 14.42%, and P4 15.26%. The highest to contain are T0 and T2 due to presence of the fatty acid in the media. According to [24] linoleic acid is considered a type of fatty acid not being produced by BSFL itself. This fatty acid is highly influenced by the content contained in the substrate. The composition of PUFA include Linoleic acid, EPA, and DHA are essential fatty acid type omega-3. In table 2. It is visible that DHA (0.59 – 1%) and EPA (<0.01%) content are found in small amounts. T2 displayed the highest value of DHA as compared to the rest of the treatments. DHA content of BSFL in all treatments are smaller in comparison to [21] that obtained 5% DHA in ensiled muscles media. Such could occur due to the difference in fatty acid content in the substrate. [25] Suspected that there is a positive correlation between the percentage of fatty acid in the media and in the larvae's body. According to [31] the low content of EPA and DHA occur due to the increasing quantity of SFA. It was proven that the highest SFA content is T1(81.25%) producing DHA content (0.59%) that is the lowest compared to the rest.

Fatty acid of BSL closely resembles the fatty acid content of palm oil and coconut oil. Therefore, it is possible to utilise fatty acid of BSFL in the related industry [26]. The presence of BSFL fatty acid makes plant based protein fatty acid also obtainable from animals. [27] found that fatty acid in BSFL is suitable for production of high quality biodiesel. From the mentioned matter. BSFL fat seemingly has good potential for being utilized in production of feed, food, and fuel

3.2 Fibre Composition of BSFL

Fibre composition of BSFL on each treatment is displayed in Table 3. On the table, it is displayed that the fibre compositions in BSFL differ according to the media of growth implemented. NDF content on this study is 20,19 – 29.11% where the content of NDF at the highest was held by T1 (29,11%) which was organic food waste. This takes place because in Table 1. it is displayed that the fibre composition in the food waste media was the highest compared to the rest. Food waste contains abundant plant and animal protein based, presence of vegetables in the food waste could increase the fibre content in the media and influence BSFL nutritional content. ADF in each treatment namely T0 11.66%, T1 17.55%, T2 10.80%, T3 10,43%, T4 11.75%. In a study of [26] content of NDF and ADF of BSFL each are respectively 35,53% and 29,27%. In this study it is also mentioned that the lower the content of crude fibre, NDF, and ADF in BSFL could indicate the declining chitin content. ADF content is a fraction of fibre that is soluble in acid detergent composed of cellulose, lignin, and silica that could indicate that the feed has low digestibility [27]. Therefore, the utilisation of BSFL for ruminant feed as well as non-ruminants should decide NDF and ADF content that are low. According to [28] crude fibre in insects represents chitin content due to chitin (linear polymer unit b-(1-4) N-Acetyl-D- glucosamine) has structure similar to that of cellulose (linear polymer unit b-(1-4) N- acetyl-D-glucosamine) and in fraction of ADF insects have proven to contain nitrogen.

Table 3. BSFL Fiber Composition

Code	NDF (%)	ADF (%)	HEMICELLULOSE (%)	CELLULOSA (%)	LIGNIN (%)
T0	20.52	11.66	8.86	3.27	0.16

T1	29.11	10.55	11.56	10.49	0.75
T2	20.97	10.80	10.17	2.51	0.20
T3	20.19	10.43	9.76	3.75	0.14
T4	22.22	11.75	10.48	8.58	0.46

NDF – Neutral Detergent fibre; ADF – Acid Detergent Fibre

In table 3. The hemicellulose content in BSFL is 8,86 – 11,56%. Hemicellulose is a fraction of crude fibre easiest to degrade, obtained from lowering of NDF with ADF. The high content of hemicellulose in the current study as in T1, 11.56% and lowest in T0, 8.86%. Cellulose and lignin content that are a fraction of fibre more challenging and difficult to digest into simpler carbohydrates. Lowest content of cellulose was on T2, 2,51%. Lowest in lignin content was T3, showing 0,14%. The decreasing content of lignin in feed, the higher the digestibility of the feed. That occurs because lignin is one of the fibre compulsions that is difficult to break down into simpler sugar. To maggots, lignin is fibre that is very difficult to turn into fatty acid [31]. In the digestive system of BSFL, there are lignin digesting microorganisms. These microorganisms will convert lignin into cellulose that will be used by maggots for energy [30]. Lignocellulose converts lignin into simple sugars in the form of enzymes that are produced by BSFL [2]. A number of studies showed that providing crude fibre could decrease the protein and fat in maggots. Maggot is suspected to not have enzymes for decomposing fibre. However, hosts fungi and microbes in the back of its middle digestive system with pH of 6-8 that could degrade crude fibre [31].

4 Conclusion

After analysis of fatty acid profile and fibre composition, it is discovered that the composition of fatty acid and fibre are influenced by the media of growth, although the exerted difference is relatively small. According to the potential of application, the content of fatty acid in BSFL could be utilised for alternative to protein based plant oil in food, feed, as well as biofuel production.

References

1. Badan Pusat Statistik. Sosial dan Kependudukan. Publikasi Statistik: Jakarta. Diakses melalui https://www.bps.go.id/ pada tanggal 1 September 2023.
2. Santi., A. T. B. Astuti, dan J. Pasamboang. Jurnal Ilmu Pertanian. 5, 2 : 91-93 (2020)
3. Yuliana L. Efektor. 9, 2 : 286-295 (2022)
4. Purnamasari, L., W. Muhlison, dan I. Sucipto. The 2nd Conference of Applied Animal Science. 105-111 (2021)
5. Masir, U., A. Fausiah, dan Sagita. Jurnal Ilmu Pertanian. 5, 2 : 87-90 (2020)
6. Magee K., Halstead J., Small R., Young I. Sustainability. 13 : 1–17 (2021)
7. Fonseca, K. B. B., M. Dicke, and J. J. A. Loon. Journal of insect as Food and Feed. 3(2) : 105-120 (2017)
8. Wang, Y.S and Shelomi,M. Foods. 6 : 91 (2017)
9. Cickova, H., G. L. Newton., R. C. Lacy, and M. Kozanek. Waste Manag. 35 (1) : 68-80 (2015)
10. El-Hack, M. E. A., M. E. Shafi., W. Y. Alghamdi., S. A. Abdelnour., A. M. Shehata., A. E. Noreldin., E. A. Ashour., A. A. Swelum., A. A. Al-Sagan., M.

Alkhateeb., A. E. Taha., A. E. Abdel-Moneim., V.Tufarelli, and M. Ragni. Agriculture. 10 : 1-31 (2020)

11. Gougbedji A, Agbohessou P, Lalèyè PA, Francis F, Caparros Megido R. Journal of Agriculture and Food Research. 4 (2021)
12. Majid, Z. A. N. M., L. Racmawati, and N. A. Rohmanna. Jurnal Teknologi Industri Pertanian. 33, 1: 41-49 (2023).
13. RAMOS-BUENO, R.P., GONZÁLEZ-FERNÁNDEZ, M.J., SÁNCHEZ-MUROS-LOZANO, M.J., GARCÍA- BARROSO, F., GUIL-GUERRERO, J.L., 2016. EUR. FOOD RES. TECHNOL. 242, 1471–1477 (2016)
14. D. N. Adli, S. Sadarman, A. Irawan, A. Jayanegara, T. M. Wardiny, T.R. Priambodo, A.P.A Yekti. Ital. J. of Anim. Sci. 22, (2023).
15. Fernando, I., S.A. Siddiqui, W.S. Nugraha, B. Yudhistira, D.N. Adli, A.A. Nagdalian, A.V. Blinov, aM.B. Mario. J. of Insect as Food and Feed. 9 (2023)
16. Tomberlin, J.K., Sheppard, D.C. J. Entomol. Sci. 37, 345–352 (2002)
17. Ushakova, N.A., Brodskii, E.S., Kovalenko, A.A., Bastrakov, A.I., Kozlova, A.A., Pavlov, D.S. Dokl. Biochem. Biophys. 468, 209–212 (2016)
18. Spanghers, T., M. Ottoboni., C. Klootwijk., A. Ovyn., S. Deboosere., B. De-Meulenaer., J. Michiels., M. Eeckhout., P. De-Clercq, and S. De-Smet. J Sci Food Agric. 97: 2594–2600 (2016)
19. Hoc B, Francis F, Carpentier J, Mostade L, Blecker J. Journal of the Saudi Society of Agricultural Sciences. 20 :155–163 (2020)
20. El-Deen, S. N., K. V. Rozen., H. Elissen., P. V. Wikselaar., I. Fodor., R. V. D. Weide., E. F.H. D. Hil., A. Z. Far, and T. Insects. 14, 204 (2023)
21. Ewald, N., Vidakovic, A., Langeland, M., Kiessling, A., Sampels, S. and Lalander, C. Waste Management The Authors. 102: 40–47 (2020)
22. Shumo, M., M. Isaac., Osuga., M. Fathiya., Khamis., M. Chrysantus., Tanga., K. M. Komi., Fiaboe., S. Subramanian., S. Ekesi., A. van-Huis, and C. Borgemeister. Scientific Report. 9 : 1-13 (2019)
23. Lawal, K. G., R. R. Kavle., T. O. Akanbi., M. Mirosa., and D. Agyei. Future Foods (2021)
24. Alifian, M. D. Disertasi IPB University (2023)
25. Liu, X., X. Chen., H. Wang., Q. Yang., K. Rehman., W. Li., M. Cai., Q. Li., L. Mazza, and J. Zhang. PLoS ONE. 12 : 1–21 (2017)
26. Matthaus, B., Piofczyk, T., Katz, H., Pudel, F. Eur. J. Lipid Sci. Technol. 121(2019)
27. Surendra, K.C., Olivier, R., Tomberlin, J.K., Jha, R., Khanal, S.K. Renew. Energy 98, 197–202 (2016)
28. Nafisah, A., Nahrowi., R. Mutia dan A. Jayanegara. Annual Basic Science International Conference. 1-5 (2019)
29. Sudirman., Suhubdy, S. D. Hasan., S. H. Dilaga., dan I. W. Karda. Jurnal Ilmu dan Teknologi Peternakan Indonesia. 1, 1 66-70 (2015)
30. Finke, M. D. 26, 105-115 (2007)
31. Kim, W. Journal of Asia-Pasific Entomology. 14, 11–14 (2011)
32. Li, W. Biotechnology for Biofuels. 8, 117 (2015)

33. Gold, M., J.K. Tomberlin., S. Diener., C. Zurbrugg., A. Mathys. Waste Management. 82, 302-318 (2018)

Associations between polymorphisms in the growth hormone locus and reproductive efficiency in dairy cattle in Pujon

A. Dzikrullah[1], *Agus* Budiarto[2], and *Sucik* Maylinda[2*]

[1]Post Graduate Student in Faculty of Animal Science, Universitas Brawijaya, Malang 65145, Indonesia
[2] Faculty of Animal Science, Universitas Brawijaya, Malang 65145, Indonesia

Abstract. Selection using genetic markers (MAS) is carried out in an effort to accelerate livestock breeding programs, through molecular studies individual superior livestock can be detected earlier so that the selection program can be directed and controlled. This study aims to analyze genetic diversity at the growth hormone locus that is thought to have a relationship with reproductive and production traits in Friesian Holstein Crossed (FHC) dairy cattle through the polymerase chain reaction - restriction fragment length polymorphisms (PCR-RFLP) method. The study was conducted in Pujon subdistrict using 53 FHC cattle samples. The results showed that at the GH gene locus only one genotype was found, AA with an allele frequency of 1.00. Based on these results, it was concluded that the GH gene in FHC cattle is monomorphic. The information in this study can be used as a reference in formulating livestock breeding programs.
Keywords: Growth Hormone Polymorphism, monomorphic, Marker Assisted Selection (MAS)

1 Introduction

Dairy cattle are a livestock commodity that is widely cultivated in the tropics of Indonesia. Recorded in the data of the Director General of PKH in 2022, East Java province is one of the provinces with the largest dairy cattle population with a population of 314,385 cattle. Dairy cattle are generally raised in highland areas, with a semi-intensive maintenance system. The increase in livestock population every year indicates an increase in the level of milk consumption and processed products. This increase needs to be supported by livestock breeds that have high genetic quality through controlled breeding, to produce breeds with high fertility rates and milk production above the group average. Implementation of controlled breeding is currently done by utilizing molecular genetics-based selection or MAS (Marker Assisted Selection) gene markers which are considered quite effective in identifying phenotypic traits for improving reproductive traits such as service per conception (S/C), days open and calving interval (CI) as well as livestock production such as carcass and milk. Through MAS identification, it is possible to determine the effect of candidate genes on livestock performance based on allele detection.

Growth Hormone (GH), known as somatotropin hormone, is one of the candidate genes derived from a peptide series of about 190 amino acids synthesized by somatotroph cells in the anterior pituitary [1]. GH plays an important role in metabolism, lactation control, mammary gland development, and fertility in cattle [2]. In a previous study by [3] it was shown that there is a positive relationship between GH gene polymorphisms and growth and

* Corresponding author: sucik@ub.ac.id

lactation performance in dairy cattle. At least four single nucleutide polymorphisms (SNPs) have been found in exon 10, which encodes the cytoplasmic domain of GHR. SNPs are located at positions 257 (A/G), 229 (T/C), 200 (G/A), 76 (T/C). Positions 200 (G/A) and 257 (A/G) induce the amino acid substance alanine to threonine, and serine to glycine. GH effects the growth, reproduction and productivity of livestock through interactions between specific receptors (GHR) on the surface of target cells. As a result, there are changes in GH activity in the target tissue due to the influence of increased signaling pathways and binding capacity. The purpose of this study was to analyze genetic diversity (genetic polymorphism) at the growth hormone locus that is thought to have a relationship with reproductive and production traits in Peranakan Friesian Holstein dairy cattle.

2 Materials and Methods

This study was conducted on 53 Friesian Holstein Crossbreed dairy cattle in lactation period II - IV with milking frequency twice per day, average body weight reached 425 ± 59.3 kg and milk production 11.13 ± 4.43 l/day kept in a smallholder farm located in Pandesari village, Pujon, Malang Regency, East Java 65391 coordinates 7°51'41.7 "S 112°28'36.6 "E with an altitude of 1200 meters above sea level with temperatures ranging from 18 oC.to 25 oC. and average humidity 74%. The research was conducted in two stages, the first stage included linear body measurements obtained from measuring body weight, body length, height, and chest circumference using digital scales, measuring rulers and measuring tapes. Daily milk production measurements were taken in the morning and evening, and reproductive data were obtained from the KOPSAE office database through the inseminator. The second stage of the research involved taking blood samples from dairy cattle using vacuntainer tubes containing EDTA anti-coagulant. Blood sampling through the jugular vein as much as 3 mL then transferred blood samples to the laboratory and stored in a refrigator at 20°C.

2.1 Amplification of DNA fragments of GH gene locus

Amplification of specific DNA fragments of the GH gene of FHC dairy cattle with PCR product length determined based on the attachment of primers designed from the position of 1537 bp to 2013 bp with a product length of 476 bp, thus obtaining DNA primers consisting of, GH-forward:

5`- GTTGTTGGATGGCAGTGG -3` and GH-reverse: 5`- CTTCTCCAAGCCTGTAGG - 3`, with gene bank acese code: JQ711182. The total volume for DNA amplification was 15 µL with a composition according to Promega product requirements which included 7.0 µL Gotaqgreen, 6.4 µL FNW (Free Nucleous Water), 0.3 µL forward and reverse DNA primers respectively and 1 µL genomic DNA sample. The mixture was processed in a BioRad T100 Thermal Cycler PCR machine with the following settings: Initial denaturation of 95oC for 5 minutes, followed by denaturation process at 95oC for 45 seconds, anneling at 58oC for 45 seconds, and extension at 72oC for 45 seconds repeated for 35x cycles, continued with the final extension process at 72oC for 5 minutes and ended with cooling down process at 12oC for 2 minutes (Rahayu, Sumitro, Susilawati and Soemarno., 2006).

The DNA amplification results (Figure 1) obtained were then subjected to RFLP process using restriction enzymes Msp1 with cutting site `5-CC | GG- 3` and Alu1 with cutting site `5-AG | CT- 3`. The total volume for RFLP using Msp1 and Alu1 was 25 µL each consisting of 1 µL of PCR product DNA sample, 21 µL of FNW (Free Nucleous Water), 2.5 µL of Buffer 10-NE and 0.5 µL of restriction enzymes (Msp1 and Alu1). Then incubate the RFLP sample at 37 oC for 15 minutes. Visualization of RFLP results is then seen at the electrophoresis stage using a 100 Volt MupidEx electrophoresis machine with agarose gel media that has been colored with bromide.

2.2 Data analysis

The visualization obtained was then continued with the calculation of genotype and allele frequencies to test the Hardy-Weinberg equilibrium (Warwick, 1990). Calculation of the magnitude of Growth Hormone gene polymorphism in PFH dairy cattle based on % PIC (Polymorphic Information Content) (Budak et al, 2003) as follows:

$$\mathbf{PICi = 1 - \sum p2i\,j}$$

Where PICi is the polymorphic information content at the i-th locus, pij is the frequency of the j-th allele and the i-th locus.
Data in the form of reproductive efficiency traits such as S/C, DO and CI as well as data on BW, LD, PB and TB, milk production were analyzed in relation to Growth Hormone gene polymorphisms using the One Way Anova method assisted by Minitab software version 17 (2023)..

3 Results and discussion

The results of data analysis using Minitab version 17 show the results as in Table 1. Because the age distribution is quite large, age grouping is carried out, namely: Age group 1 is age 2-3.5 years, age group 2 is age 3.5 - 4.5 years, and age group 3 is age > 4.5 years. Based on the results of the study mentioned that body weight (BW) was significantly influenced ($P < 0.05$) by age group, as well as in Table 2. Presented body weight at various lactation periods.

Table 1. Body weights in FHC cattle at different age groups.

Age group	N	Mean ± standard deviation (kg)*)
1	13	367.00 ± 26.53 [a]
2	13	398.15 ± 31.85 [b]
3	7	391.6 ± 45.6 [b]

[1] Different superscripts in the same column indicate significantly different ($P < 0.05$)

Table 2. Body weight in FHC cattle in different lactation groups

Lactation period	N	Mean ± standard deviation (kg)*)
1	14	369.57 ± 27.25 [a]
2	9	393.11 ± 31.33 [b]
3	5	379.20 ± 17.25 [b]
4	3	401.00 ± 50.23 [c]
5	2	438.50 ± 74.25 [c]

[2]Different superscripts in the same column indicate significantly different ($P < 0.05$)

3.1 Reproductive Efficiency of FHC dairy cattle in different age groups and lactation periods

The results of statistical analysis show that the reproductive efficiency of female dairy cattle is not influenced by age group. This can be seen in Table 3. Meanwhile, Table 4 shows that the reproductive efficiency of female cattle (S/C, DO and CI) is not influenced by body weight group. In this case the body weight is grouped into 4, namely body weight group 1 = 360-400 kg, body weight group 2 = 400-440 kg, body weight group 3 = 440-480 and body weight group 4 = > than 480 kg. . Table 4. Presents Reproductive Efficiency grouped into 4 Body Weight groups.

Table 3. Reproductive efficiency of FHC dairy cattle in different age groups

Variables	Age group	N	Mean± Standard deviation
	1	19	1.58 ± 0.51
S/C	2	21	1.38 ± 0.50
	3	13	1.23 ± 0.44
DO	1	19	101.05 ± 11.38
	2	21	96.62 ± 11.60
	3	13	95.00 ± 10.87
CI	1	19	381.05± 11.38
	2	21	376.62± 11.60
	3	13	375 ± 10.87

Table 4. Reproductive efficiency of FHC dairy cattle in 4 body weight groups

Variables	Age group	N	Mean± Standard deviation
	1	3	1.00 ± 0.00
S/C	2	25	1.28 ± 0.46
	3	2	1.5 ± 0.71
	4	3	1.33 ± 0.58
DO	1	3	102.00 ± 10.54
	2	25	99.48 ± 11.90
	3	2	91.00 ± 5.66
	4	3	105,67 ± 10.60
CI	1	3	382.00 ± 10.54
	2	25	379.48± 11.90
	3	2	371.00 ± 5.66
	4	3	385.67 ± 10.60

3.2 Allele frequencies of GH-Msp1 and GH-Alu1 loci

Growth Hormone gene polymorphism analysis in FHC cattle in this study has been carried out by PCR-RFLP method with visualization of DNA amplification using 1.5% and 2% agarose gel presented in Figure 1 with PCR visualization results showing a single band pattern at 476 bp.

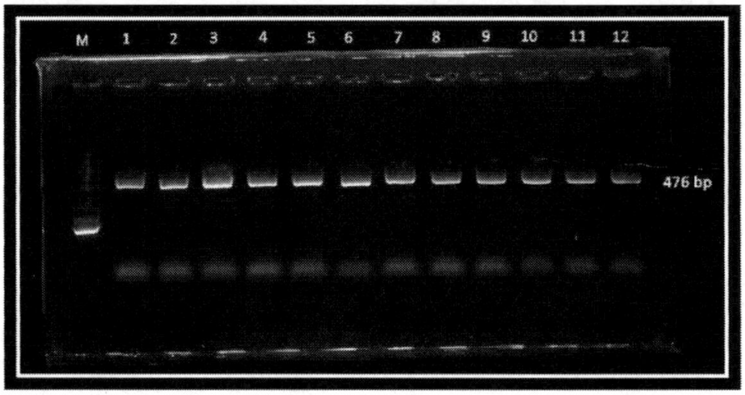

Pic 1 Visualization of GH gene amplification on 1.5% agarose gel

3.3 GH-Msp1 gene locus

PCR-RFLP visualization of the GH-Msp1 gene locus with a base length of 476 bp produced 1 single band at position 276 bp presented in Figure 2. Based on this visualization, one A allele was obtained while the G allele or other alleles were not found. This is due to the absence of mutations in the GH-Msp1 gene locus, resulting in the same or unvaried banding pattern.In general, the GH-Msp1 gene locus in FHC dairy cattle in Pujon sub-district is monomorphic with an allele frequency (A) of 1.0 because the GH gene found 100% has the AA genotype. Similar results were also found by []stating that only AA genotype was found for the GH-Msp1 gene locus in subpopulations of Balinese and PO cattle breeds. The uniformity of the GH gene locus in livestock is due to a directed selection process or controlled breeding, so the possibility of inreedig is low

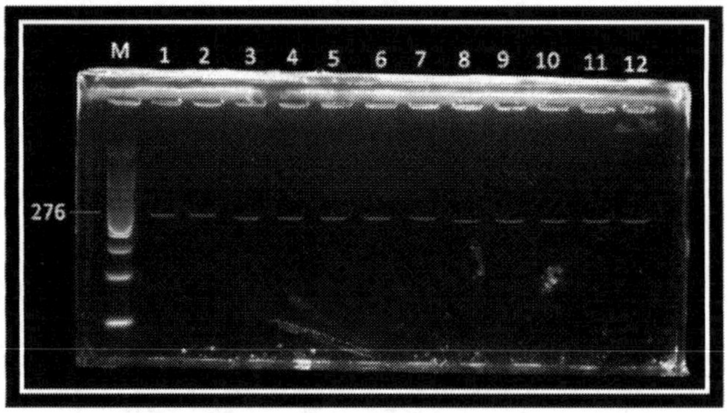

Pic 2 Visualization of PCR-RFLP with Msp1

3.4 GH-Alu1 gene locus

Presented in Figure 3 the results of cutting the GH gene using Alu1 show monomorphic results by producing 3 banding patterns at positions 130 bp, 244 bp and 476 bp. Based on this visualization, one A allele was obtained while the other alleles were not found. This is due to the absence of mutations in the GH-Alu1 gene locus, resulting in the same or unvaried banding pattern. Calculation of genotype and allele frequency at the GH-Alu1 gene locus from 53 FHC cattle obtained an allele frequency (A) of 1.0 because the GH gene found 100% has the AA genotype.

Monomorphic results at the GH gene locus were also obtained in the research of [5] which stated that in Ongole breeders only one BB genotype was found at the GH gene locus at positions 223 bp and 104 bp. Added by [6] which states that the GH-Alu1 gene locus is generally monomorphic because only the LL genotype is found in Balinese cattle at position 211 bp. It can be concluded that the frequency of genotypes and alleles in FHC cattle from generation to generation remains constant, this is possible due to several aspects such as, no mutation occurs, the number of individual cattle in the population is very large so that marriages in the population occur randomly, with this random marriage, the frequency of alleles and genotypes will be constant from generation to generation and each gene will have the same viability and fertility[7]. On the other hand, this result is different with the result of [8] in the same breed that is Local Friesian Holstein (Grati cattle) in Pasuruan, the population was polymorphic by MSp1 restriction [8].

4 Conclusion

Based on the results of the study, it was concluded that the GH (Growth Hormone) gene in Friesian Holstein Crossbreed cattle in the Pujon sub-district area was found to be monomorphic, so it could not be used in further correlation analysis between the marker gene locus and certain phenotypic traits, including reproductive traits and milk production. Further identification needs to be done using more specific gene locus markers to confirm the polymorphism status in a population.

References

1. Cattini, P.A., Bock, M.E., Jin, Y., Zanghi, J.A. and Vakili, H., *Growth Hormone & IGF Research*, **42**, 58-65 (2018)
2. Lucy, M., *Reproduction in Domestic Animals*, **43,** 31-39 (2008)
3. Sumiati, S.S., Arif, D. and Bazilah, A.T. Prosiding Seminar Nasional Industri Peternakan Bogor, 2017, **1**, 78-85 (2017)
4. Amiri, S., B. Jemmali, M.A. Ferchichi, H. Jeljeli, R. Boulbaba and A.B. Gara. Arch. Anim. Breed. **61**, 481 – 489 (2018)
5. Anggraeni, A., Talib, C., Asmarasari, S. A., Herawati, T., & Andreas, E. JITV. 22 (4): 165-172. (2017)
6. Renaville, R; Hammadi, M and Portetelle, D. Domest Anim Endocrinol. **23**, 351-36. (2002)
7. Ciptadi, G., Budiarto, A. and Oktanella. Brawijaya University Press (2019)
8. S. Maylinda. Int. J. Biotechnol. Mol. Biol. Res. **2**, 117 – 120 (2011)

Content of dry matter, crude fiber, and ash in fermented rice bran with palmyra sap at different length of fermentation

Theresia Nur Indah Koni[1*], *Regelinda* Soi[1], and *Tri Anggarini Yuniwati* Foenay[2]

[1]Feed Technology Study Program, Department of Animal Science, Politeknik Pertanian Negeri Kupang. Jl. Prof. Herman Yohanes, Kelurahan Lasiana, Kupang, East Nusa Tenggara, 85011 Indonesia

[2]Production Study Program, Department of Animal Science, Politeknik Pertanian Negeri Kupang, Jl. Prof. Herman Yohanes, Kelurahan Lasiana, Kupang, East Nusa Tenggara, 85011 Indonesia

Abstract. Rice bran is one of waste products from rice processing that can be used as nonruminant feed ingredients, but high crude fiber contained limited its utilization. Reducing crude fiber content requires processing such as fermentation. Fermentation requires carbohydrates such as palmyra sap. The purpose of this research was to determine the changes in the content of dry matter, crude fiber and ash of rice bran added with palmyra sap at different fermentation periods. This research used a Complete Random Design with four treatments and five replications. The four treatments are DN0: without fermentation, DN3: 3 days fermentation, DN6: 6 days fermentation, and DN9: 9 days fermentation. The parameters measured were the content of dry matter, crude fiber, and ash. Data were analyzed using Analysis of Variance and continued with Duncan's Multiple Range Test. The results showed that fermentation periods decreased the crude fiber and ash content but had no significant effect on the dry matter content of rice bran. The 3 days fermentation period resulting the lowest crude fiber content.

1 Introduction

One of the common agricultural byproducts used as animal feed is rice bran. Rice bran has been widely utilized by most farmers in Indonesia as animal feed. Rice bran has significant potential as an energy source for livestock. It is a byproduct of rice milling in the production of rice [1]. The nutrient content of rice bran is as follows: crude protein 11.9%, crude fat 12.1%, calcium 0.1%, phosphorus 1.3%, and metabolic energy 2200 kcal/kg [2]. Rice bran contains crude fiber ranging from 12.4% to 27.8% [3], and according to Lamid et al. [4], rice bran contains 6.9% phytic acid. Pujaningsih et al. [5] reported that this phytic acid binds minerals such as calcium and phosphorus. Additionally, phytic acid can also bind to proteins, reducing protein digestibility. Both crude fiber and phytic acid can be reduced through processing methods such as fermentation [6].

* Corresponding author: indahkoni@gmail.com

Ermalia et al. [7]reported that the fermentation of rice bran using Lactic Acid Bacteria (LAB) starter and a storage period of 12 days can reduce the phytic acid content from 6.70% to 2.07% and increase the coefficient of dry matter digestibility from 63.06% to 69.72%. Fermentation is a biochemical process involving microorganisms. One of the purposes of fermentation is to enhance the digestibility of feed ingredients [8]. The principle of fermentation involves activating the growth of necessary microorganisms, leading to the formation of new products. Anaerobic fermentation can create an acidic environment that supports the growth of lactic acid bacteria. The acidic conditions during the fermentation process can be modified using various easily fermentable carbohydrate sources as additives [9]

One of the additives used is palmyra sap or "nira lontar." Nira lontar is a liquid tapped from the male flowers of the lontar palm tree, containing sugar content ranging from 10-15% [10]. Nira lontar contains fructose at 4.0%, glucose at 3.5%, and sucrose at 3.6%[11]. The high sugar content in this substance serves as a food source to stimulate the growth of lactic acid bacteria capable of breaking down the components of crude fiber during the fermentation process. The fermentation duration is one of the most crucial factors influencing the improvement in the nutritional quality of feed ingredients. Research by Koni et al. [12] showed that fermenting banana peels with palm sap for 6 days reduced the dry matter content from 86.69% to 82.38%, crude fiber from 13.65% to 11.55%, and ash content in banana peels from 12.69% to 13.29%. Furthermore, it increased the crude protein content from 5.73% to 6.72%.

This research was aimed to assess the content of dry matter, crude fiber, and ash in rice bran with the addition of palmyra sap (nira lontar) during different fermentation durations.

2 Materials and methods

2.1 Materials

The experiment was conducted at Politeknik Pertanian Negeri Kupang, East Nusa Tenggara Province, Indonesia. The materials used in this study consist of rice bran, palmyra sap (nira lontar), and distilled water. Palmyra sap (nira lontar) was obtained lontar palm tree tappers in Lasiana, Kupang City. Aquades was used for pH measurement and rinsing the electrodes in the pH meter 5.

2.2 Research procedure

This research is divided into three parts: the preparation phase, the fermentation phase, and the analysis phase of dry matter, crude fiber, and ash content.

1. **The Stage of preparation**. The preparation phase was carried out for rice bran, palmyra sap (nira lontar), and the containers used as silos. The preparation procedure is as follows: Rice bran was sifted to separate the bran from the husk. The rice bran was analyzed for its dry matter content in the Laboratory, which was found to be 91.59%, with a moisture content of 8.41%. The rice bran was then placed into envelopes and sterilized using an autoclave at a pressure of 121 atm for 15 minutes. The palmyra sap (nira lontar) purchased directly from the tappers of lontar palm trees in Lasiana was immediately pH-tested using a pH meter after being tapped from the lontar palm tree for 2 hours. The containers were subsequently sterilized through a steaming process for 15

minutes.

2. **Fermentation Phase**. The fermentation of rice bran refers to the procedure as described by Koni et al. [13], where all the materials that were fermented were prepared to have a moisture content of 35%, while the use of palmyra sap (nira lontar) was at 20% of the moisture content of the rice bran. The fermentation was carried out following these steps: Rice bran was weighed and and then placed in a tray. Palmyra sap (nira lontar) was measured at 20% of the dry matter and placed in a provided container. Subsequently, the water content was measured, and it was mixed into the palm sap, which was then blended with the rice bran until homogeneous. Two grams of the resulting mixture were taken for pH measurement. The pH measurements were based on Bernardes et al. [14] and involved taking 2 grams of fermented rice bran, mixing it with 10 ml of distilled water, stirring, and then measuring the pH. Subsequently, the mixture from step 2 was placed in a container (jar) while being compacted. The surface of the mixture was covered with a clear plastic sheet and the jar lid was closed. The outer part of the jar lid was sealed with tape and stored in a place that was not exposed to direct sunlight. Fermentation was conducted in accordance with the specified treatment duration, which included 0, 3, 6, and 9 days. For the DN0 (without fermentation) following the mixing of fermentation materials, pH measurement was conducted, followed by weighing, and then the mixture was placed in a 60°C oven for 48 hours. Rice bran was harvested after fermentation according to the treatment. After being dried for 48 hours, the silage was then removed from the oven and weighed to determine the dry weight. The fermented rice bran was then finely ground using a blender and weighed again. Samples of the fermented rice bran were collected, comprising 10% of the silage weight, then packaged and labeled for proximate analysis in the laboratory following [15] guidelines
3. **The Stage of Analysis** for Dry Matter, Crude Fiber, and Ash Content. The procedure for analyzing dry matter, crude fiber, and ash refers to the AOAC [15] guidelines.

2.3 Experimental design

This research is an experimental study using rice bran fermented with 20% palmyra sap at different durations. The study was designed using a Completely Randomized Design (CRD) with 4 treatments and 5 replications. The treatments conducted were as follows: DN0: 0 days of fermentation, DN3: 3 days of fermentation, DN6: 6 days of fermentation, DN9: 9 days of fermentation.

2.4 Statistical analysis

The nutrient content data obtained in this study were analyzed using Analysis of Variance (ANOVA), and if any significant treatment effects were observed, Duncan's Multiple Range Test [16].

3 Results and discussion

3.1 The effect of fermentation duration on the ph of rice bran added with palmyra sap

Based on the observations of the physical quality of the fermented rice bran, there was no presence of mold. This is likely due to perfect compaction, reducing the entry of oxygen. According to Despal et al.[17], lactic acid production lowers the pH, thereby inhibiting the growth of mold.

Table 1. The initial pH and final pH of fermented rice bran with the addition of palmyra sap at different fermentation durations.

Treatment	Initial pH	pH after fermentation
DN0	6.25	6.25
DN3	6.02	4.85
DN6	6.27	4.27
DN9	6.30	4.09

Note: DN0: without fermentation, DN3: 3 days fermentation, DN6: 6 days fermentation, and DN9: 9 days fermentation

The measurement results show a decrease in pH after fermentation. The pH range obtained in this study falls within the "very good" category. This is likely because silage can reach an anaerobic condition, which suppresses fungal growth and leads to a decrease in pH. The pH values decreased in the 3, 6, and 9-day treatments. This occurred due to the presence of organic acids such as lactic acid produced by lactic acid bacteria (LAB) during the fermentation process of the rice bran. Rice bran with the addition of palmyra sap produced a higher lactic acid content than rice bran without sap addition, which was fermented for 6 days [13]. The production of lactic acid contributes to the pH reduction. The lactic acid content generated during the fermentation process acts as a preservative, thus reducing the growth of spoilage microorganisms. According to Despal et al. [17], lactic acid bacteria are capable of lowering silage pH.

3.2 The effect of fermentation duration on the dry matter content of rice bran added with palmyra sap

The data on the different effects of fermentation duration on the dry matter content of rice bran added with palmyra sap can be seen in Table 2. Based on the analysis of variance, the fermentation duration did not have a significant effect ($P>0.05$) on the dry matter content of rice bran added with palmyra sap. This is likely because the fermented material was prepared with the same dry matter content, which was 65%. Consequently, the dry matter content remained unchanged after fermentation. This is because there was no damage or degradation to the dry matter, resulting in no decrease in dry matter content. The treatment of adding rice bran to cabbage vegetable waste silage had a significant effect ($P<0.05$) on increasing the dry matter content while[1]

Table 2. Effects of fermentation duration on the dry matter, crude fiber, and ash content of rice bran added with palmyra sap.

Treatment	Parameter		
	Dry Matter (%)	Crude Fiber (%)	Ash (%)
DN0	39.29±1.69	34.02 ± 3.47c	16.13±0.10b
DN3	47.26±1.83	26.82 ± 1.79a	15.20±0.69a
DN6	42.57±1.84	27.88 ± 2.91b	16.90±0.04c
DN9	44.31±4.10	28.85 ± 2.31b	16.97±0.21c
P value	0.309	0.003	0.003

Note: DN0: without fermentation, DN3: 3 days fermentation, DN6: 6 days fermentation, and DN9: 9 days fermentation. a b c, means ± standard deviation followed by different letters in the same column indicate significant differences ($p<0.05$).

3.3 The influence of fermentation duration on the crude fiber content of rice bran added with palmyra sap

The data regarding the influence of fermentation duration on the crude fiber content of rice bran added with palmyra sap is presented in Table 2. The analysis of variance results showed a highly significant ($P<0.01$) effect of fermentation duration on the crude fiber content of rice bran. Fermentation durations of 3, 6, and 9 days were observed to reduce the crude fiber content. This is likely because microorganisms such as lactic acid bacteria in the fermentation process of rice bran added with palmyra sap produce cellulase enzymes, which can reduce crude fiber. Chayaningsih [18] stated that several lactic acid bacteria were isolated from palmyra sap in Kupang, East Nusa Tenggara, including Leuconostoc mesenteroides, Leuconostoc pseudomesenteroides, Lactobacillus plantarum, and Lactobacillus fermentum. [19] mentioned that lactic acid bacteria are capable of producing enzymes that degrade fiber, such as cellulase and xylanase. The results of Safitri et al. [20] indicated that the fermentation of rice bran using EM4 can reduce the crude fiber content of rice bran compared to non-fermented rice bran.

Based on the results of the Duncan analysis, the DN0 treatment had a higher ($P<0.05$) crude fiber content than the DN3, DN6, and DN9 treatments. The DN3 treatment significantly differed ($P<0.05$) from the DN6 and DN9 treatments, but There was no significant difference ($P>0.05$) between the DN6 and DN9 treatments. In the 3-day fermentation duration, the crude fiber content was the lowest compared to the other treatments. This is likely because a 3-day fermentation period is the optimum time for microorganisms to produce cellulase enzymes, allowing the crude fiber to be broken down into easily digestible compounds. At 6 and 9 days, it may be due to a reduction in nutrients in the substrate, leading to a decrease in microorganism activity and limited fiber degradation. According to the findings of Safitri et al.[20], a fermentation duration of 36 hours can reduce the crude fiber content in rice bran fermented with EM4

The reduction in crude fiber is suspected to be due to the microorganisms present in palm sap being able to digest the crude fiber in rice bran. This study's results are consistent with the findings of [20], where the average crude fiber content of banana peels fermented with palm sap decreased from 18.7% to 11.55%.

3.4 The effect of fermentation duration on the ash content of rice bran added with palmyra sap

The data regarding the effect of fermentation duration on the ash content of rice bran added with palmyra sap is presented in Table 2. The analysis of variance results indicate that fermentation duration significantly affects ($P<0.05$) the ash content of rice bran added with palmyra sap. It can be observed that the ash content decreases in the 3-day treatment. This is likely because lactic acid bacteria utilize mineral contents such as calcium and phosphorus from palmyra sap and rice bran during the fermentation process, resulting in a lower ash content. According to [20], when inorganic matter (ash) decreases, it is likely that the content of organic matter containing nutrients increases. In the 6 and 9-day treatments, there is an increase in ash content due to the degradation of organic matter during the fermentation process, as a result of microbial activity.

The results of the Duncan test indicate that the 3-day fermentation duration has a significantly lower ($P<0.05$) ash content compared to the other treatments. The 3-day treatment has a significantly lower ($P<0.05$) ash content than the 6 and 9-day treatments. There is no significant difference ($P>0.05$) between the 6 and 9-day treatments. The low ash content in the 3-day fermentation duration is likely due to microorganisms using nutrients, including ash, for their growth, leading to a decrease in ash content in the substrate. Meanwhile, in the 6 and 9-day treatments, there is an increase in ash content because a longer fermentation duration results in a reduction in organic matter caused by microbes. Beneficial microbes are sourced from additives in the combustion process, and their inorganic elements are retained [21].

4 Conclusions

Based on the results and discussion, the fermentation duration did not affect the dry matter content but had a significant impact on the crude fiber and ash content in rice bran with palmyra sap addition. Considering the lowest crude fiber content observed, a 3-day fermentation period appears to be the most effective treatment for rice bran supplemented with palmyra sap.

This research was financially supported by DIPA PNBP under the scheme of Competitive Applied Research organized by the Research Center of Politeknik Pertanian Negeri Kupang. Therefore, the research team would like to thank for the exceptional support during the planning and development of this research work.

References

1. S. Superianto, A. E. Harahap, and A. Ali, J. Sain Peternak. Indones. **13**, 172 (2018)
2. H. Hartadi, S. d Reksohadiprodjo, and A. D. Tilman, *Tabel Komposisi Pakan Untuk Indonesia*, III (Universitas Gadjah Mada, Yogyakarta, 1993)
3. A. D. Tillman, H. Hartadi, S. Reksohadiprojo, S. Prawirokusumo, and S. Lebdosoekojo, *Ilmu Makanan Ternak Dasar* . (Gadjah Mada University Press, Yogyakarta, 1991)
4. M. Lamid, N. N. T. Puspaningsih, and O. Asmarani, J Appl Env. Biol Sci **4**, 377 (2014)
5. R. I. Pujaningsih, J. Indones. Trop. Anim. Agric. **29**, 100 (2004)
6. C. H. Kim, S. B. Park, J. J. Jeon, H. Kim, S. H. Kim, E. C. Hong, and H. K. Kang, Korean J. J. Poult. Sci. **44**, 235 (2017)
7. A. A. U. Ermalia, O. Sjofjan, and I. H. Djunaidi, Bul. Anim. Sci. **40**, 113 (2016)

8. S. Munira, L. O. Nafiu, and A. M. Tasse, J. Ilmu Dan Teknol. Peternak. Trop. **3**, 21 (2016)
9. R. Utomo, S. P. S. Budhi, and I. F. Astuti, Bul. Peternak. **37**, 173 (2013)
10. E. Irmayuni, Nurmila, and A. Sukainah, J. Pendidik. Teknol. Pertan. **4**, 170 (2018)
11. E. Naiola, Biodiversitas **9**, 165 (2008)
12. T. N. I. Koni, T. A. Y. Foenay, and H. Y. Chrysostomus, J. Ilmu Dan Teknol. Peternak. **9**, 381 (2022)
13. T. N. I. Koni, T. A. Y. Foenay, and A. Jehemat, Livest. Res. Rural Dev. **34**, 88 (2022)
14. T. F. Bernardes, J. R. S. Gervásio, G. De Morais, and D. R. Casagrande, J. Dairy Sci. **102**, 9039 (2019)
15. AOAC (Association of Official Analytical Chemists), *Official Methods of Analysis of the Association of Official Analytical Chemists*, 18th ed. (Association of Official Analytical Chemist, Washington, DC., 2005)
16. V. Gasperz, *Teknik Analisa Dalam Penelitian Percobaan*, Edisi III (Tarsito, Bandung, 2006)
17. Despal, I. G. Permana, S. N. Safarina, and A. J. Tatra, Media Peternak. **34**, 69 (2011)
18. H. . Chayaningsih, Identifikasi Bakteri Asam Laktat Dari Nira Lontar Serta Aplikasinya Dalam Mereduksi Salmonella Typhimurium Dan Aspergillus Flavus Pada Biji Kakao, Sekolah Pasca Sarjana IPB, 2006
19. Y. Yanuartono, S. Indarjulianto, H. Purnamaningsih, A. Nururrozi, and S. Raharjo, J. Sain Peternak. Indones. **14**, 49 (2019)
20. Y. Safitri, Zulfan, and H. Latif, J. Ilm. Mhs. Pertan. Unsyoah **1**, 781 (2016)
21. H. Y. Chrysostomus, T. N. I. Koni, and T. A. Y. Foenay, J. Trop. Anim. Vet. Sci. **10**, 91 (2020)

The impact of *Sargassum crassifolium* brown seaweed flour with lower salt content in diets on internal organ of broilers

Maria Endo Mahata[1*], *Rohfranco Tua* Cibro[2], *Sepri* Rezki[1], *Zurmiati* Zurmiati[1], and *Yose* Rizal[1]

[1] Lecturer at Departement of Nutrition and Feed Industry Technology, Faculty of Animal Science, Universita Andalas, Padang 25163, Indonesia.
[2] Undergraduate Student at Departement of Nutrition and Feed Industry Technology, Faculty of Animal Science, Universita Andalas, Padang 25163, Indonesia.

Abstract. The research aims to determine the effect of *Sargassum crassifolium* brown seaweed flour in diet,on the internal organs of broiler. Salt content of brown seaweed was decreased by immersion in flowing water for 15 hours. This research used 100 DOC broilers of the Lohman strain. In pre- starter (0-1 week of age) period were given commercial diet for adaptation, and at the age of 2 to 6 weeks are given diet containing *Sargassum crassifolium* brown seaweed flour. The cages used were 20 box cages, and each cage was placed by 5 broilers. The experiment using a Completely Randomized Design (CRD) with 4 different treatments of *Sargassum crassifolium* brown seaweed flour (0%, 6%, 12% and 18%) in the diet, and each treatment was repeated 5 times. Measurements were percentage of liver, of pancreas, gizzard, small intestine, and small intestine length (cm). The results showed that the inclusion of the brown seaweed *Sargassum crassifolium* flour in the diet had no significant effect ($P>0.05$) on percentage of liver, small intestine, and the length of the small intestine, but it had a very significant effect ($P<0.01$) on percentage of pancreatic and gizzard. Based on the research results, it can be concluded that the brown seaweed *Sargassum crassifolium* can be given up to a level of 18% in the diet without having a negative effect on the percentage of liver,small intestine and length of the small intestine, and increased the percentage of gizzard and pancreas of broilers.

1 Introduction

Indonesia is known as one of the largest seaweed exporting countries in the world. [1] states that Indonesia's seaweed production will reach 9.12 million tons in 2021. According to [2] that seaweed is divided into three large classes, namely *Rhodophyceae* (red algae), *Phaeophyceae* (brown algae), and *Chlorophyceae* (green algae). Sargassum is a seaweed belonging to the *Phaeophyceae* class that grows in Indonesia and is distributed in many Indonesian marine environments. Sargassum is widely exported because of its

* Corresponding author: maria@ansci.unand.ac.id

polysaccharide content, namely alginate. Alginate is the main component of brown seaweed sap (*Phaeophyceae*) and is an important compound in cell walls, which consists of two types of monomers, namely β-D-Mannopyranosyl Uronate and α-L-Gulopyranosyl Uronic Acid. Alginate is the most used marine polysaccharide, and it is explored for old and new functions as food additive or a biomedical essential polymer in wound dressings [3].

The brown seaweed *Sargassum crassifolium* has not been widely used as a poultry feed ingredient, even though the nutrients and bioactive substances it contains include alginate [4], fucoxanthin 1.64 mg/g [5], and fucoidan 0.87% of the dry weight *of Sargassum crassifolium* [6] is really needed by poultry to meet their nutritional needs and support their health. The bioactive compounds alginate, fucoidan and fucoxanthin found in brown seaweed have been reported to suppress fat absorption and inhibit cholesterol synthesis, and are useful for reducing fat in the body of livestock.

According to [7] brown seaweed has active components in the form of phenolic compounds, steroids, alkaloids and triterpenoids which function as anti-bacterial, anti-virus, anti-cancer, and anti-fungal. [8] also stated that brown seaweed also contains bioactive compounds such as phlorotannin ranging from 0.74-5.06% which functions as an antimicrobial.

Previous research shows that including 10% *Sargassum crassifolium* seaweed in broiler diet has a negative influence on the performance of broiler chickens because *Sargassum crassifolium* seaweed contains quite high salt, that is 11.21% and causes diarrhea in broiler chickens, high mortality rates, and liver problems [9], and according to [10] the tolerance salt content in poultry diet is 0.3% and 0.4-1.0% [11].

Furthermore, [4] reported that the salt content of the brown seaweed *Sargassum crassifolium* can be reduced by soaking it in running water for 15 hours and the salt content decreased from 11.21% to 2.9% or a decrease of 74.13%, and the content of nutrient and metabolic energy that are: dry matter 85.91%, crude protein 8.89%, crude fat 1.55%, ash 17.20%, crude fiber 12.73%, BETN 59.63% , Ca 3.80%, P 1.29%, 2,243 Kcal/kg, and contains 0.39% lysine, 0.14% methionine + cystine, 0.06% tryptophan, and 0.45% threonine. Seaweed is reported to have biofilter capabilities that can absorb compounds in waters and also heavy metals, and can be used as a pollution bioindicator as well as biocontrol in marine waters. [12] reported that seaweed has the ability to absorb heavy metals at quite high concentrations. *Sargassum* spp. seaweed is more tolerant and resistant to heavy metals compared to green and red seaweed [13]. The metal absorption mechanism by *Sargassum* spp is reported to be through functional groups in alginate and fucoidan compounds which are cell wall component of seaweed [14].

It is feared that the biofilter's ability to absorb heavy metal compounds contained in the brown seaweed *Sargassum crassifolium* could interfere with the development of internal organs (liver, gizzard, pancreas and small intestine) if used as broiler feed. Apart from that, the content of bioactive compounds alginate and fucoidan which are polysaccharides are also expected to affect the internal organs of broilers.

There have not been many reports about the impact of *Sargassum crassifolium* brown seaweed which has had its salt content reduced as a broiler feed. Therefore, the aim of this research was to see the impact of giving *Sargassum crassifolium* brown seaweed flour with low salt content on the internal organs of broilers.

2 Materials and Methods

This study used 100 DOC broiler strains Lohman MB-202 Platinum from a mixture of males and females with an average initial weight of 45.2 ± 2.82 g/head. The cages used in this research were box types, as many as 20 units measuring 60 x 50 x 50 cm. Each cage unit contains five broilers equipped with eating and drinking utensils and a 60-watt incandescent lamp, which functions as heating and lighting, in each cage.

The diet consists of ground corn, rice bran, fish meal, soybean meal, meat bone meal, corn gluten meal, methionine, top mix, coconut oil, and *Sargassum crassifolium* brown seaweed flour. The composition of feed, nutrient , and metabolic energy of experimental diet can be seen in Table 1.

This research used a completely randomized design (CRD) consisting of 4 levels of the brown seaweed *Sargassum crassifolium* in broiler diets, that is, 0 (A), 6 (B), 12 (C), and 18 % (D), each repeated five times. Parameters measured were the percentage of liver, pancreas, gizzard, small intestine weight, and small intestine length (cm).

2.1 The percentage of liver

The percentage of liver weight was obtained by weighing the liver (g) of 6-week-old broilers that had been slaughtered for each treatment compared with the live weight (g) and multiplying by 100%.

2.2 The percentage of gizzard

The percentage of gizzard weight was obtained by weighing gizzards (g) on 6-week-old broilers that had been slaughtered for each treatment and compared with their live weight (g) and multiplying by 100%.

2.3 The percentage of pancreas

The percentage of pancreas weight was obtained by weighing the pancreas (g) in 6-week-old broilers slaughtered for each treatment, comparing it with the live weight (g), and multiplying it by 100%.

2.4 The percentage of small intestine.

The weight percentage was obtained by weighing the intestines of 6 week old broiler chickens that had been slaughtered for each treatment, and compared with their live weight and multiplying by 100%.

2.5 The Length of the small intestine

It is measured by measuring the length of the small intestine in 6-week-old broilers slaughtered for each treatment.

2.6 Data analysis

In this study, the data are presented as mean. Significance between the mean values was determined statistically by ANOVA in 4 treatments with five replications, and results are reported as mean.

Tabel 1. Composition of feed, nutrient content (%) and metabolic energy (Kcal/Kg) of experimental diets

Feed ingredient,nutrient (%), and Metabolic energy (Kkal/Kg)	Experimental diet (%)							
	Starter phase (2 weeks)				Grower phase (3-6 weeks)			
	A	**B**	**C**	**D**	**A**	**B**	**C**	**D**
Ground corn	49.00	45.00	41.00	37.00	56.50	52.50	48.50	44.5
Rice bran	13.75	11.25	8.75	6.25	15.00	12.50	10.00	7.50
Fish meal	15.00	15.00	15.00	15.00	8.25	8.25	8.25	8.25
Soybean meal	14.25	14.75	15.25	15.75	9.00	9.50	10.00	10.50
Meat bone meal	3.00	3.00	3.00	3.00	4.00	4.00	4.00	4.00
Corn gluten meal	4.50	4.50	4.50	4.50	6.75	6.75	6.75	6.75
Top mix	0.05	0.05	0.05	0.05	0.05	0.05	0.05	0.05
Methionine	0.20	0.20	0.20	0.20	0.20	0.20	0.20	0.20
Coconut oil	0.25	0.25	0.25	0.25	0.25	0.25	0.25	0.25
Sargassum crassifolium	0.00	6.00	12.00	18.00	0.00	6.00	12.00	18.00
Total	**100.00**	**100.00**	**100.00**	**100.00**	**100.00**	**100.00**	**100.00**	**100.00**
Feed ingredient,nutrient (%), and Metabolic energy (Kkal/Kg)	Experimental diet (%)							
	Starter phase (2 weeks)				Grower phase (3-6 weeks)			
	A	**B**	**C**	**D**	**A**	**B**	**C**	**D**
Crude protein	22.96	23.05	23.13	23.22	20.05	20.14	20.22	20.31
Crude lipid	3.54	3.35	3.16	2.97	3.59	3.40	3.21	3.01
Crude fiber	3.63	4.08	4.53	4.98	3.56	4.01	4.46	4.91
Calcium	1.03	1.07	1.46	1.67	0.91	1.13	1.34	1.55
Available Phospor	0.79	0.76	0.47	0.72	0.72	0.70	0.68	0.66
Lysin	1.64	1.65	1.66	1.67	1.15	1.16	1.17	1.18
Methionine	0.79	0.79	0.79	0.79	0.70	0.70	0.70	0.69
Alginate	0.00	2.61	5.22	7.82	0.00	2,61	5.22	7.82
Fucoidan	0.00	0.05	0.10	0.16	0.00	0.05	0.10	0.16
Fucoxanthin	0.00	0.01	0.02	0.03	0.00	0.01	0.02	0.03
NaCl	-	0.17	0.35	0.52	-	0.17	0.35	0.52
Metabolic energy (Kkal/kg)	2.820	2.789	2.758	2,727	2.907	2.876	2.846	2.815

3 Results and discussion

3.1 The percentage of liver

The statistical analysis showed that the inclusion of *Sargassum crassifolium* brown seaweed flour in a broiler diet had no significant effect (P>0.05) on the liver weight percentage of 6-week-old broilers. In Table 2, the percentage of liver weight with different levels of *Sargassum crassifolium* brown seaweed flour in the diet ranging from 1.58 to 1.86% of live weight. [15] stated that broiler liver weight ranged from 2.67% and 2.76% of live weight; the percentage of broiler's liver weight given *Sargassum crassifolium* in their diet in this study was still within the normal range.

Table 2. Mean of percentage of liver

Sargassum crassifolium percentage in broiler diet	Mean of percentage of liver
0	1.58
6	1.86
12	1.78
18	1.70
Standard error	0.08

There was no significant effect of *Sargassum crassifolium* on the percentage of broiler liver weight, showing that the diet containing *Sargassum crassifolium* does not contain dangerous toxic compounds and can still be accepted by the broiler's body, so that the liver's performance is not disturbed by *Sargassum crassifolium* in diet. The absence of changes in consistency and the reddish-brown color of the broiler's liver in this experiment also supports this. According to [16] Normal chicken liver color is brownish-red, if the liver is poisoned, the color of the liver will turn yellow. The salt content factor of *Sargassum crassifolium* brown seaweed, reduced from 11.21 to 2.9%, also supports broilers' liver health. This is in line with the statement of [17] that the sodium content in poultry feed above 0.5% is considered toxic.

The brown seaweed *Sargassum crassifoilum* also contains the bioactive compound fucoidan. According to [18] the fucoidan compound found in the brown seaweed *Sargassum crassifoilum* can be an antioxidant. Fucoidan can protect liver cells from free radicals. Thus, the fucoidan content found in *Sargassum Crassifolium* brown seaweed can protect broiler livers from the influence of other active substances, which can cause liver disorders in broilers. In this study, the fucoidan content in each level of *Sargassum crassifolium* (0, 6, 12 and 18%) in broiler diets was 0.05%, 0.10%, and 0.16%, respectively.

3.2 The percentage of gizzard

The statistical analysis showed that the inclusion of *Sargassum crassifolium* in the broiler diet had a significant effect ($P<0.01$) on the gizzard weight percentage of broilers aged six weeks. In Table 3 the percentage of gizzard increased with increasing of *Sargassum crassifolium* brown seaweed at level 12 and 18 % by 2.20 and 2.21 %.

Table 3. Mean of percentage of gizzard

Sargassum crassifolium percentage in broiler diet	Mean of percentage of gizzard
0	1.72^{b}
6	1.74^{b}
12	2.20^{a}
18	2.21^{a}
Standard error	0.08

An increasing percentage of gizzard indicates disturbances in the food digestion process. The texture of *Sargassum crassifolium* seaweed is coarser compared to the consistency of other ingredients such as ground corn, rice bran, fish meal, soybean meal, Corn Gluten Meal, and Meat Bone Meal, which are used as feed ingredients for the experimental diet. Increasing the level of use *Sargassum crassifolium* in broiler diet changes the texture of the diet to become coarser. The consistency of the diet will affect the gizzard as a digestive organ whose role is to pulverize (grind) the diet before it is passed to the small intestine. The rough texture of the diet causes the gizzard to work harder to crush the diet, thus affecting its weight. According to [19], gizzard weight will be positively correlated with feed particle size, further explaining that coarse corn particles can cause an increase in gizzard weight by 26-41%.

This study found that the weight percentage of broiler gizzard organs ranged from 1.72 – 2.21%. The gizzard weight percentage found was still higher than the broiler gizzard weight reported by [20], ranging from 1.35 to 1.41%, but is still within the gizzard weight range reported by [21], that is 1.6 – 2.3%.

3.3 The percentage of pancreas

The results of statistical analysis showed that the inclusion of *Sargassum crassifolium* in the broiler diet had a very significant effect (P<0.01) on the percentage of pancreatic weight of broilers aged six weeks.

Table 4. Mean of percentage of pancreas

Sargassum crassifolium percentage in broiler diet	Mean of percentage of pancreas
0	0,19[b]
6	0,28[a]
12	0,25[a]
18	0,26[a]
Standard error	0,02

Giving brown seaweed *Sargassum crassifolium* in diet at levels of 6,12, and 18% caused an increase in pancreatic weight in broilers. [22] explained that increasing the weight of the pancreas is a form of adaptation to meet the increased need for digestive enzymes. The brown seaweed *Sargassum crassifolium* contains the bioactive compound alginate, which cannot be digested by poultry because it does not produce the alginate lyase enzyme [23]. According to [24], brown alga (brown seaweed) cell walls are majorly comprised of alginate and fucose-containing sulfated polysaccharides.

In this study, the inclusion of *Sargassum crassifolium* 6, 12, and 18% in diet would increase the alginate content in broiler rations by 2.61, 5.22, and 7.82%, respectively. The increased alginate content will affect the digestion and absorption process in the broiler intestine because the alginate lyase enzyme is not produced to digest alginate. This condition also affects the weight of the pancreas organ due to disruption of the digestive and absorption processes in the small intestine. According to [25], the experiments on mice that consumed alginate in the form of Na alginate disrupted changes in the function of pancreatic fluid secretion, they suppressed the process of digestion and absorption in the intestine. It was further explained that the rat's body will compensate by increasing the number of cells (hyperplasia) and the size of pancreatic cells (hypertrophy) to increase the secretion of digestive juices.

This study found that the weight percentage of broiler pancreatic organs ranged from 0.19 to 0.28%. The percentage of pancreatic weight found is still in the range of broiler pancreatic weight reported by [21] is 0.25 – 0.40%.

3.4 The percentage of small intestine

The results of the statistical analysis showed that the inclusion of *Sargassum crassifolium* in broiler diet had no significant effect (P>0.05) on the percentage of small intestine weight of broilers aged at six weeks.

Table 5. Mean of percentage of small intestine

Sargassum crassifolium percentage in broiler diet	Mean of percentage of small intestine
0	2.75
6	3.61
12	2.99
18	3.23
Standard error	0.24

Table 5 shows that the average weight of broiler small intestines with different levels of *Sargassum crassifolium* in the diet ranged from 2.75 to 3.61% of live weight. Besides alginate, brown seaweed contains other bioactive compounds, such as phlorotannin. [26] reported that the phlorotannin content in brown seaweed ranged from 0.74 - 5.06%. According to [27] phlorotannin is a polyphenolic compound with antimicrobial activity. The activity of phlorotannin as an antimicrobial can maintain good intestinal health and development; thus, increasing *Sargassum crassifolium* levels in the broiler diet can maintain intestinal weight equivalent to the control diet (0% *Sargassum crassifolium*).

Brown seaweed is reported also contain bioactive flavonoid as much as 0.17% [28]. The flavonoid compounds can protect the mucosal walls of the small intestine, according to [29] the flavonoid compounds in herbs play an important role in protecting the mucosal walls of the small intestine. In this study, the inclusion of *Sargassum crassifolium* up to 18% in the broiler diet could maintain the weight of the small intestine equivalent to the weight of the small intestine of broilers that were not given *Sargassum crassifolium* in the diet (0%). This shows that the bioactive flavonoid compounds in *Sargassum crassifolium* can maintain broiler intestinal health, so increasing the level of *Sargassum crassifolium* in broiler diet can maintain the health and development of the broiler's small intestine, and protecting the mucosal walls of the small intestine from non-beneficial microbesimp and improve the absorption of nutrients in animal feed. Optimal nutrient absorption will increase nutrient digestibility, including protein digestibility. [30] stated increased protein digestibility can affect the weight of the small intestine. [31] mentioned, protein plays a role in cell formation, replacing dead cells and forming body tissue. The body tissue cells that are formed include small intestinal epithelial cells. The more small intestinal epithelial cells there are, the wider the surface of the small intestine, and the greater the number of villi, so the weight of the small intestine will be heavier [32].

In this study, increasing the *Sargassum crassifolium* brown seaweed in broiler diet did not show a difference in the percentage of small intestine weight of broilers who were not given *Sargassum crassifolium* brown seaweed in their diet. It is suspected that the dose of the bioactive compounds phlorotannin and flavonoids contained in the brown seaweed *Sargassum crassifolium* is still low, so it does not show a difference in the percentage of small intestine weight. Beside that, there was no effect on the percentage of small intestine weight due to the administration of *Sargassum crassifolium* brown seaweed after optimal growth of the small intestine (6-week-old broilers). According to [33], after hatching, the weight of the small intestine increases faster than that of other organs. Based on research conducted), intestinal weight increased fourfold after hatching until four days of age. Growth began to decline at 7 to 21 days of age, so that the weight of the broiler small intestine observed in this study could no longer increase in intestinal weight because the intestine no longer experienced related growth, and the broiler's age has reached six weeks.

3.5 The length of small intestine

The results of statistical analysis showed that the provision of *Sargassum crassifolium* in broiler diet had no significant effect (P>0.05) on the length of the small intestine of broilers aged six weeks. Table 5 shows that the average length of the small intestine of broilers with different levels of *Sargassum crassifolium* inclusion in the diet ranged from 180.52 to 200.58 cm. The average length of the small intestine is longer when compared to the results of research by **[34]** regarding the addition of white ginger flour to broiler chickens, ranging from 184 cm to 199 cm.

Table 6. Mean of the length of small intestine

Sargassum crassifolium percentage in broiler diet	Mean of the length of small intestine (cm)
0	180.52
6	200.58
12	189.32
18	194.70
Standard error	6.44

In this study, it was predicted that the length of the small intestine is influenced by the content of bioactive compounds flavonoids and phlorotannins. Giving brown seaweed *Sargassum crassifolium* containing bioactive compounds flavonoids and phlorotannins can protect the mucosal walls of the small intestine and can increase the number of small intestinal villi. [35] explained that the high growth of small intestinal villi is closely related to the potential of the small intestine to absorb nutrients. The higher the small intestinal villi, the greater the effectiveness of nutrient absorption through the small intestinal epithelium. A longer small intestine is an indication of a larger area for digestion and nutrient absorption [36]. In this study, the bioactive compounds phlorotannin and flavonoids contained in the brown seaweed *Sargassum crassifolium* did not affect the length of the small intestine of broiler chickens. This is also because the time for giving *Sargassum crassifolium* brown seaweed has passed the optimal growth time for the small intestine (broilers are six weeks old). According to [37] optimal growth occurs between 4 and 8 days of age, with an increase of almost six times, and after that, there will be a decline in growth.

Apart from that, the crude fiber content in the brown seaweed *Sargassum crassifolium* is still safe, below 7% (Table 1). High crude fiber will stimulate the growth of the small intestine because there is an increase in its performance in digesting coarse fiber. In most poultry research, insoluble dietary fiber has been used as nutrient diluent due to the lack of enzymes to digest β 1–4, β 1–3, and β 1–6 linkages found in such non-starch polysaccharides [38] which have been regarded as impairing performance when used in high amounts due to a slowing down and dilution of nutrient intake [39].

4 Conclusion

The brown seaweed *Sargassum crassifolium* can be given up to a level of 18% in the diet without having a negative effect on the percentage of liver, small intestine and length of the small intestine, and increased the percentage of gizzard and pancreas of broilers.

This reseach was funded by the Profesor Research Grant, Universitas Andalas, Padang.(PTU-KRPIGB-UNAND, batch I, 2023. Contract number: T/9/UN16-19/Pangan-PTU-KRPIGB-Unand/2023, 04 April, 2023.

References

1. Indonesia. D. https://dataindonesia.id/agribisnis-kehutanan/detail/sulawesi-selatan-produksi-rumput-laut-terbesar-di-indonesia-2021. Data accessed November 24, 2023, at 11.30. (2023).
2. Inem.O. Agrikan: Jurnal Agribisnis Perikanan Volumen 6, Pages 47-54 (2013)
3. Leonel Pereira and João Cotas Book Chapter, https://www.intechopen.com/chapters/68305 (2019).
4. Mahata. M.E. Laporan penelitian pendanaan 70 % skema terapan unggulan klaster riset, publikasi guru besar (PTUKRPIGB Unand). Kontrak No: T/9/UN 6.19/Pangan-PTU-KRPIGB-Unand/2023 (2023)
5. Susanto. E, Fahmi. A.S, Abe.M, Hosokawa.M, & Miyashita.K Lipids, Aquatic Procedia, 7, 66-75 (2016)
6. Sinurat. E, R. Peranginangin, E. Saepudin. Jurnal Pascapanen dan Bioteknologi Kelautan dan Perikanan, Volume 6, Nomor 2, Halaman : 131-138 (2011).
7. Nagappan. H, Pee PP. Kee SHY, O.W. JT, Yan .S.W, Chew L.Y, Kong K.W. Journal Science Technology. Volume 26, Number 2, Page:11-219 (2017)
8. Samee H. Z.X Li, H .Lin, J .Khalid,Y.C Guo. Journal of Zhejiang University Science B. Volume 10, Number 2, Page:147-153. (2009)
9. Mahata. E. M, Y. L. Dewi, M.O.Sativa, S. Reski, Hendro, Zulhaqqi, A. Zahara. http://repo.unand.ac.id/44708/ (2015)
10. A.A.El-Deek,Mervat.S.El-Deen, Safaa. M.Hamdy,M.A.Asar,H.M.Yakout,Y.Attia. Egypt. Poult. Sci. Volume: 29, Issue: IV, Page: 907-921 2009)
11. Notes. Can. J. Anim. Sci. 65:773-775 (1985)
12. Yatris R.T , E.Y. Herawati1 , Y. Kilawati. .Exp. Life Sci. Vol. 9 No. 2 (2019)
13. Diaz .M.A.H, F.D.L.Chavira, M. L. Lares, A.C.Barragán, A. S.Valencia, Applied Geochemistry,Volume 22, Issue 7, Pages 1380-1392 (2007)
14. Bijang. C.M, J.Latupeirissa, M. Ratuhanrasa. Indo. J. Chem. Res, Volume.6, Issue. 1, Page:26-37 (2018).
15. Putnam P. A. Handbook of Animal Science ,San Diego: Academic Press (1991)
16. D. Damara, I. K. Berata, I. B. K. Ardana, N. L. E. Setiasih, and I. N. Sulabda. Indones. Med. Veterinus, Volume. 10, Issue 5, Pagep. 714–724 (2021)
17. Baloš. M.Z, S. Jakšić, K.Slobodan, M.Kapetanov's. Archives of Veterinary Medicine Volume 9, Issue1, Page:31-42 (2016)
18. Sinurat. E, E. Marraskuranto. Squalen Bulletin of Marine and Fisheries Postharvest and Biotechnology, Volume 7, Issue 3, Page : 131-138 (2012)
19. Medion,https://www.medion.co.id/faktor-yang-mendorong-perkembangan-gizzard-pada-broiler/ . 2023 Accessed in30 Agustus, 2023, 09.10 a.m)
20. Hernandez F., J. Madrid, V. Farcia, J. Orengo and M.D. Megias. Poultry Sci.,83:169-74. (2004)
21. Sturkie, P. D. Avian Physiology. 4 th Edition. Spinger-Verlag, New York (2020)

22. Lisnahan.C.V.,Wihandoyo.W, Zupriza. Z, Harimurti.S. Jurnal Ilmiah Peternakan Terpadu, Volume 6, Issue 2, Page: 128-133 (2018)
23. Subaryono. Bulletin of Marine and Fisheries Postharvest and Biotechnology Volume 11,Issue 2,Page: 45–52 (2016)
24. Beuder. S, S. A. Braybrook. Seminars in Cell & Developmental Biology. Volume 134, Page: 103-111. https://doi.org/10.1016/j.semcdb.2022.03.003 (2023)
25. Ikegami, S., Fumie, T., Hironobu, H., Noboru, T., F. Eiichi. N., Satoshi, I. J.Nutr. 120:353-360. (1990).
26. Samee H, ZX Li, H Lin, J Khalid and YC Guo. Journal of Zhejiang University Science Volume10, Issue 2, Page:147-153. (2009)
27. Chojnacka, K., A. Saeid, Z. Witkowska and L. Tuhy. The Open Conference Proceedings Journal. Page : 20-28. (2012)
28. Pangesti, U. T., Natsir, M. H., & Sudjarwo, E. Journal of Tropical Animal Production,Volume 17, Issue 2, Page :58-65 (2017)
29. Setiawan, H., N. U. Listiatie dan M. Zulfikar. 2018. J. Veteriner. Volume 19, Page:554-567 (2018
30. Fard, S. H., M. Toghyani and S. A. Tabeidian .The J. of Int. Recycl Org. Waste Agriculture. 3:141-146 (2014)
31. Ketaren, P. P. Wartazoa.Volume 4, Page:172-180 (2010)
32. Satimah S.D, Yunianto dan F Wahyono. J Sainpet Indonesia. Volume 14, Issue 4, Page :396 – 403.(2019)
33. Katanbaf, M. N., Siegel, P. B., & Dunnington, E. A. Theoretical and Applied Genetics, 76, 540-544 (1988)
34. Anggraini, A. D., Widodo, W., Rahayu, I. D., & Sutanto, A. Prosiding Seminar Nasional Teknologi Peternakan dan Veteriner .Page: 615-621 (2020)
35. Lenhardt, L. and S. Mozes. 2003. Acta Vet. Brno. Volume 72, Page: 353 – 358 (2003)
36. Wang, X., Y. Z. Farnell, E. D. Peebles, A. S. Kiess, K. G. S. Wamsley and W. Zhai. Poultry Science 95: 1332 – 1340 (2016)
37. Ravindran, V., & Abdollahi, M. R. Animals, Volume 11, Issue 10, Page: 2795 (2021)
38. Raza.A, Bashir S. Tabassum. R. Heliyon. Volume;5, Issue: 4,e01437 (2019)
39. Heywang B.W. Poult. Sci ;29:804–811. doi: 10.3382/ps.0290804 (1950)

Qualitative performance of participated cow for progeny test on smallholder dairy farms

Tri Eko Susilorini[1,*], *Rini Dwi* Wahyuni[2], *Hanum* Muarifah[1], *Rizka* Larasati[3], *Abimanyu* Wicaksono[3], and *Donny Wahyu* Indriatmoko[3]

[1] Department of Animal Production, Faculty of Animal Science, Universitas Brawijaya, Malang 65145, Indonesia
[2] Department of Animal Nutrition, Faculty of Animal Science, Universitas Brawijaya, Malang 65145, Indonesia
[3] Master Student of Animal Science, Faculty of Animal Science, Universitas Brawijaya, Malang 65145, Indonesia

Abstract. Dairy cow health status affects to milk production and contributes to the profit in many types of dairy management. Locomotion score has been used to determine the qualitative performance score of dairy cows. Investigations on dairy cow located in Wagir Malang were obtained to assess the qualitative performance of participated cow. Selecting participated cow to assure the potential of superior bull from his pedigree in progeny test. There were 85 dairy cows belongs to the smallholder dairy farmer contributed to this research. The result showed that Chest Width 20.99 ± 2.63 cm; Body Depth 79.58 ± 4.79 cm; Rump Angle -2.59 ± 2.43°; Thurl Width 16.46 ± 1.71 cm; Rear Leg Side View 146.82 ± 6.08°; Rear Leg Rear View 15.96 ± 4.19°; Font Foot Angle 42.94 ± 4.81° and Rear Foot Angle 40.12 ± 5.09°. It can be concluded that the qualitative performance of participated cow shows moderate performance based on the standard performance of FH. It is necessary to select the participated cow to improve milk yield of its pedigree and qualitative performance score can be used as the basis of breeding program.

1 Introduction

Dairy cattle in Indonesia, primarily raised by smallholder farmers, belong to the Friesian Holstein (FH) breed and its descendants, with the aim of obtaining milk production. Milk production is a result of the interaction between genetic and environmental factors. To achieve optimal milk production, one approach is to improve genetic quality, intending to enhance the genetic makeup of the population in subsequent generations through a directed selection program. Selection is not limited to males with a progeny test program but should also be conducted on participated female cows.

The performance of participated cows used in progeny tests varies widely. Therefore, conducting proper evaluations of participated cows is an effective way to improve the genetic quality and milk production of dairy cows. Performance assessment can be carried out based

* Corresponding author: triekos@ub.ac.id

on two characteristics: quantitative and qualitative performance. Quantitative performance includes vital statistics of the cow and milk production, while qualitative performance encompasses angularity, rump angle, side-view foot positioning, and rear-view foot positioning.

According to Chemunique [1], qualitative performance is crucial for selecting dairy cow dams because animal comfort is a factor influencing decreased milk production, feed consumption, animal weight loss, and increased reproductive failure and mastitis. The research of Zhang et al. [2] found that rump angle significantly influences milk production in dairy buffaloes. Qualitative performance assessment can be conducted based on the National Holstein Friesian Association standards [3].

The objective of this research is to analyze the quality of qualitative performance of participated cows used as dams in the progeny testing program in the Wagir District of Malang Regency.

2 Materials and methods

2.1 Data information

Eighty-five dairy cows in their lactation period were used in this research and selected as participated cows for the progeny testing program of the Livestock Department in 2023. These dairy cows are owned by smallholder farmers in the Wagir District of Malang Regency, Indonesia. The research method is a survey. The observed variables and their measurement methods are presented in Table 1.

2.2 Statistical Analysis

The data obtained was analyzed descriptively and the results were compared with guidelines from the National Holstein Friesian to determine the performance of participated cows.

3 Results and Discussion

The descriptive statistics of type traits on participated cows were presented in Table 2. Qualitative performance is crucial for participated cows because it is linked to the health status of the livestock. The observation results indicate that all participated cows fall into the intermediate or moderate category. Chest width is associated with the ability of cows to consume feed. In the research, the chest width of participated cows was found to be consistent with the findings of Shahid et al. [4] who reported that indigenous cows in Northern Azad Jammu & Kashmir had an average chest width of 34.1 ± 0.7 cm.

Regarding all qualitative performance aspects of participated cows, the results were almost the same as the study conducted by Kumar et al. [5] on Sahiwal dairy cows, where all were categorized as intermediate. It is further noted that the condition of cows in the intermediate category still needs improvement as it is related to milk production and cow reproduction. Choosing better qualitative performance in participated cows were able to enhance functionality and longevity in milk production.

Table 1. Observed variable and how to measure it [3]

Qualitative performance	**Codes**	**Specifications**	
Chest Width	CW	The distance between the top of the inside section of the front legs, measured diagonally	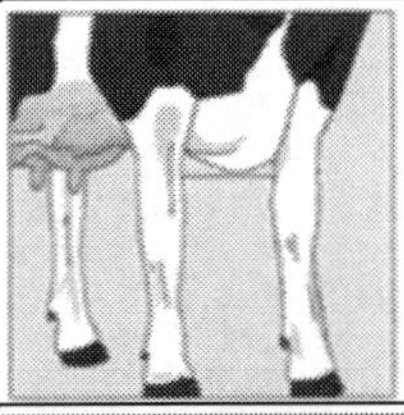
Body Depth	BD	The distance between the pin bone and the shoulder's tip	
Thurl Width	TW	From side to side, gauge the pelvic region	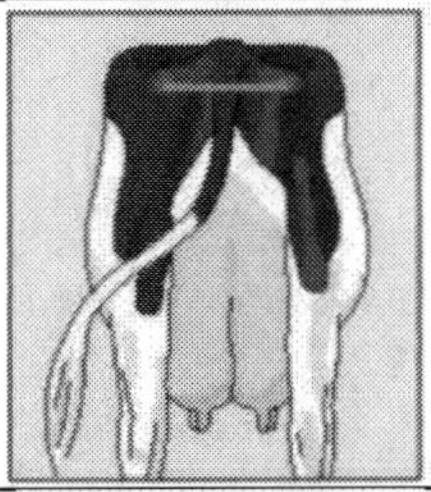
Rump Angle	RA	The slope from hooks to pins, when the cow is standing, is measured by pins high/low.	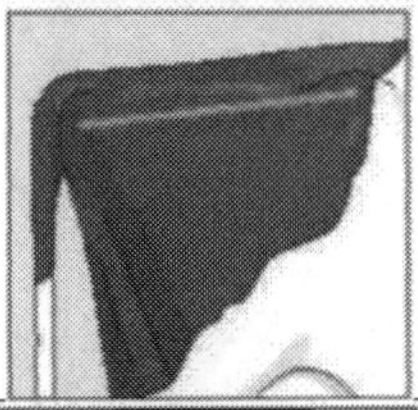
Rear Legs Side View	RLSV	Angle determined at the hock's front	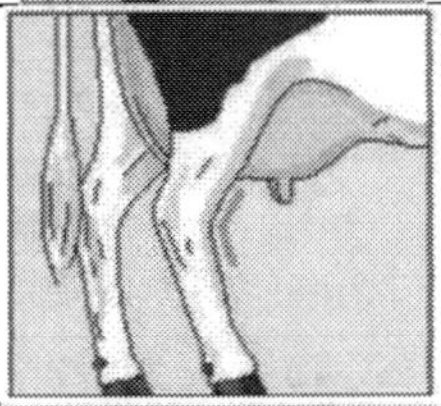
Rear Legs Rear View	RLRV	When viewed from behind, the rear legs' feet' direction	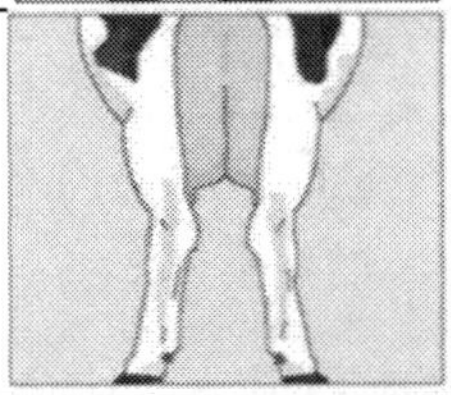
Font Foot Angle	FFA	The angle at the front of the front hoof is measured from the ground to the right hoof's hairline.	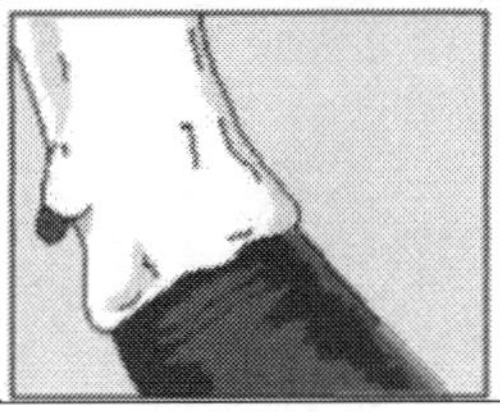
Rear Foot Angle	RFA	The angle at the front of the back hoof is measured from the ground to the right hoof's hairline.	

Table 2. Information type traits of participated cows

Traits	Maximum	Minimum	Mean	Score	Category
Chest Width (cm)	34	17	20.99	5.3	Intermediate
Body Depth (cm)	91	65	79.58	5.5	Intermediate
Thurl Width (cm)	20	12	16.46	4.5	Intermediate
Rump Angel (cm)	4	-7	-2.59	4.4	slight slope
RLSV (Rear Legs Side View) (°)	50	20	146.82	4.9	Intermediate
RLRV (Rear Legs Rear View) (°)	25	5	15.96	4.3	Intermediate
Front Foot angel (°)	50	30	42.94	4.6	Intermediate
Rear Foot Angel (°)	55	30	40.12	4.1	Intermediate

4 Conclusion

It could be concluded that the category of the participated cow was intermediate and had good qualitative performance.

This publication is part of Professor Grants 2023. We are thankful to Faculty of Animal Science Universitas Brawijaya for the financial support. The authors extend their sincere thanks to the students who involved in collecting data.

References

1. Chemunique, The importance of cow comfort – Cows with healthy feet eat, milk and breed better – Identifying dairy cow lameness. 2011.
2. X.X. Zhang, Z.G. An, K.F. Niu, C. Chen, T.Z. Ye, A. Shaukat, L.G. Yang, Animal, **16**, 100653 (2022)
3. Dexter Cattle Society, A Useful Guide to Linear Assessment of Dexter Cattle, no. January (Dale, Holme House, 2020)
4. B. Shahid, K. Z. Abbasi, M. I. Khan, and F. S. Awan, J. Anim. Plant Sci. **32**, 670–679 (2022)
5. S. Kumar, V.V. Varada, P.S. Banakar, N. Tyagi, R. Chouraddi, R.H. Mallapa, A.K. Tyagi, Anim. Biotechnol. **34**, 1612-1625 (2023)

Effect of dietary acidifier on intestinal pH of New Zealand White rabbit

Heli Tistiana[1*], *Eko* Widodo[1], and *Hermanto* Hermanto[1]

[1]Faculty of Animal Science, Universitas Brawijaya, 65145 Malang, Indonesia

Abstract. The microflora condition in the small intestine of weaning rabbits hasn't developed yet. The use of acidifiers in rabbit feed aims to suppress harmful microorganisms and digestive disorders, especially in weaning rabbits. This research used different levels of acidifier as feed additives for weaning rabbits given on pellet feed. The acidifier addition level was control (T_0), 0.1% (T_1), 0.2% (T_2) and 0.3% (T_3). The parameter measured was the pH of the different parts of the small intestine. The research used 64 weaning rabbits (35 days old) which were kept for 6 weeks. Parameter measured was the pH of the parts of the small intestine consisting of the duodenum, jejunum and ileum. This research used a randomized group design with 4 treatments and 8 replications. The results showed that the addition of acidifiers had no significant difference effect ($P>0.05$) on pH of the duodenum and ileum, but there was a trend to decrease pH in all parts of intestine. While, the effect of acidifiers showed significant differences ($P<0.05$) on pH of jejunum. The conclusion is that the use of acidifier with a level of 0.3% gives the best effect to decrease of intestine pH of New Zealand White Rabbits.
Keywords: acidifier, pH, small intestine, rabbit

1 Introduction

Rabbit farming in Indonesia shows constant development after the Covid 19 case. Rabbits farming are able to survive because these livestock use source feedstuff from local resources. The carrying capacity of feed sources in Indonesia is an advantage in managing livestock businesses. It's just that the obstacles experienced by breeders are digestive problems that often occur on farms. This condition often even proceeds to cases of death, especially in small rabbits. This problem of digestive disease in young rabbits is caused by many factors, one of which is from feed.

Feed is the main source of nutrition for livestock. Providing feed that suitable with their nutritional. High quality of feed will make livestock develop properly. The problem is that rabbits usually lack fiber intake in their feed, so this is the most common cause of gastrointestinal disorders in rabbits. The condition of the digestive ecosystem is mainly related to the immune regulatory system. Microbial control is carried out to improve digestive efficiency and health. This microbial regulation of the digestive tract can overcome digestive problems in weaning rabbits [1].

* Corresponding author: tistiana_heli@ub.ac.id

The condition of microorganisms in the digestive tract has different resistance to intestinal pH, this can be the cause of digestive disorders. [1] The digestive ecosystem is involved in several major physiological functions and particularly in digestive and immune system regulation. Therefore, microbiota control can be done to improve digestive efficiency and digestive health in animal husbandry. [2] Intestinal connection with components of this microflora from birth. But it is still limited in how it affects normal development and physiology

Rabbits generally have a small intestine pH value ranging from 6.4-7.4 [3]. At this pH, the condition of the microflora in the intestines of weaning rabbits that are not yet stable can easily cause diarrhea and death in young rabbits. The high pH of the rabbit intestine results in not being able to stabilize the bacteria that enter the digestion, so this is a problem for breeders. Farmers will overcome this problem by giving antibiotics, but since antibiotic was restricted or prohibited, farmers experience problems. Therefore, it is necessary to find a solution with the use of other feed additives in the form of organic acids or acidifiers. The addition of organic acids to the diet of rabbits is also important because weaned rabbits do not have enough capacity to acidify their stomach content [4].

Antimicrobial feed additives used worldwide in animal husbandry to improve the economy and ecology of animal production. It can be increasing growth rates, reducing feed expenditure and reducing the risk of disease [5].

This research to determine the effect of the use of acidifiers in feed on the intestine pH of New Zealand White rabbits. The expectation use of acidifiers can reduce the pH value in the digestive system. pH on intestine reach acidic conditions so that it can control and increase the number of nonpathogenic microbes (BAL) in the digestive system.

2 Materials and Methods

The research material used 64 New *Zealand* White rabbits. The rabbits uses weaning rabbit with an age of 1 month. The in vivo experiments was carried out for 6 weeks. The feed ingredients used are corn, rapeseed meal, soybean meal, pollard, grass, coconut meal, corn gluten feed, mineral mix and molasses. The feed ingredients are formulated and processed into one to become pellet-shaped basal feed. The treatment feed used basal feed added with Pros G. Brand of acidifier containing formic acid (350 gm), lactic acid (58 gm), propionic acid (78 gm), citric acid (99.50 gm) and silocon dioxide (18.50 gm).

The research method used field experiments, namely *in vivo tests on* New Zealand White rabbits. The experimental design used was a randomized design complete with 4 treatments and 8 replication. The treatments tested are as follows:

T0 = basal diet

T1 = basal diet + 0,1 % *acidifier* (PROS Acid G)

T2 = basal diet + 0,2 % *acidifier* (PROS Acid G)

T3 = basal diet + 0,3 % *acidifier* (PROS Acid G)

The variable observed in this research was the pH of the parts of the small intestine consisting of the duodenum, jejunum and ileum. pH measurement is done by removing digesta from the small intestine and placed in a tray or container, then digesta is measured using a pH meter [6]. The data obtained during the study were then analyzed using ANOVA (*Analysis of variance) fingerprints*, if there is a real or very real difference in influence between treatments, it will be continued with the Duncan Multiple Distance Test.

Table 1. Feed Ingredient and Nutrition Content

Feed Ingredient	%
Corn	18
Rice bran	15
Pollard	24
Copra meal	3
Spybean meal	12
Rapeseed meal	7
Corn Gluten Feed	7.8
corn pericarp	5
Brachiaria Decumbens	6
Molases	2
premix	0.2
Total	100
Nutrition Content	
Crude Protein (%)	19.05
Crude Fiber (%)	16.85
Crude Fat (%)	2,73
NDF (%)	58,18
ADF (%)	31,40
Selulose (%)	16,90

3 Results and Discussion

The results on the effect of adding *acidifiers* in pellet feed on the pH of the small intestine of *New Zealand White rabbits* are presented in table 2. The statistical analysis showed that the addition *of acidifiers* to the feed did not have a different effect (P>0.05) on the pH value of the duodenum. The highest duodenum pH value is shown by T_1 (6.77±0.30) while the lowest duodenal pH value is shown by T_3 (6.64 ±0.19). In Figure 1 can be seen that the duodenum pH trend increases at T_1, then at T_2 and T_3 decreases along with the increase in the level of acidifier.

Table 2. Average pH value of parts of the small intestine

Variables	Treatments			
	T0	T1	T2	T3
pH Duodenum	6.73±0.28	6.77±0.30	6.71±0.37	6.64±0.19
pH Jejunum*	$7.02{\pm}0.29^{ab}$	$6.76{\pm}0.19^{a}$	$7.11{\pm}0.46^{b}$	$6.91{\pm}0.43^{ab}$
pH Ileum	7.41±0.37	7.37±0.33	7.36±0.30	7.23±0.44

Information: T0 = without acidifier, T1 = addition of 0.1% acidifier; T2=addition of 0.2% acidifier, T3=addition of 0.3% acidifier; * $^{a–b}$ means within the same row without the same superscripts are significantly different ($p < 0.05$)

The effect of the treatment showed a significanty different (P<0.05) in the pH of the jejenum. The highest pH value of the jejunum was shown by the T_2 (7.11±0.46); where the To, T_2 and T_3 treatments show not different from the same notation. In figure 1 can be seen that the trend of jejenum pH decreases, but at T_2 it increases. And at T_3 was decreases again. The lowest jejunum pH was shown by T_1 (6.76±0.19).

The addition of acidifier on feed did not difference effect between treatments (P>0.05) in the ileum. The highest ileal pH value was shown by T_0 (7.41±0.37) while the lowest ileal pH value was shown T_3 (7.23±0.44). In figure 1, it can be seen that the pH trend of the ileum decreases along with the increase level of acidifier.

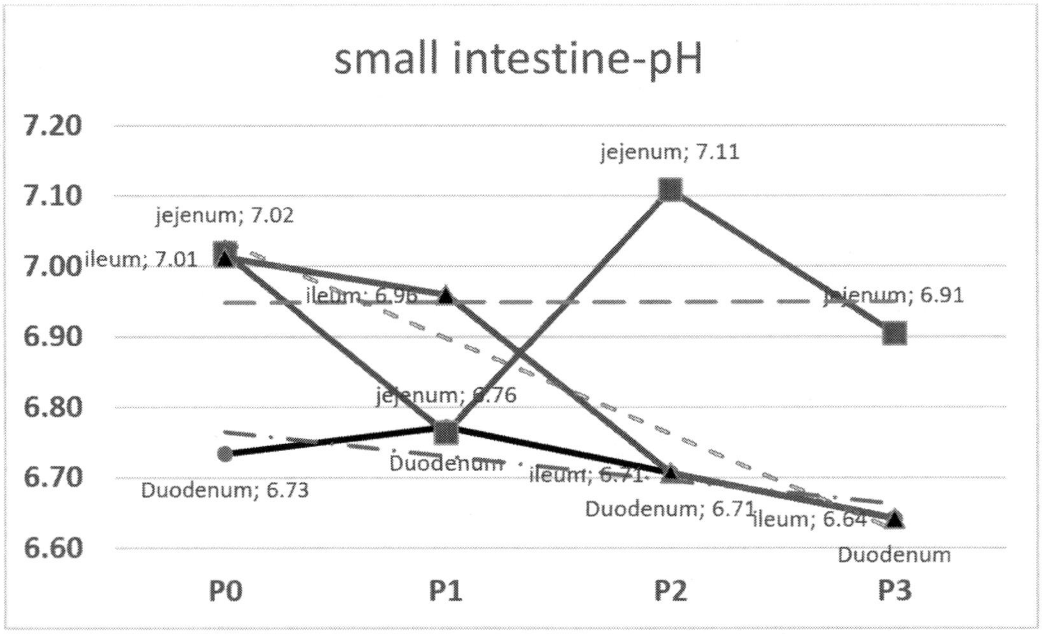

Fig. 1. Trends in the pH of the duodenum, jejenum and ileum in each treatment

Treatment level 0.3% acidifier (T_3) to pellet feed gave the best results in lowering the pH value of the small intestine of New Zealand White rabbits. The pH value of the duodenum at T_3 was 6.64±0.19, jejunum pH 6.91±0.43 and ileal pH 7.23±0.44. The pH value obtained can still be good and normal condition. [3] The pH of the small intestine in guinea pigs and rabbits is in the range of pH values around 6.4-7.4. [7] Further added that explain detail in rabbits, the pH value of the diduodenum was between 6.0-6.6 and the jejunum pH value was 6.6-7.0 and around the distal ileum value was 7.4. The use of *acidifier* as much as 0.3% in pellet feed was still safe and tolerable, because the final pH obtained does not drop dramatically and exceeds the normal limit of rabbit intestinal pH in general. In addition acidifier, pH conditions that are not too acidic will prevent *acidosis* in livestock.

Overall result from treatment of adding acidifier in rabbit feed can be seen in Table 2. The addition of acidifiers to pellet feed results in a decrease in pH in each treatment, this happens because acidifiers act as organic acids that have a role to acidify the digestive tract, especially in small intestine. According to [8] a decrease pH in the digestive tract due to acidifier administration occurs when acid breaks down in the digestive tract and liberates hydrogen ions (H +) ; then more hydrogen ions (H +) are released by the acidifier from organic acids, the pH in the digestive tract was getting more acidic. The acidic atmosphere in the digestive system provides several advantages, including suppressing the development of pathogenic bacteria that inhibit the digestive process and increasing the amount of BAL in the intestine so that indirectly the use of acidifiers will maintain microbial balance and even increase microbes in the digestive system. The good condition on digestive system will facilitate absorbing nutrients. The intestine will be optimal in digestion of feed and it will be more healthful .

4 Conclusion

The conclusion of this research was the adding of *acidifier* with level 0.3% in feed can have the best effect on reducing intestinal pH of New *Zealand White Rabbit*.

References

1. S. Combes, L. Fortun-Lamothe, L.Cauquil, T. Gidenne. Engineering the rabbit digestive ecosystem to improve digestive health and efficacy. *Animal*, *7*(9), pp.1429-1439 (2013)
2. G. Tellez, S.E. Higgins, A.M. Donoghue, B.M. Hargis. Digestive physiology and the role of microorganisms. *Journal of Applied Poultry Research*, *15*(1), pp.136-144 (2006).
3. H.A. Merchant, E.L. McConnell, F. Liu, C. Ramaswamy, R.P. Kulkarni, A.W. Basit, S. Murdan. Assessment of gastrointestinal pH, fluid and lymphoid tissue in the guinea pig, rabbit and pig, and implications for their use in drug development. *European Journal of Pharmaceutical Sciences*, *42*(1-2), pp.3-10 (2011).
4. K.U. Amaefule, I.I. Mbagwu,, N.E. Inyang. Performance, nutrient utilization and intestinal environment of weaned rabbits fed diets supplemented with organic acids in the humid tropics. *Nigerian Journal of Animal Science*, *13*, pp.59-69 (2011).
5. M.J. Uddin, K.M.S. Islam, A. Reza, R. Chowdhury. Citric acid as feed additive in diet of rabbit-effect on growth performance. *Journal of the Bangladesh Agricultural University*, *12*(452-2016-35621), pp.87-90 (2014).
6. W.S.M. Emma, O. Sjofjan, E. Widodo. Karakteristik usus halus ayam pedaging yang diberikan asam jeruk nipis dalam pakan. *J. Veteriner*, *14*(1), pp.105-110 (2013).
7. Z. Hong, Y. Xu., J.F. Yin, J. Jin, Y. Jiang, Q. Du. Improving the effectiveness of (−)-epigallocatechin gallate (EGCG) against rabbit atherosclerosis by EGCG-loaded nanoparticles prepared from chitosan and polyaspartic acid. *Journal of agricultural and food chemistry*, *62*(52), pp.12603-12609 (2014).
8. S. Imam, L.D. Mahfudz, N. Suthama. Perkembangan mikrobia usus ayam broiler yang diberi pakan stepdown protein dengan penambahan asam sitrat sebagai acidifier. *Jurnal Litbang Provinsi Jawa Tengah*, *16*(2), pp.191-200 (2018).

Importance-Performance Analysis (IPA) of sustainable broiler agribusiness initiatives: The plasma farmers perspective

Nanang Febrianto[1*], *Priyo* Sugeng Winarto[1], *Puji* Akhiroh[1], *Hafidz* Abdullah[1], and *Budi* Hartono[1]

[1]Faculty of Animal Science, Universitas Brawijaya, Veteran Rd, Malang 65145, Indonesia

Abstract. Plasma farmer satisfaction is pivotal in the broiler partnership cooperation system, serving as a key indicator of the extent to which partnership principles are effectively upheld. This study assessed plasma farmers' satisfaction level regarding broiler partnership services' performance in Malang. The research sample comprised 87 plasma farmers selected through purposive sampling. The data analysis employed the Importance Performance Analysis (IPA) method to discern the positioning of service attributes in quadrant analysis. The Customer Satisfaction Index (CSI) method was also employed to gauge user satisfaction. The study revealed a noteworthy CSI value of 0.87%, signifying a high degree of contentment among farmers with the core company's performance and services. Delving further into the analysis of 24 partnership attributes using the IPA assessment, several vital observations emerged. Firstly, two attributes were identified as possessing substantial importance yet exhibiting subpar performance. Secondly, seven attributes demonstrated partnership performance that necessitates ongoing maintenance. Thirdly, six attributes exhibited partnership performance that warranted maintenance but with relatively lower priority. Lastly, nine attributes showed an excessive level of performance.

1 Introduction

Broilers are nonruminant livestock that is raised to produce meat. Broiler livestock businesses contribute to the fulfillment of people's animal protein needs, especially chicken meat. The problems experienced by broiler farms are the increasing costs of production facilities (feed, DOC, vaccines, and medicines), which are not matched by an increase in the selling price of chicken and price stability. limited funding, availability of inputs, and fluctuations in the selling price of chicken. This condition makes farmers increasingly helpless, so one solution to minimize this problem is running a broiler farming business with a partnership pattern. The partnership business pattern is a livestock business involving a relationship between companies producing inputs and local farmers bound by mutually beneficial cooperation.

The core company engaged in broiler partnerships provides facilities and infrastructure provided by the partnership as production support such as DOC, feed,

* Corresponding author: nanangfeb@ub.ac.id

medicine, vaccines, vitamins, Field Extension Officers, and marketing back at harvest with a contract price agreement. So that farmers only provide cages, equipment, and labor in raising chickens. The implementation of the partnership is related to the performance services provided to the plasma that provide satisfaction to plasma farmers [1]. [2] added that plasma farmer satisfaction is an essential factor to consider in the partnership business development process. The level of satisfaction of plasma farmers with the core company will positively impact the continuity of the partnership business. Plasma farmers who are satisfied with the core company tend to maintain their cooperation with the company. This study aims to analyze farmers' satisfaction levels with the core company's contract farming model.

2 Materials and Methods

The research was conducted in Malang Regency. The research location was determined considering that Malang Regency is one of the districts with the largest broiler population in East Java Province in 2021, with a population of 25,591,600 birds [3]. The data collected is primary data and secondary data. Primary data was obtained by interviews with respondents using questionnaire, documentation, and observation. Secondary data was taken from the statistical data collection agency and the local animal husbandry office. The assessment of the level of satisfaction with the company's performance attributes is divided into several variables and indicators, as follows:

Table 1. Assessment variables and indicators

No	Variable	Indicator
1	Partner Acceptance Procedure	Partner Services Partner acceptance procedure Marketing assurance
2.	Contract price of production facilities	DOC Contract Price Feed Contract Price Vaccine Contract Price
3.	Quality of production facilities	Quality of DOC Quality of feed Quality of vaccine
4.	Delivery schedule of production facility item	DOC delivery schedule Feed delivery schedule Vaccine delivery schedule
5.	The role of Technical Service (TS)	Maintenance management technical guidance Frequency of guidance time by TS Supporting facilities used by TS Response to complaints Courtesy TS
6.	Broiler chicken harvesting	Implementation of production standards Timeliness of harvest Appropriateness of chicken selling price
7.	Post-harvest handling	Payment Incentive/bonus DOC bonus amount Mortality bonus

Source: Primary data, 2023

Determination of the score or value of the level of satisfaction on each attribute using a Likert scale calculation of 1 - 5 with the lowest score information farmers consider the attribute is not important or not satisfied and the highest score is considered very important and very satisfied. Data analysis used to determine the level of satisfaction of plasma farmers is the Importance Performance Analysis (IPA) method and the Customer Satisfaction Index (CSI) [4].

3 Result and Discussion

3.1 *Calculation of Customer Satisfaction Index (CSI)*

Measurement of overall breeder satisfaction is obtained by calculating the Customer Satisfaction Index (CSI) to obtain the CSI value, the average value of the importance level (Mean Importance Satisfaction / MIS), and the average value of the performance level (Mean Satisfaction Score / MSS) are required. The assessment was carried out on 24 indicators that became the company's service attributes. The calculation (CSI) is in the Table 2.

Table 2. Results of the calculation of the Costumber Satisfaction Index (CSI)

No	Attributes	Average Importance (MIS)	Average Satisfaction (MSS)	*Weighted Factor* (WF)	*Weighted Score* (WS)
1.	Partner Services	4.28	4.03	0.041	0.18
2.	Partner acceptance procedure	4.70	4.02	0.045	0.21
3.	Marketing assurance	4.28	4.08	0.041	0.18
4.	DOC Contract Price	4.23	4.10	0.041	0.17
5.	Feed Contract Price	4.67	4.13	0.045	0.21
6.	Vaccine Contract Price	4.28	4.10	0.041	0.18
7.	Quality of DOC	4.15	3.98	0.040	0.16
8.	Quality of feed	4.28	4.08	0.041	0.18
9.	Quality of vaccine	4.15	4.07	0.040	0.16
10.	DOC delivery schedule	4.40	4.13	0.042	0.19
11.	Feed delivery schedule	4.50	4.07	0.043	0.19
12.	Vaccine delivery schedule	4.47	4.07	0.043	0.19
13.	Maintenance management technical guidance	4.20	4.07	0.040	0.17
14.	Frequency of guidance time by TS	4.38	4.07	0.042	0.18
15.	Supporting facilities used by TS	4.40	4.02	0.042	0.19
16.	Response to complaints	4.28	4.00	0.041	0.18
17.	Courtesy TS	4.70	4.08	0.045	0.21
18.	Implementation of production standards	4.28	4.07	0.041	0.18
19.	Timeliness of harvest	4.23	4.13	0.041	0.17
20.	Appropriateness of chicken selling price	4.67	4.08	0.045	0.21
21.	Payment	4.28	4.08	0.041	0.18
22.	Incentive/Bonus	4.15	4.03	0.040	0.16
23.	DOC bonus amount	4.28	4.05	0.041	0.18
24.	Mortality bonus	4.15	4.02	0.040	0.16
	Total	104.6	97.6	1.000	4.4
	Total Weight Value (WT)				4.4
	CSI value = (WT : 5) x 100%				0.87

Source: Primary data, 2023

Based on the assessment conducted by plasma farmers on the performance of the core company, the CSI score is 0.87 percent. This value is in the range of 0.81 - 1.00, which indicates that overall, farmers are delighted with the performance. Although the satisfaction value is on a satisfied scale, farmers still have unmet expectations, amounting to 0.13%. Farmer satisfaction scores obtained from several attributes of company performance that have the highest scores are on the contract price of DOC and drugs and vaccines, DOC delivery schedule, and timeliness of harvest. In addition to the satisfaction value, a factor in the high CSI score, is the average of the breeder's interest value in the company's performance attributes. The company attributes that have a high importance value are the partner acceptance procedure, feed contract price, feed delivery schedule, TS (Technical Service) courtesy, and the suitability of the chicken selling price. [5] added that the customer satisfaction index value of 81.20% or 0.812 means that farmers are satisfied with the company's performance.

3.2 Importance Performance Analysis (IPA)

Importance Performance Analysis (IPA) is used to assess plasma farmers' performance on partnership companies' performance. Through this analysis, it will be known what things need to be improved to improve their performance and achieve plasma farmer satisfaction. Based on the study's results, the average value of the importance score was 4.35. This value means that plasma farmers consider overall partnership attributes in the core company very important. Based on the performance level assessment, the results show that the average performance score of the overall partnership attribute is 4.07. This score means that overall, plasma farmers are satisfied with the performance provided by the core company. Customer satisfaction reflects how far the company has responded to the wishes and market expectations [6].

The partnership attributes are then divided into four quadrants that reflect each attribute's importance and performance conditions. The IPA matrix consists of four quadrants, namely quadrant I (Top priority), quadrant II (Maintain achievement), quadrant III (Low priority), and quadrant IV (Excessive). A cut point separates the quadrants, which comes from the average importance and performance value. The partnership attributes contained in each quadrant can be seen in Figure 1.

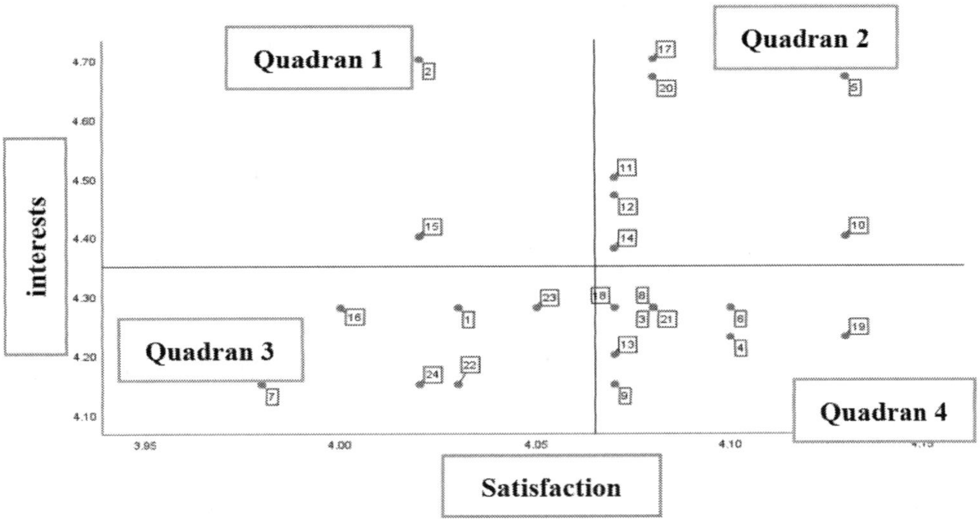

Figure 1. Importance Performance Analysis Quadrant

3.2.1 Quadrant I (Top Priority)

Quadrant I contain company attributes with a high importance value but still low performance, which is interpreted as a company's weakness. The variables in this quadrant must be improved again; the company continuously improves to meet breeders' expectations. The attributes in this quadrant consist of the acceptance procedure to become a partner and the supporting facilities used by TS (Technical Service). Attributes that are the weaknesses of the company in quadrant I must be evaluated for the performance of the services provided, directly reviewing the partner breeders so that the company can improve its services.

3.2.2 Quadrant II (Maintain Achievement)

Breeders consider Attributes in quadrant II important, and the company's performance can provide satisfactory service. These attributes show the company's superiority from the breeder's point of view and must maintain its performance. The Attributes in quadrant II are the company's strengths, especially in service to breeders. Companies need to maintain the performance of the attributes in this quadrant. Variables in quadrant II include feed contract price, DOC delivery schedule, feed, medicine and vitamins, frequency of guidance time by TS (Technical Service), and courtesy of TS (Technical Service).

3.2.3 Quadrant III (Low Priority)

Company attributes in quadrant III breeders consider that the level of importance is low, and the company's performance is still considered unfavourable by breeders. This is consistent with the opinion of [7] states that the company must improve performance so that the attributes in quadrant III do not move to quadrant I, which is a weakness. The partnership attributes in this quadrant are partner service, doc quality, response to complaints, provision of additional bonuses/intensive, doc bonus amount, and mortality bonus.

3.2.4 Quadrant IV (Excessive)

Attributes in this quadrant have low importance for breeders, but the company's performance is so good that it is considered excessive. The performance that the partnership attributes in quadrant IV have achieved must be maintained and does not need to be improved anymore because it will only cause a waste of resources. The attributes included in this quadrant are marketing assurance, doc contract price, drug and vitamin contract price, feed quality, drug and vitamin quality, maintenance management technical guidance, application of production standards, timeliness of harvest, and speed of payment of results.

4 Conclusion

The conclusion of this research showed that plasma farmers are delighted with the performance of partnership companies in the Malang district based on the calculation of the Customer Satisfaction Index (CSI) value of 0.87 and the calculation of Importance Performance Analysis (IPA), which provides advantages from the core company related to services to farmers.

References

1. N. Azmi, B.A. Nugroho, S. Azizah. Journal of Tropical Animal Science and Technology. **6**, 27-33. (2019)

2. Palmarudi, Kasim. J. Animal Science and Technology. **2**, 50-59. (2012)
3. Statistics East Java. Surabaya. (2022)
4. Martilla, James. **14**, 77-79. (1977)
5. C. Subiyanto, C. Cepriadi, E. Sayamar, Jurnal Faperta. **3**, 1-14. (2016)
6. E.D.S. Agustia, D. Haryono, A. Nugraha. **8**, 418-422. (2020)
7. N. Istiqamah, D. Suherman, B. Zain. **8**, 35-48. (2019)

The quality of spent hen chicken sausage with the addition of tomato paste (*Solanum lycopersicum*) as natural food colorant

Cintya Dienardila Anisa[1], *Herly* Evanuarini[1*], and *Imam* Thohari[1]

[1]Faculty of Animal Science, Universitas Brawijaya Malang, 65145, Indonesia

Abstract. Spent hen chicken meat are a meat that from 70 weeks old chicken that usually has a hard texture so further processing are needed to improve the quality. Spent hen chicken meat can be processed as sausage but chicken sausage usually have a pale color. Tomato have a natural coloring agent namely lycopene. Tomato paste can be used as a coloring agent and improve the quality of chicken sausage. The aim of this research was to determine the best percentage of tomato paste addition on chicken sausage based on physicochemical quality. The material used were spent hen chicken meat, tapioca, vegetable oil, tomato paste, seasoning, herbs and spices, skim milk, and ice cube. The research method is a laboratory experimental with a completely randomize design with 4 treatments and 4 replications. Tomato paste was added to sausage formulation 0%, 5%, 10%, and 15% of the total ingredients used. The measured variables were L a*b* color, cooking loss, water holding capacity, and moisture content. The results showed that the treatments gave very significant effect ($P<0.01$) on L a*b* color, cooking loss, water holding capacity, and moisture content. Tomato paste addition on sausage can increase redness and yellowness but decrease lightness, increase cooking loss and moisture content, but decreased water holding capacity. It can be concluded that the best treatment on this reesearch was chicken sausage with 5% tomato paste.

1 Introduction

Spent hen chicken meat are a meat that from a 70 weeks or more old chicken. Spent hen chicken have a low productivity. Usually laying hen have egg productivity up to 94-95% but spent hen chicken only have 50% of egg productivity so the meat can be used. But, people usually didn't like the meat from spent hen chicken because it has a hard texture and easier to get rancid [1]. Spent hen chicken meat h*ave a component that almost the same with usual chicken meat but have a higher fat and moisture content. Spent hen chicken meat can processed as food diversification with restructured meat technology to improve the quality and change people's mindset. Restructured meat are a technology that using a low quality carcass and smaller piece of meat to improve the value and making a new product using a binder, filler, and emulssion [2]. Sausage are the example of restructured meat product.

* Corresponding author: herlyfptub@ub.ac.id

Sausage are an oil in water emulssion made by grinding meat with the additional ingredient and put it on casing. Chicken sausage usually have a pale color so it needed a food colorant to increase the product appearance. Color are the first aspect for consumer to buy something [3]. Nitrite is a food colorant that usually being used for sausage. Nitrite has a bright red color but it can cause a carcinogenic when used excessively [4]. Tomato are ingredients that can be used as a natural food colorant and didn't have a side effect for consumer. Tomatoes contain a vitamin, glucose, carbohyrdrate (3.1%), protein (1.1%), ash (0.5%), fat (0.2%) and moisture content (95.1%). Tomato also has an antioxidant ability because of carotenoid [5]. Lycopene are the part of carotenoid who has an antioxidant ability and also a food colorant agent. Lycopene can be found on ripped tomatoes. Lycopene are soluble to fat and can gave a red and yellow color. Tomatoes contain a 0.85mg – 13.6mg of lycopene. Processed tomatoes usually have a higher lycopene content than a fresh tomato. Tomato paste are the example of processed tomatoes [6]. Several previous studies also use a natural food colorant to improve the quality of chicken sausage. Addition of beetroot extract and red dragon fruit can improve the physicho-chemical quality of chicken sausage. It also can improve the redness on sausage that attract panelist and it because the betalains and betacyanin pigment that improve the redness of sausage [7] [8]. The use of tomato paste can be used as natural food colorant on spent hen chicken sausage to improve the quality. This research aims to determined the best percentage addition of tomato paste on spent hen chicken sausage based on L a*b* color, cooking loss, water holding capacity, and moisture content.

2 Materials and methods

2.1 Materials

Materials used on this research are spent hent chicken sausage with the optional ingredients including spent hen chicken meat breast (60%), crushed garlic (3%), salt (2%), tapioca (6%), pepper (0.5%), nutmeg powder (0.25%), ginger powder (0.25%), sugar (2%), mushroom powder (1%), vegetable oil (10%), ice (10%), powdered skim milk (5%), and tomato paste for the treatments (0%, 5%, 10%, 15%). Materials for analysis are distilled water. The tools used on making sausage are meat grinder, spoon, knife, plastic glove, stove, bowl, pot, cutting board, thermometer, scale, and sausage casing made from celullose diameter 21 mm. The tools used on analysis are oven 105°C, petri dish, waterbath, exicator, colorider, 35 kg weight, and glasses.

2.2 Methods

The method of this research is a laboratory experimental with a completely randomize design with 4 treatments and 4 replications. The experimental design are presented in Table 1.

Table 1. Experimental design and research treatment

Treatments	Description
Control	Sausage with no addition of tomato paste
S1	Sausage with the addition of 5% tomato paste
S2	Sausage with the addition of 10% tomato paste
S3	Sausage with the addition of 15% tomato paste

2.3 Tomato Paste Preparations

The tomatoes used for this research was purchased from the traditional market in Malang. First, wash the tomatoes using water then cut the tomatoes. Blanch the tomatoes at 40 °C for 5 minutes and then put it on the cold water so it easier to peel. Peel of the tomatoes skin, put them on the blender and grind the tomatoes. After grinding, filter the pulp so the seed are separated from the juice. Evaporated the tomato juice until it thicken. Tomato paste was ready to used.

2.4 Sausage Preparation

Clean the spent hen chicken meat before processing, remove the skin, bones, and fat. Chop the meat and put it on the meat grinder. Add the salt and ice then minced for a while. Put the other ingredients, tapioca, garlic, pepper, nutmeg, ginger, sugar, mushroom powder, vegetable oil, and skim milk, then minced it again. For the treatment, put the tomato paste (5%, 10%, 15) and minced them until it homogenous. Put the sausage batter in celullose casing using sausage stuffer, give the space about 7 cm and tied it with thread so the sausage can be divided. Boiled the sausage with 80°C for 20 minutes and then put it on the cold water for 10 minutes so the sausage can be hardened. Spent hen chicken meat can be used for further analysis.

2.5 Data Analysis

Data were calculated using Microsoft Excel 2010 then were analysis with analysis of variance (ANOVA). If there were a significant effect, it will continued by Duncan's Multiple Range Test (DMRT).

2.6 Characterization of Sausage

*2.6.1 L a*b* color*

Color analysis was using a colorider. Calibrated the colorider with black and white plate before using it for analysis. Cut the center part of sausage and place it on a smooth surface. Put the colorider near the samples and click the button. The values of lightness, redness, and yellowness will be shown on the colorider screen.

2.6.2 Cooking Loss

Cooking loss was determined using a 80°C waterbath. Put the 10 g sausage batter on the vacuum plastic bag (A). Boiled the sample using a waterbath at 80°C for 30 minutes. Cooled the sample and dry it with tissue. Weighed the cooked sample (B). Calculated the cooking loss using the formulation =

$$cooking\ loss\ (\%) = \frac{A-B}{A}\ x\ 100\% \quad (1)$$

2.6.3 Water Holding Capacity

Water holding capacity was determined using a 35 kg weight. Put the 0,3 g sample on the filter paper Whatmann 42 and then place it on two glass plates. Place the 35 kg weight on the top glass plates and wait for 5 minutes. Draw the water loss on milimeter block for counting diameter of wet area and calculated WHC using formulation =

$$(MgH2O) = \frac{diameter\ of\ wet\ area\ (cm2)-8}{0{,}0948} \quad (2)$$

$$wet\ sample\ area = \frac{mgH2O}{sample\ weight} x\ 100\% \quad (3)$$

$$WHC\ (\%) = Moisture\ content\ (\%) - wet\ sample\ area(\%) \quad (4)$$

2.6.4 Moisture Content

Moisture content was determined with thermogravimetri method. Put the petri dish on the 105°C oven for 24 hours. After that, put the petri dish on the eksikator for 30 minutes then weighed it using analytical scale (A). Put the 2 gram sampel on the petri dish and weighed it (B). Dry the sample on the 105°C oven for 24 hours, then put it on the eksikator for 30 minutes. Weighed the petri dish + sample after dried (C). Calculated the final weight using the formulation =

$$Moisture\ content\ (\%) = \frac{(B-C)}{(B-A)} x\ 100\% \quad (5)$$

3 Results and Discussion

3.1 L a*b* color

Addition of different percentage of tomato paste (0%, 5%, 10%, and 15%) gave highly significant effect (P<0.01) on sausage color. The average L a*b* color of chicken sausage with the addition of tomato paste shown on Table 2. The difference color of spent hen chicken sausage with the addition of tomato paste shown on Figure 1.

Table 2. Average of L a*b* color of chicken sausage with the addition of tomato paste

Treatments	Lightness (L)	Redness (a*)	Yellowness (b*)
SC	67.57 ± 0.22^b	1.06 ± 0.06^a	14.14 ± 0.15^a
S1	67.12 ± 0.03^b	2.13 ± 0.08^b	22.27 ± 0.42^b
S2	66.94 ± 0.77^b	3.81 ± 0.07^c	23.20 ± 0.06^c
S3	65.80 ± 0.06^a	4.89 ± 0.03^d	24.74 ± 0.23^d

Notes : superscripts on the column (^{a,b,c,d}) shown a highly significant effect (P<0.01)

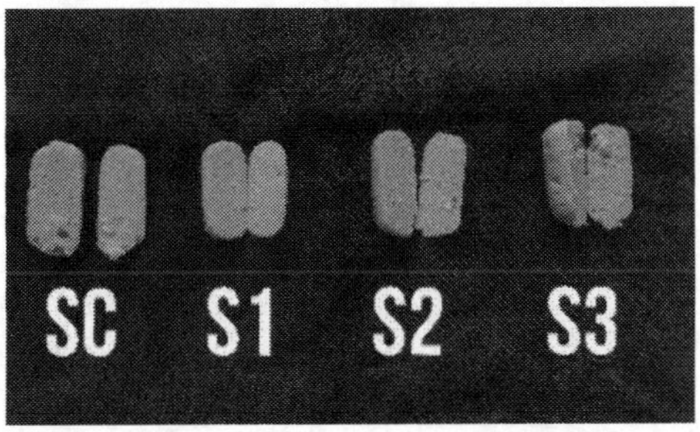

Fig. 1. Difference color of chicken sausage with the addition of tomato paste

3.1.1 Lightness

Lighntess of chicken sausage keep decreasing with the addition of tomato paste. Addition of a component on a meat can decrease the color of food product. Maillard reaction also the cause of decreased lgitness. Maillard reaction can be happening when cooking a sausage addition of tomato paste with a high temperature so ascorbate acid in tomatoes denaturated [9]. Maillard reaction are a glucose – amino reaction. When it getting a high temperature, the bond of glucose – amino can create a melanoidin which gave a browning color on food [10].

3.1.2 Redness

Redness of chicken sausage increased with the addition of tomato paste. This is because of natural food colorant pigment in tomato namely lycopene. Lycopene are a red pigment which part of carotenoid. The lycopene stability are higher when the tomatoes getting processed into a tomato paste, tomato pomace, or tomato flour. Lycopene content on tomato paste are nearly 184.29 g' [11]. Processed tomato into paste can change the isomer trans to cis that makes a more soluble lycopene when put it on meat batter. The change of trans into cis are happening when lycopene get a high temperature when cooking the tomato paste [9].

3.1.3 Yellowness

Yellowness of chicken sausage also increased with the addition of tomato paste. This is because of tomato pigment, beta-caroten and xantofil, which give a orange-yellowness color [12]. Beta-caroten and xantofil also have a cis isomer which can turn into trans when tomato are processed into paste. Processed tomato into paste can make beta-caroten and xantofil more soluble and homogenous in the chicken sausage so it has a yellowish color [13].

3.2 Cooking loss

Addition of different percentage of tomato paste (0%, 5%, 10%, and 15%) gave highly significant effect (P<0.01) on cooking loss. The average cooking loss of spent hen chicken sausage with the addition of tomato paste shown on Table 3.

Table 3. Average of cooking loss of chicken sausage with the addition of tomato paste

Treatments	Cooking loss (%)
SC	1.59 ± 0.20^{a}
S1	2.01 ± 0.06^{b}
S2	2.36 ± 0.09^{c}
S3	2.75 ± 0.10^{d}

Notes : superscripts on the column (^{a,b,c,d}) shown a highly significant effect (P<0.01)

Cooking loss are an analysis to find how much protein can hold water after meat batter getting cooked. Cooking loss can affect texture from meat product [14]. Cooking loss of chicken sausage with the addition of tomato paste are increasing as the percentage of tomato paste increase. The reason of this are the moisture content. When moisture content increase, it's possible that pH decrease when cooking loss increase [15]. Protein can also affect cooking loss. The higher cooking loss on the chicken sausage means that tomato paste can't stop syneresis. It because tomato paste component have a low ability to be a binder. Binder usually has a higher protein so it can make a good texture and emmulsion [16] but tomato paste contain higher carbohydrates (4.50%) than protein (1.44%).

3.3 Water Holding Capacity

Addition of different percentage of tomato paste (0%, 5%, 10%, and 15%) gave highly significant effect (P<0.01) on water holding capacity. The average water holding capacity of spent hen chicken sausage with the addition of tomato paste shown on Table 4.

Table 4. Average of water holding capacity of chicken sausage with the addition of tomato paste

Treatments	Water Holding Capacity (%)
SC	36.85 ± 0.76^{d}
S1	35.36 ± 0.22^{c}
S2	33.43 ± 0.43^{b}
S3	31.54 ± 0.54^{a}

Notes : superscripts on the column (^{a,b,c,d}) shown a highly significant effect (P<0.01)

Water Holding Capacity (WHC) are an analysis to find how much sample can bind the water on food products. WHC can affect the texture, especially juiciness, on sausage. The higher WHC means that the sample can bind water very well [17]. WHC on chicken sausage decrease as the percentage of tomato paste increase. It's because tomato paste contain high water. Tomato paste are an acid product because of ascorbat acid and it can affect WHC. When an acid component meet the protein it can lead to protein denaturation. Protein denaturation can make the water on sample didn't bind as much as usual so the water keep leaking on sample that means the sample didn't have a great binding capability [18].

3.4 Moisture Content

Addition of different percentage of tomato paste (0%, 5%, 10%, and 15%) gave highly significant effect (P<0.01) on moisture content. The average moisture content of spent hen chicken sausage with the addition of tomato paste shown on Table 5.

Table 5. Average of moisture content of chicken sausage with the addition of tomato paste

Treatments	Moisture Content (%)
SC	44.42 ± 0.81^{a}
S1	48.64 ± 0.51^{b}
S2	51.11 ± 0.37^{c}
S3	53.26 ± 0.23^{d}

Notes : superscripts on the column (^{a,b,c,d}) shown a highly significant effect (P<0.01)

Moisture content on chicken sausage are increase as the percentage of tomato paste increase. It's because tomato paste have a higher moisture content that are almost 94% [19]. Moisture content can indicate the stability and shelf life of product. The higher moisture content can improve the growth of microorganism [20]. The increased of moisture content also because the low fiber on tomato paste. Tomato paste have 0.35% fiber that can't bind the water that much. Pectin, tomato fiber, usually have the ability to bind the water in meat products but when it get an unstable – high temperature, it can lead to pectin degradation which mean pectin didn't have strong ability to bind water anymore [21].

4 Conclusion

The addition of 5% tomato paste on spent hen chicken meat are the best treatment on this research from the L a*b* color, cooking loss, water holding capacity, and moisture content.

References

1. M. S. El-Tarabany, O. A. Ahmed-Farid, S. M. El-Bahy, M. A. Nassan, A. S. Salah, Frontiers in Veterinary Science, **9**, 1-9 (2022)
2. S. A. Rather, F. A. Masoodi, J. A. Rather, R. Akhter, A. Gani, T. A. Ganaie, Food Chemistry, **359**, 1-9 (2021)
3. M. Neville, A. Tarrega, L. Hewson, T. Foster, Food Science and Nutrition, **5**, 852-864 (2017)
4. M. M. Silva and F. C. Lidon, Emirates Journal of Food and Agriculture, **28**, 366-373
5. T. A. Castro, B. S. Leite, L. S. Assuncao, T. d. J. Freitas, N. B. Colauto, G. A. Linde, D. M. Otero, B. A. S. Machado, C. D. F. Ribeiro, Molecules, **25**, 1-24 (2021)
6. S. R. Ghadage, K. A. Mane, R. S. Agrawal, V. N. Pawar, The Pharma Innovation Journal, **8**, 1245-1248 (2019)
7. E. A. Dincer, K. K. E. Caglar, M. Erbas, International Journal Food Engineering, **17**, 75-82 (2021)
8. F. M. Manihuruk, T. Suryati, I. I. Arief, Media Peternakan, **40**, 47-54 (2017)
9. M. U. Nasir, S. Hussain, S. Jabbar, Science Letters, **3**, 1-5 (2015)
10. M. Murata, Glycoconjugate Journal, **38**, 283-292 (2020)
11. D. Gorecka, A. Wawrzyniak, A. Jedrusek-Golinska, K. Dziedzic, J. Hamulka, P. L. Kowalczewski, J. Walkowiak, Open Chemistry, **18**, 752-756 (2020)
12. Arifulloh, I. Oktavianawati, I. N. A. Winata, Berkala Saisntek, **4**, 15-18 (2016)
13. H. E. Khoo, K. N. Prasad, K. W. Kong, Y. Jiang, A. Ismail, Molecules, **16**, 1710-1738 (2011)
14. S. M. V. Mejia, A. Shaheen, Z. Zhou, D. McNeill, B. M. Bohrer, Meat Science, **156**, 85-92 (2019)
15. M. V. Melendres, J. P. Camou, N. G. Torrentera Olivera, E. A. Almora, D. G. Mendoza, L. A. Reyes, H. G. Rios, Meat Science, **97**, 54-61 (2014)
16. K. Kyriakopoulou, J. K. Keppler, A. J. van der Gootm Foods, **10**, 1-29 (2021)
17. S. H. V. Cornet, S. J. E. Snel, J. Lesschen, A. J. van der Goot, R. G. M. van der Sman, Journal of Food Engineering, **290**, 1-9 (2020)
18. B. P. Ismail, L. Senaratne-Lenagaia, A. Stube, A. Brackenridge, Feeature Article, **10**, 54-63 (2020)
19. V.N. Latpate, Dr. AK. Singh, B. B. Renge, Journal of Pharmacognosy and Phytochemistry, **8**, 1254-1257 (2019)
20. M. V. Zambrano, B. Dutta, D. G. Mercer, H. L. Maclean, M. F. Touchie, Trends in Food Science and Technology, **88**, 484-496 (2019)
21. R. Gao, Z. Wu, Q. Ma, Z. Lu, F. Ye, G. Zhao, Foods, **10**, 1-21 (2021)

The effectiveness of cassava waste and skim milk as a filler of phytogenic and probiotic blends to inhibit the pathogenic bacteria and aflatoxin

Nindy Amelia Zalfa[1], *Yuli* Frita Nuningtyas[2], *Feri* Eko *Hermanto*[2], *Fajar* Shodiq Permata[3], and *Muhammad* Halim Natsir[2*]

[1]Master Student, Faculty of Animal Science, Universitas Brawijaya, Malang 65145, Indonesia
[2]Lecturer, Faculty of Animal Science, Universitas Brawijaya, Malang 65145, Indonesia
[3]Lecturer, Faculty of Veterinary Medicine, Universitas Brawijaya, Malang 65145, Indonesia

Abstract. This research aims to determine the effectiveness of cassava waste and skim milk as a filler of phytogenic and probiotic blends to inhibit phtytogenic bacteria and aflatoxin. The phytogenics and probiotics blends were used in the yellow type with the composition of Galangal *(Kaempferia rhizome)*, Temulawak *(Curcuma xanthorrihiza roxb)*, Red Ginger *(Zingiber officinale)*, Turmeric *(Curcuma domestica val)*, Actinomycetes, lactic acid bacteria, photosynthetic bacteria, tempeh yeast, and fermentation fungi *(Aspergillus)*; furthermore, the green type consists of Sambiloto (*Andrographis paniculata*), Betel *(Piper betle)*, Moringa *(Moringa Oliefera)*, Papaya *(Carica papaya)*, actinomycetes, lactic acid bacteria, photosynthetic bacteria, tempe yeast, and fermentation fungi *(Aspergillus)*. Each pytobiotic and probiotic in yellow and green types were filled with cassava waste and skim milk, then dried in an oven at 50 °C for 24 hours. The ratio between the combination of phytogenics and probiotics with the filler is 1:1. The inhibition was divided into four types, consisting of positive control, negative control, cassava waste, and skim milk. The inhibitory pathogenic bacteria and fungi used the Well Method. The data were analyzed using a complete randomized design. If there were a significantly different result, then the analysis of the Duncan Multiple Range Test (DMRT) would be continued. The results showed that control positive in the yellow and green type has the higher inhibitory pathogenic on salmonella, escherichia coli, and aflatoxin. However, the filler casava waste has higher inhibitory salmonella, escherichia coli, and aflatoxin compared to skim milk. The conclusion of this research showed that cassava waste is effective as a filler for phytogenic and probiotics as an inhibitory pathogenic bacteria and aflatoxin.

* Corresponding author: emhanatsir@ub.ac.id

1 Introduction

Salmonella bacteria have a significant impact on poultry production and cause systemic disease in broilers. *Salmonella* contamination of poultry products, such as eggs, is a major concern for food safety and human health. *Salmonella* can pass through feces, contaminating egg surfaces and posing a risk to consumers. *Salmonella* has a wide host range, can colonize the gastrointestinal tract of poultry, and transfer antibiotic resistance genes to other bacteria, thus contributing to the prevalence of drug-resistant microbes in commercial poultry. Therefore, it is important to find prophylactic options that can reduce the number of bacteria and increase poultry production. *Escherichia Coli* is a poultry pathogen that can cause systemic disease in both broilers and laying hens, causing high mortality rates. *Escherichia Coli* bacteria often contain many virulence factors and antimicrobial resistance, making them difficult to control. These bacteria can interact with heterophiles and macrophages of broilers, thus causing severe infections. It is important to find effective measures to reduce the colonization of *Escherichia Coli* and improve poultry health [21]. The poison produced by fungi, namely aflatoxin, can also be harmful to poultry production. This can inhibit the formation of liver proteins and lipid metabolism in the liver, causing a swollen and fatty liver. Aflatoxin can also lower cholesterol levels in broiler's serum and affect the role of hepatocytes. In addition, aflatoxin can contaminate feed and reduce its quality, thus negatively impacting poultry performance [1].

Probiotics can be used as an alternative to prevent the growth of *Salmonella* and *Escherichia Coli* bacteria. Such ingredients can increase the immunity of broilers and improve the vaccination response, thereby increasing protection against bacterial infections. Probiotics have been shown to stimulate the production of natural antibodies in broilers, which can help fight *Escherichia Coli* and *Salmonella* In addition, probiotics have the potential to reduce the number of bacteria and increase overall poultry production. It is important to continue to research and develop these strategies in order to effectively control *Escherichia Coli* and *Salmonella* in poultry and ensure optimal poultry health and food safety [21]. Phytogenic are natural and residue-free compounds derived from plants that have been used as feed additives for livestock production. They include various bioactive components or substances of plant origin such as terpenoids, alkaloids, glycosides, and phenolics. Phytogenic also has antimicrobial properties, fighting intestinal pathogenic microorganisms while maintaining a population of beneficial microflora. In addition, phytogenic has an immunostimulant effect on humoral and cellular immunity and has antioxidant properties. They are used as growth-promoting feed additives, antimicrobial agents, and immunostimulants in the poultry industry [7].

A combination of phytogenic and probiotic products. Type H phytogenic and probiotic (Green) with the composition of Sambiloto *(Andrographis paniculata),* Betel *(Piper betle)*, Moringa *(Moringa Oliefera)*, Papaya *(Carica papaya)*, actinomycetes, lactic acid bacteria, photosynthetic bacteria, tempeh, and fermented fungi *(Aspergilus)* and type K phytogenic and probiotic (Yellow) with the composition of Kencur *(Kaempferia rhizome)*, Temulawak *(Curcuma xanthorrihiza roxb)*, Red Ginger *(Zingiber officinale)*, Turmeric *(Curcuma domestica val)*, Actinomycetes, lactic acid bacteria, photosynthetic bacteria, tempeh yeast, and fermentation fungi *(Aspergilus)*. The bioactive content of phytogenic and probiotics contains antibiotic and antimicrobial compounds that are expected to inhibit the growth of *Salmonella*, *Escherichia Coli*, and aflatoxin fungi in the form of mash. The purpose of this study was to determine the best filler that will be combined with probiotics and phytogenic to inhibit *Salmonella*, *Escherichia Coli*, and aflatoxin fungi.

2 Material and Method

The method is carried out through 2 stages, namely, making a mixture of feed additives with fillers two kinds of filler, and the second is the inhibitory power test. The mixture was carried out in the Laboratory of Animal Feed Nutrition and the inhibitory pathogenic in the Epidemiology Laboratory of the Faculty of Animal Science, Universitas Brawijaya, Malang.

2.1 Mixing of Feed Additive with Fillers

The production of mash-shaped phytogenic and probiotics has 2 types of fillers used, namely cassava waste and skim milk. Mixing phytogenic and probiotic with each type of filler with the ratio of 1: 1 which is then homogenized by stirring. Then the mixture of phytogenic and probiotic mixture dough with each filler is applied thinly to the baking sheet. After spreading evenly, the dough is put in a 50°C oven for 24 hours. After that the dough will dry completely and take the mongering dough from the pan and puree it with a mixer.

2.2 Inhibition Test

2.2.1 Media Production

MHA (Mueller Hinton Agar) media is made by weighing 15.2 grams then dissolved into 400 mL aquades with a magnetic stirer by heating to boiling. Then the media is sterilized using an autoclave with heating to 121°C. Sterilized Mueller Hinton Agar media is poured into sterile petri dishes and allowed to stand at room temperature until solidified.

2.2.2 Bacteria Culture

SA (Sodium Agar) media weighed 0.4 grams then dissolved into 20 mL of aquades homogenized with a magnetic stirer, heat until boiling. Prepare 9 mL of peptone then sterilize both using an autoclave until the temperature is 121°C. Bacteria are taken from pigeon feces which will then be spread to SA (Sodium Agar) media isolated using selective media. Take 1 gram of pigeon feces, dissolve it with 1 mL of sterile aquades then fortex until homogeneous. Take 1 mL of feces using a reconstituted micropipette and place it in a sterile petri dish. Then pour sterile NA media into a petri dish containing feces and homogenize it. Let stand until the substrate hardens at room temperature. After hardening, put in the incubator with a temperature of 37 °C for 24 hours. So that colonies are formed.

2.2.3 Bacteria Isolation

Each bacterial isolation medium weighed, namely SSA (Salmonella Shigela Agar) for *Salmonella* 1.26 grams, EMBA (Eosin Methylen Blue Agar) for *Escherichia Coli* bacteria 0.75 grams, and MEA (Malt Agar Extract) for aflatoxin 0.96 grams each dissolved into aquades 20 mL and homogenized at once by heating to boiling using a magnetic stirer. Then sterilized using an autoclave to a temperature of 121 ^{0}C. Then pour into a sterile petri dish and let it stand until it hardens. Bacteria emerging from NA media were taken using swabs then stricked into each selective media. Put in the incubator with a temperature of 37°C for 24 hours.

2.2.4 Inhibition Test

MHA media weighed 22.8 grams dissolved into 600 mL of homogenized aquades and heated to boiling using a magnetic stirrer. Take a small amount of each colony of *Salmonella*, *Escherichia Coli*, and Aflatoxin bacteria using a swab. Then tricked into a cup containing MHA media. Form a wells by perforating the media with a kock borer then fill the wells hole with a phytogenic and probiotic sample of 0.1 mL. Then put in an incubator with a temperature of 37°C for 24 hours. Then measure the clear zone around the wells using a caliper.

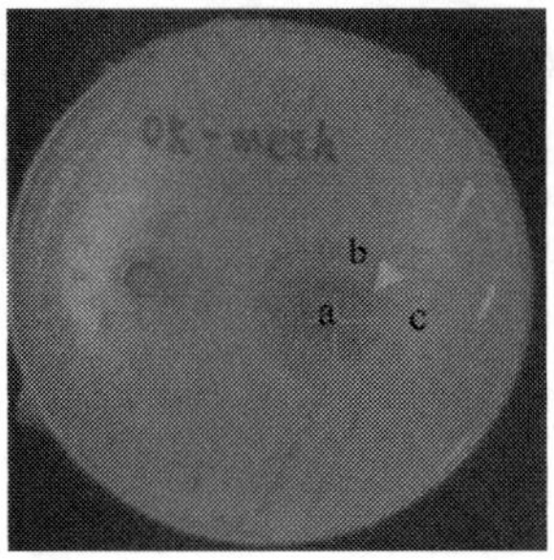

Figure 1. Measurement Inhibition Zone

Information :
a : horizontal diameter
b : vertical diameter
c : diameter of wells
The diameter of the inhibitory zone can be calculated using the following formula:

$$Inhibitory\ zone = \left(\frac{a+b}{2}\right) - c \qquad (1)$$

2.3 Statistical Analyse

The experiment was conducted in duplicate and the values were written in the average ± standard deviations tabulated in Microsoft Excel. Then continued statistical analysis was carried out using a one-way-analysis of variance (ANOVA) (SPSS version 27). Then continued with Duncan's analysis (LSD) on a statistically significantly different treatment (P<0.05).

3 Results and Discussion

Testing the inhibitory zone of *Salmonella*, *Escherichia Coli*, and aflatoxin fungi on a combination of phytogenic with probiotics using the wells method. There are 2 types of phytogenic and probiotic with different phytogenicsotic content, namely yellow phytogenic and probiotic and green phytogenic and probiotic. Phytogenic and probiotic in the form of mash using cassava waste filler and skim milk. Table 1 has presented the results of inhibitory power tests of both types of phytogenic and probiotic and two types of fillers.

3.1 The Effect of Phytogenic and Probiotic Yellow Type on The Growth Inhibition of Salmonella sp, Escherichia Coli, and Aflatoxin

Testing the inhibitory zone of *Salmonella*, *Escherichia Coli*, and aflatoxin on a combination of phytogenic with probiotics using the wells method. There are 2 types of phytogenic and probiotic with different phytogenicsotic content, namely phytogenic and probiotic yellow type and phytogenic and probiotic green type. Phytogenic and probiotic in the form of mash using cassava waste filler and skim milk. Table 1 has presented the results of inhibitory power tests of both types of phytogenic and probiotic and two types of fillers.

Table 1. Result of Inhibition Test *Salmonella sp.*, *Escherichia coli*, and Aflatoxin

Bacteria	Yellow Type				Green Type			
	Control +	Control -	Cassava Waste Filler	Skim Milk Filler	Control +	Control -	Cassava Waste Filler	Skim Milk Filler
Salmonella	35.25 ± 0.78	0.00 ± 0.00	5.76 ± 0.23	2.86 ± 0.09	34.85 ± 0.21	0.00 ± 0.00	2.86 ± 0.09	0.75 ± 0.07
E. coli	34.07 ± 1.15	0.00 ± 0.00	3.83 ± 0.81	1.64 ± 0.58	33.88 ± 0.88	0.00 ± 0.00	7.98 ± 0.39	0.86 ± 0.41
Aflatoxin	33.07 ± 1,12	0.00 ± 0.00	0.50 ± 0,14	0.26 ± 0.02	33.02 ± 1.20	0.00 ± 0.00	0.50 ± 0.35	0.13 ± 0.04
Average	33.13 ± 0.91	0.00 ± 0.00	3.36 ± 2.66	1.59 ± 1.30	33.92 ± 0.92	0.00 ± 0.00	3.78 ± 3.82	0.58 ± 0.39

Phytogenic and probiotic yellow type found the diameter of the inhibitory zone of the highest value was produced in positive control (33.25 ± 0.91), onggok filler (3.36 ± 2.66), skimmed milk filler (1.59 ± 1.30), and negative control (0.00 ± 0.00). The positive control gave the highest inhibitory zone diameter value due to the administration of antibiotics in the form of ciprofloxacin which can significantly inhibit the growth of *Salmonella sp* bacteria, *Escherichia Coli* bacteria, and aflatoxins. Cassava waste filler given a lower value than positive control but higher than skim milk filler and negative control. This is due to the huge starch content ranging > 65.90% which is the main form of complex carbohydrates in the product [6]. Starch can also be converted into glucose which can be a source of energy for probiotic bacteria present in bioherbs such as actiomycetes, LAB (Lactic Acid Bacteria), and fermented fungi (Aspergilus). While the value of the inhibitory zone in skim milk filler is lower than the positive control and cassava waste filler but higher than the negative control, this is due to the fact that skim milk does not contain significant amounts of glucose [21]. This also causes the value of the inhibitory zone in skim milk fillers to be lower when compared to onggok fillers which contain a lot of starch, an energy source of several probiotics.

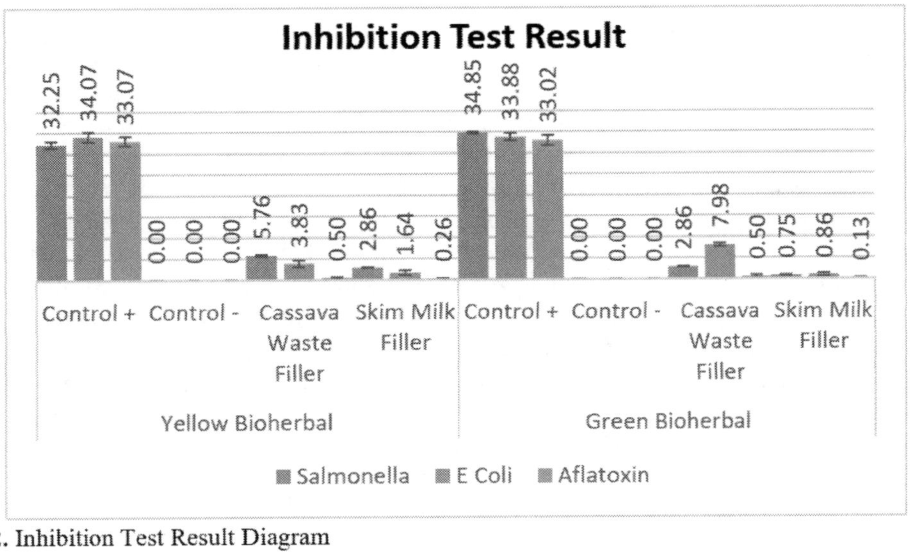

Fig. 2. Inhibition Test Result Diagram

The formation of inhibitory zones in yellow phytogenic and probiotic types against *Salmonella sp.* and *Escherichia Coli* is caused by several bioactive compounds it contains, namely the *Zingiber officinale* plant, namely ginger which can inhibit bacterial growth, including compounds such as Allicin, Allistatin, Ajoene, Gingerol, Gingerdiol, and Gingerdione. These compounds have therapeutic properties and play an important role in metabolism and antibiotic activity [12]. *Curcuma xanthorrhiza Roxb* contains curcumin which effectively inhibits bacterial growth and biofilm formation [27]. The antibacterial effect of *Curcuma domestica* is provided quickly financial and F. Discover ne domestica may be due to cell wall differences from Gram-positive and Gram-negative bacteria, allowing easier infiltration of hydrophobic compounds and increasing inhibitory effects on bacteria [17]. The purity of curcumin in extracted *Curcuma domestica* was found to be quite high compared to pure curcumin samples. The celery extraction with a concentration showed a higher inhibitory power compared to the use of antibiotics against *Salmonella* and *E Coli* bacteria [31]. The maximum weight percentage of curcumin is reported at 5% [18]. The probiotics in bioherbs also help inhibit *Salmonella sp.* bacteria. and *E Coli*. Actinomycetes are considered a source of bioactive compounds and are capable of producing various classes of antibiotics with diverse chemical structures and mechanisms of action. Actinomycetes produce different classes of antibiotics, including aminoglycosides, peptides, ansamycins, B-lactams, tetracyclines, macrolides, lincosamides, epoxides, and aminocoumarins. These antibiotics work on specific prokaryotic targets without damaging the eukaryotic system [8]. Actinomycetes contain about 75% of antibiotics, mainly antibacterial, produced by. This antibiotic has shown high effectiveness against the inhibition of Gram positive and Gram negative bacteria [16]. Lactic acid bacteria have several mechanisms that contribute to their ability to inhibit bacterial growth. One of the mechanisms is the production of antimicrobial compounds, such as organic acids (lactic, citric, acetic, fumaric, and malic acid), hydrogen peroxide , CO2, diacetyl, ethanol, reuterin, acetaldehyde, acetoin, ammonia, bacteriocin, bacteriocin, bacteriocin-like inhibition. In addition, lactic acid bacteria compete with pathogenic microorganisms for nutrition, which further contributes to antimicrobial effects [3]. Rhizopus oryzae can produce organic acids such as lactic acid, which has been shown to have antibacterial effects.

The formation of inhibitory zones in the yellow type phytogenic and probiotic against aflatoxin is caused by antifungal compounds in *Curcuma xanthorrhiza* called xanthorrhizol. It has been reported to reduce the germination of conidia of various fungi, including A. flavus, A. fumigatus, A. niger, F. oxysporum, R. oryzae, and T. mentagrophytes. Curcumin and demothoxycurcumin compounds inhibit fungal growth by damaging various cellular processes, such as damaging cell membranes, inhibiting cell wall synthesis, and disrupting fungal enzymes and proteins [4]. Actinomycetes isolated from hypersaline soil have shown antifungal activity against various fungi. Overall, actinomycetes are considered a good source of antibacterial and antifungal compounds. Antimicrobial compounds produced by actinomycetes can inhibit the growth and development of fungi by targeting specific cellular processes or structures [26]. In LAB, especially Lactobacillus plantarum, it has been shown to reduce aflatoxin production. In one study, Lactobacillus plantarum showed antifungal and antimycotoxigenic activity against *Aspergillus flavus* and *Fusarium verticillioides*, which are aflatoxin-producing fungi [22]. Yeast helps control aflatoxin through its binding capacity. The yeast cell wall contains glucomannan compounds that can bind to aflatoxin thereby preventing the harmful effects of aflatoxin [2]. Antiaflatoxin in *Aspergillus* involves evaluating the antifungal properties of various phytochemicals, such as ascorbic acid, gallic acid, caffeine, and quercetin. This phytochemical has been tested against two strains of *Aspergillus*, namely *Aspergillus flavus* and *Aspergillus parasiticus*. The results showed that the four phytochemicals succeeded in reducing the growth and activity of the two test mushrooms. Quercetin showed the highest inhibitory effect on both strains of *Aspergillus* [28].

3.2 The Effect of Phytogenic and Probiotic Green Type on The Growth Inhibition of Salmonella sp, Escherichia Coli, and Aflatoxin

Green type phytogenic and probiotic found the diameter of the inhibitory zone from the highest values produced in positive control (33.92 ± 0.91), onggok filler (3.78 ± 3.82), skimmed milk filler (0.58 ± 0.39), and negative control (0.00 ± 0.00). The positive control gave the highest inhibitory zone diameter value due to the administration of antibiotics in the form of ciprofloxacin which can significantly inhibit the growth of *Salmonella sp* bacteria, *Escherichia Coli* bacteria, and aflatoxins. Cassava waste filler given a lower value than positive control but higher than skim milk filler and negative control. This is due to the large amount of starch content in it which certainly affects probiotics, namely increasing their antioxidant content when put into biocomposite films and making energy sources for probiotic metabolism [1]. While the value of the inhibitory zone in skim milk filler is lower than the positive control and cassava waste filler but higher than the negative control, this is due to the fact that skim milk does not contain significant amounts of glucose [21]. This also causes the value of the inhibitory zone in skim milk fillers to be lower when compared to onggok fillers which contain a lot of starch, an energy source of several probiotics.

The formation of inhibitory zones in green phytogenic and probiotic types against *Salmonella sp.* and *Escherichia Coli* due to the content of phytogenic and probiotic bioactive compounds, such as in *Andrographis paniculate* there are antimicrobial compounds that can inhibit bacteria through various mechanisms. For example, andrographolide compounds can affect the Quorum Sensing System (QSS), the communication system between bacteria, thereby reducing biofilm production and inhibiting virulence factors. It can also restore antibiotic sensitivity in P. aeruginosa by reducing efflux pump expression. Other compounds in *Andrographis paniculata* have been shown to damage bacterial cell membranes, inhibit bacterial enzymes, and impair bacterial DNA replication and transcription [13]. Antimicrobial compounds found in *Piper betle*, such as hydroxychavicol, eugenol, chavibetol, and allylpyrocatechol, can inhibit bacteria through various mechanisms. For example, hydroxychavicol has been shown to disrupt bacterial cell membranes, leading to cell death. Eugenol has been found to inhibit bacterial cell wall synthesis and interfere with bacterial DNA replication. Chavibetol has been shown to inhibit the growth of some strains of bacteria, including Staphylococcus aureus, by disrupting bacterial cell membranes. Allylpyrocatechol has been found to inhibit the growth of several strains of bacteria, including Escherichia and Pseudomonas aeruginosa, by interfering with bacterial DNA replication and transcription. Overall, the antimicrobial activity of *Piper betle* is likely due to a combination of these mechanisms, as well as other factors such as concentration and mode of delivery of the compound [11]. In addition, Rhizopus oryzae may produce enzymes that may have antimicrobial properties that may inhibit pathogenic bacteria [19]. *Aspergillus* acts as an antibacterial agent through the production of secondary metabolites called antimicrobial compounds. The compound is produced as a fungal defense mechanism to inhibit bacterial growth in its environment [23]. Actinomycetes are known to be a rich source of natural products, including antibiotics. Mycelial bacteria produce most of the antibiotics that help in inhibiting the growth of *Salmonella sp.* and *Escherichia Coli* [30]. The antibacterial properties of their potential LAB (Lactic Acid Bacteria), which can be attributed to the production of organic acids, hydrogen peroxide, and bacteriokine. These substances help inhibit the growth of harmful bacteria and contribute to the preservation of food. In addition, lactic acid bacteria have been investigated for their potential use as probiotics, which are beneficial bacteria that can improve gut health and boost the immune system [10].

The formation of an inhibitory zone in the green type phytogenic and probiotic against aflatoxin is due to the herbal plant *Andrographis paniculata* containing several antifungal compounds. One of its main active compounds is andrographolide, which has been extensively studied for its antifungal properties by damaging fungal cell membranes leading to leakage of cell contents and cell death. It can also inhibit the growth and reproduction of fungi by interfering with the metabolic processes of fungi. In addition, andrographolide has been found to modulate the immune system and increase the body's natural defenses against fungal infections [13]. Antifungal activity in actinomycetes that produces compounds that can inhibit the growth and reproduction of fungi, thereby preventing or treating fungal infections. The specific compound mentioned in the document is Caerulomycin A [5]. The production of organic acids and other metabolites in LAB (Lactic Acid Bacteria) creates an unfavorable environment for fungal growth. In addition, lactic acid bacteria can also produce antifungal compounds called bacteriocins, which can inhibit the growth of certain fungi [10]. Disruption of cell membranes in the antifungal compound *Aspergillus* leads to leakage of cell contents and eventually cell death. Another class of antifungal agents, polyenes such as amphotericin B, bind to ergosterol in fungal cell membranes, forming pores that disrupt membrane integrity and cause cell death. Echinocandins, such as caspofungin and micafungin, inhibit the synthesis of beta-glucan, a key component of fungal cell walls, leading to weakening of cell walls and cell death. These are just a few examples of mechanisms by which antifungal agents target and kill *Aspergillus* fungi [29].

4 Conclusion

The conclusion that can be drawn from this study is the best filler that can help in inhibiting the growth of *Salmonella, Escherichia Coli*, and aflatoxin from phytogenic and probiotic type yellow namely cassava waste and phytogenic and green probiotic type which is also cassava waste.

References

1. Abotbina, W., Sapuan, S. M., Ilyas, R. A., Sultan, M. T. H., Alkbir, M. F. M., Sulaiman, S., & Bayraktar, E. *Materials*, **15**(19), 6992. (2022)
2. Afzal, N., Hassan, S. M., Mughal, S. S., Pando, A., & Rafiq, A.. *American Journal of Chemical and Biochemical Engineering*, **6**(1), 21-26. (2022)
3. Agriopoulou, S., Stamatelopoulou, E., Sachadyn-Król, M., & Varzakas, T. *Microorganisms*, **8** (6), 952. (2020)
4. Akter, J., Amzad Hossain, M., Sano, A., Takara, K., Zahorul Islam, M., & Hou, D. X.. *Pharmaceutical Chemistry Journal*, **52**, 320-325. (2018)
5. Ambavane, V., Tokdar, P., Parab, R., Sreekumar, E. S., Mahajan, G., Mishra, P. D., ... & Ranadive, P. *Advances in Microbiology*, (2014)
6. Amini, A. PHYTOGENIC AND PROBIOTIC., & Larasati, T. D.. *Jurnal Chemurgy*, **6** (2), 70-79. (2022)
7. Chalupa-Krebzdak, S., Long, C. J., & Bohrer, B. M. *International dairy journal*, **87**, 84-92. (2018)
8. De Simeis, D., & Serra, S. *Antibiotics*, **10** (5), 483. (2021)
9. El-Ghany, A. *Journal of World's Poultry Research*, **10** (4), 571-579. (2020)
10. Gajbhiye, M. H., & Kapadnis, B. P. *Biocontrol science and technology*, **26**(11), 1451-1470. (2016)

11. Gupta, R. K., Guha, P., & Srivastav, P. P. *Acta Ecologica Sinica*, **43** (5), 721-732. (2023)
12. Helen, O. N., Job, A. O., & Oghenebrorhie, O. *Adv. Anim. Vet. Sci*, **8** (10), 1019-1027. (2020)
13. Hossain, S., Urbi, Z., Karuniawati, PHYTOGENIC AND PROBIOTIC., Mohiuddin, R. B., Moh Qrimida, A., Allzrag, A. M. M., & Capasso, R. *Life*, **11**(4), 348. (2021)
14. Hu, Z. C., Li, PHYTOGENIC AND PROBIOTIC. J., Zou, S. P., Niu, K., & Zheng, PHYTOGENIC AND PROBIOTIC. G. *Preparative biochemistry & biotechnology*, **50** (8), 745-752. (2020)
15. Jagannathan, S. V., Manemann, E. M., Rowe, S. E., Callender, M. C., & Soto, PHYTOGENIC AND PROBIOTIC. *Marine Drugs*, **19** (7), 365. (2021)
16. Jakubiec-Krzesniak, K., Rajnisz-Mateusiak, A., Guspiel, A., Ziemska, J., & Solecka, J. *Polish journal of microbiology*, **67** (3), 259-272. (2018)
17. Kebede, B. PHYTOGENIC AND PROBIOTIC., Forsido, S. F., Tola, PHYTOGENIC AND PROBIOTIC. B., & Astatkie, T. *Heliyon*, **7**(2). (2021)
18. Kiamahalleh, M. V., Najafpour-Darzi, G., Rahimnejad, M., Moghadamnia, A. A., & Kiamahalleh, M. V. *Journal of Chromatography B*, **1022**, 191-198. (2016).
19. Londoño-Hernández, L., Ramírez-Toro, C., Ruiz, PHYTOGENIC AND PROBIOTIC. A., Ascacio-Valdés, J. A., Aguilar-Gonzalez, M. A., Rodríguez-Herrera, R., & Aguilar, C. N. *International journal of food microbiology*, **257**, 110-127. (2017).
20. Mallikarjunaswamy, G. E., & Noushad, N. *Current Plant Biology*, **27**, 100217. (2021)
21. Mehrabani, S., Safavi, S. M., Mehrabani, S., Asemi, M., Feizi, A., Bellissimo, N., & Salehi-Abargouei, A. *European journal of nutrition*, **55**, 1389-1396. (2016).
22. Nazareth, T. D. M., Luz, C., Torrijos, R., Quiles, J. M., Luciano, F. B., Mañes, J., & Meca, G. *Toxins*, **12** (1), 21. (2019)
23. Pinheiro, E. A. A., Carvalho, J. M., dos Santos, D. C. P., Feitosa, A. D. O., Marinho, P. S. B., Guilhon, G. M. S. P., & Marinho, A. M. D. R. *Natural product research*, **27** (18), 1633-1638. (2013)
24. Redweik, G. A., Stromberg, Z. R., Van Goor, A., & Mellata, M. *Poultry science*, **99** (2), 752-762 (2020)
25. Saeed A, PHYTOGENIC AND PROBIOTIC., & Salam A, I. Current limitations and challenges with lactic acid bacteria: a review. *Food and Nutrition Sciences*, (2013)
26. Sarika, K., Sampath, G., Govindarajan, R. K., Ameen, F., Alwakeel, S., Al Gwaiz, PHYTOGENIC AND PROBIOTIC. I., & Ravi, G. *Saudi Journal of Biological Sciences*, **28** (6), 3553-3558 (2021).
27. Septama, A. PHYTOGENIC AND PROBIOTIC., Tasfiyati, A. N., Kristiana, R., & Jaisi, A. *South African Journal of Botany*, **146**, 728-734 (2022)
28. Tiwari, S., Gupta, N., Malairaman, U., & Shankar, J. *National Academy science letters*, **40**, 267-271 (2017).
29. Yerbanga, I. W., Diallo, S. N., Rouamba, T., Denis, O., Rodriguez-Villalobos, H., Montesinos, I., & Bamba, S. *Journal of Medical Mycology*, **33**(1), 101328. (2023)
30. Zhu, H., Sandiford, S. K., & van Wezel, G. P. *Journal of Industrial Microbiology and Biotechnology*, **41**(2), 371-386. (2014)
31. Nuningtyas, Y.F., Sjofjan, O., Djunaidi,I.H., and Natsir, M.H. *IOP Conf. Ser.: Earth Environ. Sci.*, **411**. (2019)

Effect of extraction *Andrographis paniculata* and *Moringa oleifera* on microbiological jejunal of broiler chickens

Ilham Fithrah Hasanain[1], *Yuli* Frita Nuningtyas[1], *Filoza* Marwi[1], *Feri* Eko Hermanto[1] and *Muhammad* Halim Natsir[1*]

[1]Faculty of Animal Science, Universitas Brawijaya, Malang 65145, Indonesia

Abstract. This study aims to evaluate the effect of the combined extraction of *Andrographis paniculata* and *Moringa oleifera* leaves as phytobiotics on the digestive microbiology of broiler chickens. The number of chickens used in this study was one hundred twenty-eight (128) chickens divided into 4 treatments, 4 replications where each replication consisted of 8 chickens. The treatments in this study were T0 (control), T1 (basal diet + 0.25% *Andrographis paniculata* and *Moringa oleifera* leaves), T2 (0.50% *Andrographis paniculata* and *Moringa oleifera* leaves), and T3 (0.75% *Andrographis paniculata* and *Moringa oleifera* leaves). The method used in this research is in vivo. Data analysis used a completely randomized design. If there is a significant difference, it is analysed using the Duncan Multiple Range Test (DMRT). The results of the study showed that the use of *Andrographis paniculata* and *Moringa oleifera* leaves had a significant effect on the number of lactic acid bacteria, Escherichia coli and Salmonella in the digestion of broiler chickens. T3 showed the highest number of lactic acid bacteria compared to other treatments, while T3 showed the lowest number of E. coli and Salmonella. The use of *Andrographis paniculata* 0.75% and *Moringa oleifera* leaves had a positive influence on the number of lactic acid bacteria in the digestion of broiler chickens.

1 Introduction

Poultry farming has the potential to be developed to meet food needs for animal protein sources. However, the toughest challenges for breeders are the emergence of diseases that can cause losses and the prohibition of the use of Antibiotics Growth Promoter (AGP). So it becomes an obstacle for breeders to completely abandon the use of AGP. This is because AGP can have a negative effect on the poultry digestive system. Apart from that, the use of AGP can leave residue in the bird's body and the final product is meat or eggs which will be consumed by humans. This makes the product unsafe for consumption. Several researchers have presented previously related to the alternative for AGPs including the use of enzymes, lactic acid bacteria, or even the use of prebiotics [1-3].

* Corresponding author: emhanatsir@ub.ac.id

However, coccidiosis is a disease caused by parasites from the genera Eimeria and Isospora which are included in the phylum Apicomplexa with a complex life cycle [1]. This parasite will attach to the wall of the small intestine and affect the absorption of nutrients, resulting in damage to other organs [2]. This disease is transmitted through feces, especially in birds that live in groups in large populations and high densities [3]. So coccidia will reduce poultry productivity and require quite large treatment costs. Herbal plants that have the potential to act as anticocidiosis are Moringa leaves and bitter leaves. Chemical compounds identified in the Moringa plant include Alkaloids, Flavonoids, Saponins, Terpenoids, Tannins, Steroids [4]. At the same time, the pharmacological activities found in the Moringa plant are antioxidant, antihyperuricemic, analgesic, anticancer, antihyperglycemic, anti-inflammatory, antihyperlipidemic and antimicrobial activities [4]. In Addition, bitter leaves contain orthosiphon glucose, essential oils, saponins, polyphenols, flavonoids, sapophonins, potassium salts and myonositol [5]. Based on the 2 problems above, namely replacing AGP and coccidiosis and looking at the potential opportunities for the content of bitter and Moringa leaves, the aim of this research is to utilize herbal plants as an innovative replacement for AGP as well as anti-coccidiosis in poultry. The urgency or priority of this research is to prevent coccidiosis caused by parasites using active substances from herbal plants. This is also an effort to replace antibiotic growth promoter (AGP) in poultry. The use of herbal plants is expected to reduce disease and improve the quality of products in the form of chicken meat. Products produced from poultry in the form of healthy meat will support the availability of quality food. The novelty in research on NGP herbal products from a combination of *Andrographis paniculata* and *Moringa oleifera* leaves is 1) NGP products are produced which are able to prevent the emergence of coccidiosis, 2) NGP herbal products use crude herbal extract technology and nanoparticle technology, 3) NGP herbal products are used as chicken feed additives with 2 forms, namely extraction. The purpose of this research is to produce quality herbal products to prevent or treat coccidiosis and increase chicken productivity.

2 Materials and Methods

2.1 Experimental design

The materials used in this research were 128 broiler chickens obtained from PT JAPFA Comfeed. Other ingredients used are *Andrographis paniculata* extract and *Moringa* leaf extract obtained from the Center for Research and Development of Medicinal Plants and Traditional Medicine. The research was conducted at the Sumbersean field laboratory located in Dau District, Malang Regency. The method used in this research was in vivo using 4 treatments and 4 replicated consisting of 32 broiler chickens as follows:
T0= Control without additional treatment
T1= Use 0.25% mixture of Bitter and *Moringa* extracts in liquid form
T2= Use 0.50% mixture of Bitter and *Moringa* extracts in liquid form
T3= Use 0.75% mixture of Bitter and *Moringa* extracts in liquid form
Data was analysed using a completely randomised design. Completely randomised design is a type of experimental design where treatment is given randomly to all experimental units. This can be done because the environment where the experiment is held is relatively homogeneous so that the media or place of the experiment does not have a significant influence on the observed response. If there were a significant difference, then analysis with Duncan Multiple Range Test (DMRT).

3 Results and discussion

. Lactic acid bacteria (LAB) is non-pathogenic bacteria that has an important role in the digestive tract. The bacteriocin properties of herbal ingredients can have an antagonistic effect on the growth of pathogenic bacteria, thereby reducing bacterial growth and improving the microflora in the intestinal [6]. For instance, the lactic acid bacteria, while pathogenic bacteria in the poultry digestive tract that have high mortality are, for example, Escherichia coli. Lactic acid bacteria can be grouped as probiotic bacteria which can be mixed with locally formulated feed or commercial feed. Lactic acid bacteria are able to kill pathogenic bacteria, one of which is Escherichia coli which can cause colibacilosis.
Based on the results of the analysis of the variety of lactic acid bacteria (LAB) in table 1 in digesta broiler chickens, it shows that there is no significant effect (P>0.05). The differences in lactic acid bacteria were not significant, indicating that up to a dose of 0.75% the combination of bitter and moringa extracts had no effect on chicken digesta. This is thought to occur because broiler chickens have a good digestive system because there are many positive bacteria that support the chicken's digestive system to work optimally. Apart from that [8] stated that differences in total lactic acid bacteria were influenced by differences in age and the influence of probiotic administration. The highest average number of lactic acid bacteria was obtained from treatment T3, namely 39.15 x 10^7 CFU/ml and the lowest average number of lactic acid bacteria in treatment T0 was 6.6 x 10^7 CFU/ml. The average percentage of lactic acid bacteria in broiler chickens from the lowest respectively is T0 (6.6 x 10^7 CFU/ml), T1 (8.95 x 10^7 CFU/ml), T2 (11.95 x 10^7 CFU/ml) and T3 (39.15 x 10^7 CFU/ml). T3 is the best treatment because it has the highest average number of lactic acid bacteria (LAB) compared to other treatments.
Absorption of nutrients in feed requires good digestive tract function to optimise the digestive process [9]. Intestinal health is influenced by the population of microbes or bacteria that live in it [10]. Lactic acid bacteria can be grouped as probiotic bacteria which can be mixed with locally formulated feed or commercial feed. Lactic acid bacteria are able to kill pathogenic bacteria, one of which is Escherichia coli which can cause colibacilosis [7].

Table 1. The Results of analysis various effects of *Andrographis paniculata* and *Moringa oleifera* extraction on the microbiology of broiler chicken Digista

Treatment	*Lactic Acid Bacteria*	*Escherichia Coli*	*Salmonella*
T0	6.6 ± 0.282	53.5 ± 2.828^{cd}	26.95 ± 6.85^{cd}
T1	8.95 ± 1.626	30.95 ± 12.51^{b}	1645 ± 10.39^{bc}
T2	11.95 ± 3.88	41.5 ± 6.5^{bc}	6.40 ± 2.12^{ab}
T3	39.15 ± 17.88	13.35 ± 2.89^{a}	0 ± 0^{a}

Note: The different superscripts in columns indicate significant differences (P<0.05).

Escherichia coli (*E. coli*) is a bacterium that lives in the digestive tract of warm-blooded animals, including mammals and birds. This bacterium was first isolated by Theodor in 1885. E.coli is a rod-shaped Gram-negative bacteria. E. coli cells have a length of 2.6 – 6.0 μm and a diameter of 1.1 – 1.5 μm, singly or in pairs and are non-motile with *peritrichous flagella* [11]. Based on the results of the analysis of Escherichia coli bacteria in broiler chicken digesta, it showed a significant effect (P<0.05). The lowest average number of *E. coli* bacteria was obtained from treatment T3, namely 13.35 x 10^4 CFU/ml and the highest average number of E.coli bacteria in treatment P0 was 53.5 x 10^4 CFU/ml. The average percentage of broiler chicken carcasses from the lowest respectively is T3 (13.35 x 10^4 CFU/ml), T1 (30.95 x 10^4 CFU/ml), T2 (41.5 x 10^4 CFU/ml) and T0 (53.5 x 10^4 CFU/ml). T3 is the best treatment because it contains 0.75% bitter and Moringa leaf extracts.

Escherichia coli lives in the digestive tract of poultry, especially in the jejunum, ileum and cecum. Apart from that, it is also often found in the trachea and oesophagus. Several clinical symptoms caused by pathogenic E. coli are grouped into *E. coli* which causes diarrhea, septicemia and Avian Pathogenic Escherichia coli (APEC). In broiler chickens, *colibacillosis* has a negative impact and causes death during the rearing period so that body weight at harvest is below standard with varying morbidity rates and mortality reaching 5-20% [12]. The addition of bitter and moringa extracts can reduce the presence of E. coli bacteria in the digestive tract. Moringa leaves contain flavonoids, sterols, triterpenoids, alkaloids, saponins, tannins and phenols while bitter leaves contain flavonoid chemical compounds [13-14]. Apart from that, bitter also contains alkanes, ketones, aldehydes and minerals [15].

Moringa leaves as additive feed for broiler chickens is reported to be a strong antioxidant that can protect and maintain the condition of chickens against oxidative stress, thereby providing results in the form of better growth rates and carcass quality [16]. Moringa also contains amino acids such as scordinine, methionine, lysine, and cystine which can stimulate chicken growth, increase body weight and increase energy [17]. Sambiloto has activity in maintaining beneficial gastrointestinal microflora, including lactobacilli, and inhibiting harmful bacteria, namely E. coli and Staphylococcus [18]. The flavonoid compounds in bitter and moringa act as antioxidants and can support the immune system. The flavonoids and tannins in bitter have potential as an antelemintic [19].

Broiler chickens can be contaminated with Salmonella sp. starting from livestock which is influenced by the cage and livestock environment [20]. Salmonella sp. rod-shaped, perithtic flagella for movement, has no spores, is gram negative, measures 0.5-0.8 μm in diameter and 1-3.5 μm in length. Salmonella sp. grows easily on simple media and almost never ferments lactose or saccharose and forms acid and sometimes produces gas from glucose and mannit and gives negative results in the indole reaction. The average colony size is 2-4 mm [21]. Infection with *Salmonella sp.* called Salmonellosis. Salmonellosis is an infection that can disrupt the digestive tract and can cause death in broiler chickens [22].

Based on the results of the analysis of *Salmonella sp.* in broiler chicken, digesta showed a significant effect ($P<0.05$). The average number of *Salmonella spp bacteria*. The lowest was obtained from treatment T3, namely 0.00 x 10^3 CFU/ml and the average number of *Salmonella sp.* The highest in treatment T0 was 26.95 x 10^3 CFU/ml. The total average of *Salmonella spp* bacteria. from the lowest respectively, namely T3 (0.00 x 10^3 CFU/ml), T2 (6.4 x 10^3 CFU/ml), T1 (16.45 x 10^3 CFU/ml) and T0 (26.95 x 10^3 CFU/ml). ml). T3 was the best treatment with the addition of 0.75% bitter and Moringa leaf extract. P3 has an average number of Salmonella sp bacteria. lowest compared to other treatments. This is caused by the influence of compounds contained in bitter extract and Moringa leaves.

Moringa leaf extract has antioxidant activity which is known to scavenge free radicals and provides significant protection against oxidative damage, so Moringa leaves can be used as an additive to animal feed [23]. The most prominent content of the Moringa plant is antioxidants, especially in the leaves which contain high antioxidants. Based on phytochemical tests, Moringa leaves contain tannins, steroids, triterpenoids, flavonoids, saponins, interquinones and alkaloids, all of which are antioxidants [24]. Sambiloto (*Andrographis paniculata*) is a plant that can prevent increases in cholesterol levels. Sambiloto contains phybiotics and contains an active substance known as andrographolide, which gives a bitter taste and has anti-toxic, cancer-preventing, anti-allergic, anti-inflammatory and anti-bacterial [25].

Moringa contains nutrients and compounds that are important for the body, Moringa leaves also contain phytochemical substances such as tannins, steroids, triterpenoids, flavonoids, saponins, anthraquinones and alkaloids. This compound has the ability to act as an antibiotic, anti-inflammatory, detoxification and antibacterial drug, [26]. Compounds, saponins, flavonoids, tannins and terpenoids can inhibit bacterial growth [27]. Flavonoids which are classified as phenols are antibacterial and are able to denature proteins and damage bacterial cell membranes [28].

Antimicrobial active ingredients can destroy bacterial cell membranes and break down bacteria by increasing cell wall permeability. With the presence of antimicrobials, the nutrients in the ration can be optimally absorbed by the digestive tract, thereby increasing the efficiency of ration use [29]. Apart from that, according to [30] the addition of Moringa leaf extract can increase protein levels in the digestive tract so that lactic acid bacteria can also increase and suppress pathogenic bacteria. Phytogenic can inhibit pathogenic bacteria such as *escherichia coli* and *Salmonella sp*. [31].

4 Conclusion

In conclusion, this study underlines the significance of carefully managing egg storage durations for Arab Chickens, with a particular emphasis on minimising adverse effects on growth-related parameters. The hatching eggs of Arab chickens can be safely stored for up to 4 days without any detrimental effect on their performance during the first week post-hatch. *Andrographis paniculata* and *Moringa oleifera* extract 0.75 % has a positive effect on the amount of microbiology in the digesta of broiler chickens.

The authors acknowledge financial support from research grants doctor Number 979/UN10.F05/PN/2023, chaired by Prof. Dr. Ir. M. Halim Natsir, S.Pt., MP., IPM., ASEAN Eng.

References

1. Sjofjan, O., D. N. Adli, M. H. Natsir, Y. F. Nuningtyas, I. Bastomi, F. R. Amalia. JITAA, **46** (2021)
2. Adli, D.N., O. Sjofjan, M.H. Natsir, Y.F. Nuningtyas, N. Sholikah, A. C. Marbun. Livest. Res. For Rur. Dev. **32** (7) (2020).
3. Sjofjan, O., M. H. Natsir, D. N. Adli, D. D. Adelina, L. M. Triana. In IOP Conf. Ser.: Earth and Env. Sci. **465** (2020).
4. Sawitti, M.Y., H. Mahatmi, I.N.K. Besung, Ind. Med. Vet. **2**, (2013).
5. Peek, H. W. University of Utrecht (2010).
6. Mesa-Pineda, C., J.L. Navarro-Ruíz, S. López-Osorio, J.J. Chaparro-Gutiérrez, L.M. Gómez-Osorio, Front. in Vet. Sci. **8** (2021).
7. Bhattacharya, A., P.Tiwari, P.K. Sahu, S. Kumar, J. of Pharm. & Bioall. Sci. **10**, (2018).
8. Hossain, M.D.S., Z. Urbi, A. Sule, K.M.H. Rahman, The Sci. World J. (2014).
9. Purwati, E., S. Syukur, Dharma Wanita Assoc. of West Sumatra Province, Padang (2006).
10. Kabosu, Y.H.L., U.D. Frans, I.R.D. Annytha, Ind. Vet. J. **6**, (2023).
11. Millah, F., D.P. Fatan, Y. Turrini, S. Sugiharto, Nat. Sem. on UNS Anim. Husbandry Study Pro, (2016).
12. Widodo T. S., B. Sulistiyanto, C.S. Utama, Agr. J. **15**, (2015).
13. Pertiwi D.D.R., R. Murwani, T. Yudiarti, Ind. Anim. Husband. J., **19** (2017).
14. Sumampouw, O. J, Deepublish (2019).

15. Wahyuwardani, S., S.M. Noor, M. Nurhasanah, Proceed. of the Nat. Sem. on Anim. Husband. and Vet. Tech., **504** (2015).
16. Jola, J.M.R. Londok, J.S. Mandey, J. for Sci. and Tech. **1** (2014).
17. Ikalinus, R., S.K. Widyastuti, N.L.E. Setiasih, J. of Trop. Anim. Husband **4** (2015).
18. Illah, Z.A., R.D. Ratnani, Suwardiyono, I. Hartati, Moment. **10** (2014).
19. Cwayita, W, Univ. of Fort Hare. Alice. South Africa (2014).
20. Kusmardika, D. A., J. of Healt. Sci. Phy. **1** (2019).
21. Guther, K.D., M. Ulfah, 4th Buffalo Symp. New Delhi. India (2003).
22. Indrati, R., M. Sawitri, U. Wisaptiningsih in Proceedings of the National Seminar on Animal Husbandry and Veterinary Technology. 447–455, (2019).
23. Diyana, U., E. Erina, M. Abrar, Vet. Stud. Sci. J. (2021).
24. Rizki, R. P., A. Zainul, A. Inayatul, Medicra **5** (2022).
25. Sartika, D., Susilawati, G. Arfani, J. of Ind. Tech. and Agr. Prod. **21** (2016).
26. Agoyi, E. E., E. A. Padonou., W. Amoussa., E. Achille., Assogbadjo, R. G. Kakaï., B. Sinsin. IJAAR. **6** (2015).
27. Kasolo, J. N., G. S. Bimenya., L. Ojok., J. Ochieng., J. W. Ogwal-okeng, J. of Med. Plants Res. **4** (2010).
28. Joseph, J., J. Solomon, Med. Arom. Plants **3** (2014).
29. Mardiana, L, Penebar Swadaya: Jakarta (2013).
30. Ariani, I.D. Insan Cendekia Medika. Jombang (2017).
31. Nuningtyas, Y.F., O. Sjofjan, I.H. Djunaidi, M.H. Natsir. In IOP Conference Series Earth Environmental Science. **411** (2019)

The effect of leguminous leaf meal substitution as a tannin source in complete feed on Rumen Degradable Protein (RDP), Rumen Undegradable Protein (UDP), and Digestible Energy (DE) using *in vitro* analysis

Siti Chuzaemi[1], *Mashudi* Mashudi[1], *Poespitasari Hazanah* Ndaru[1*], *Herni* Sudarwati[1], and *Lilia Putri* Agustina[2]

[1] Lecturer of Animal Science Faculty, Department of Animal Science, Universitas Brawijaya. Indonesia

[2] Student of Animal Science Faculty, Department of Animal Science, Universitas Brawijaya. Indonesia

Abstract. This study aimed to assess the impact of substituting concentrate with leguminous leaf meal as a source of tannins in complete feed on Rumen Degradable Protein (RDP), Rumen Undegradable Protein (UDP), and Digestible Energy (DE) content in vitro. The complete feeds were formulated using ingredients such as corn stover, soybean meal, copra meal, cassava meal, rice bran, pollard, molasses, minerals, calliandra meal, gamal meal, and indigofera meal. The experiment included ten treatments, each replicated three times. The treatments were as follows: T1 complete feed (60% corn stover + 40% concentrate), T2 complete feed (60% corn stover + (35% concentrate + 5% calliandra meal), T3 complete feed (60% corn stover + (35 % concentrate) + 5% indigofera meal), complete feed T4 (60% corn stover + (35% concentrate + 5% gamal meal), T5 complete feed (60% corn stover + (30% concentrate + 10% calliandra meal), T6 complete feed (60% corn stover + (30% concentrate + 10% indigofera meal), complete feed T7 (60% corn stover + (30% concentrate + 10% gamal meal), complete feed T8 (60% corn stover + (25) % concentrate + 15% calliandra meal), complete feed T9 (60% corn stover + (25% concentrate + 15% indigofera meal), complete feed T10 (60% corn stover + (25% concentrate + 15% gamal meal). The conclusion of this study reveals that the addition of 15% calliandra meal reduced RDP by 46.56% and increased UDP by 53.44%. Furthermore, the inclusion of 15% gliricidia meal led to an increase in DE value by 3.46 MJ/KgDM.

* Corresponding author: poespitasarihn@ub.ac.id

1 Introduction

Ruminant livestock are typically provided with forage, which is animal feed characterized by low protein content, ranging from 7-12% (Sutardi, 1981; Ensminger, 2001), but it contains a high level of crude fiber exceeding 18%. Corn plantations are among the forages suitable for animal feed, encompassing stems, leaves, and other parts of the corn plant with elevated water content (Ahmad et al., 2020). Athori (2023) notes that corn kernels exhibit substantial nutritional value, including 9.84% crude protein, 25.2% crude fiber, and 3.86% crude fat, accompanied by a Nitrogen-Free Extract content of approximately 59.9%. While providing forage alone is insufficient to fulfill the nutritional requirements of ruminant livestock, it is essential to supplement their diet with foods rich in protein to meet both production and breeding demands, such as concentrates. Concentrates serve as a valuable source of energy and protein, offering easily digestible nutrients that contribute to enhancing livestock productivity. Nevertheless, breeders often encounter the challenge of the relatively high cost associated with concentrate feeds. As a result, the expense of maintaining ruminant livestock is predominantly driven by the expenditure on feed, constituting approximately 70%. This challenge can be addressed by adopting alternative local feeds, with legume leaves emerging as a viable protein source.

According to Kushartono and Iriani (2004), legumes boast an average protein content of 22%. Additionally, livestock require legumes to uphold the stability of microbes in the rumen due to their high fiber content. Some legumes suitable for animal feed include *Calliandra*, *Indigofera*, and *Gliricidia*. These legumes encompass secondary metabolic compounds known as tannins. Tannins possess protein-binding properties and play a crucial role in the rumen, serving as protein protectors in the digestion of feed by ruminants. This compound has the ability to safeguard proteins in the feed, and feeding high-protein content without tannin protection can diminish feed efficiency as the protein is susceptible to degradation into ammonia (NH3) by rumen microbes. Tannins also function to control parasites, particularly nematodes, in the digestive tract of livestock, contributing to overall animal health and increased productivity.

Tannins act as protein protectors by binding feed ingredients into complex compounds that resist degradation by proteolytic bacteria. These protected proteins undergo hydrolysis in the abomasum under low pH conditions. Proteins passing through the abomasum are then enzymatically digested in the intestine into amino acids, serving as a nutritional supply for the host. Proteins escaping microbial degradation in the rumen are referred to as by-pass proteins. Tannins effectively convert Rumen Degradable Protein (RDP) into Undegradable Protein (UDP), subsequently enhancing metabolizable protein (MP). This means that the protein is absorbed by the small intestine (Jayanegara, et al., 2019).

The aim of protein protection is to enhance the quantity of protein entering the digestive tract post-rumen. Substituting concentrate feed with legume flour is anticipated to positively impact this research, as the tannin content in legume plants is capable of shielding protein from rumen microbial activity. This protection allows the protein to undergo enzymatic digestion in the small intestine, and the crude protein that evades digestion by rumen microbes (by-pass protein) can be efficiently utilized by the small intestine.

2 Materials and Methods

2.1 Location

This research was carried out in the Animal Nutrition and Feed Laboratory, Faculty of Animal Science, Universitas Brawijaya.

2.2 Materials

The materials utilized in this research include corn stover, a concentrate comprising rice bran, pollard, cassava flour, copra meal, soybean meal, molasses, minerals, urea, leaf flour, and legume leaf flour, specifically from tannin-rich sources such as *Calliandra*, *Indigofera*, and *Gliricidia*. The tools and materials employed in the production of complete feed include those for obtaining rumen fluid, proximate analysis, in vitro analysis (following the method outlined by Tilley and Terry in 1963), as well as the analysis of Rumen Degradable Protein (RDP), Undegradable Protein (UDP), and Digestible Energy (DE) protein.

2.3 Methods

The experiment included ten treatments, each replicated three times. The treatments were as follows: T1 complete feed (60% corn stover + 40% concentrate), T2 complete feed (60% corn stover + (35% concentrate + 5% *Calliandra* meal), T3 complete feed (60% corn stover + (35 % concentrate) + 5% *Indigofera* meal), complete feed T4 (60% corn stover + (35% concentrate + 5% gamal meal), T5 complete feed (60% corn stover + (30% concentrate + 10% *Calliandra* meal), T6 complete feed (60% corn stover + (30% concentrate + 10% *Indigofera* meal), complete feed T7 (60% corn stover + (30% concentrate + 10% gamal meal), complete feed T8 (60% corn stover + (25) % concentrate + 15% *Calliandra* meal), complete feed T9 (60% corn stover + (25% concentrate + 15% *Indigofera* meal), complete feed T10 (60% corn stover + (25% concentrate + 15% gamal meal).

2.4 Statistical Analysis

The data obtained in this research were analyzed using a Randomized Block Design comprising 10 treatments with 3 replications, employing analysis of variance (ANOVA). Additionally, in cases where there are substantial and statistically significant differences between treatments, Duncan's multiple range test will be conducted to ascertain the specific variations among each treatment (Steel and Torrie, 1993).

3 Results and discussion

3.1 Chemical composition of feeds used in this study

The feed is a combination of feed ingredients consumed by livestock, providing essential nutrients such as energy and protein to meet the animals' dietary requirements over a 24-hour period (Parakkasi, 1999). Prior to being formulated into complete feed, the individual ingredients undergo proximate testing at the Animal Nutrition and Feed Laboratory. The nutrient composition of these feed ingredients is outlined in Table 1 and includes corn stover, legume leaves (*calliandra*, *indigofera*, and *gliricidia*), and concentrate ingredients (rice bran, pollard, soybean meal, copra meal, cassava, molasses, and urea).

Based on the proximate analysis results, corn stover exhibits the following composition: DM 24.52%, OM 93.32%, CP 8.68%, CF 26.29%, and EE 1.36%. In comparison, the CP content in this study is higher than that reported by Mustika and Hartutik (2021), who found for corn stover: DM 31.2%, OM 80.3%, ash 7.43%, CP 7.8%, CF 23.55%, and EE 2.34%. This discrepancy is attributed to variations in the cutting age of corn stover, where an increase in age leads to a decrease in protein content and an increase in crude fiber (Indriani et al., 2022). The detailed nutritional content analysis is presented in Table 1.

Table 1. Nutritional Content of Feed

	DM(1) (%)	OM(1)	Ash(1)	CP(1) (% BK)	CF(1)	EE(1)	Tannin(2)
Pollard	87,85	94,09	5,91	15,05	7,44	3,47	
Gaplek	87,13	95,29	4,71	2,98	6,97	0,42	
Rice Bran	90,81	80,06	19,94	10,12	31,16	6,46	
Copra Meal	90,39	92,44	7,56	20,49	23,22	3,48	
Soybean Meal	90,85	92,23	7,77	43,59	5,18	2,72	
Molasses	76,02	92,82	7,28	4,26	0,86	0,22	
Urea	99,89	99,85					
Mineral	99,82						
Corn Sover	24,52	93,32	6,68	8,68	26,29	1,36	
Concentrate	90,68	89,52	10,48	12,88	13,79	2,08	
Calliandra Flour	97,35	92,72	7,28	21,32	10,77	2,15	3,06
Gliricidia Flour	95,37	90,22	9,78	22,53	11,8	5,79	6,83
Indigofera Flour	94,21	90,18	9,82	24,28	10,34	5,69	4,06

Note:
(1) Analysis results from The Laboratory of Animal Nutrition and Feed, Faculty of Animal Science, Universitas Brawijaya.
(2) Tannin analysis From The Laboratory of Chemistry and Biochemistry of Food and Agricultural Products, Universitas Brawijaya

3.2 The RDP, RUP and DE levels in diet

Referring to Table2, it can be seen that the treatment has a highly statistically significant effect ($P < 0.01$) on RDP and RUP, whereas it demonstrates a significant effect on the DE value.

Table 2. The RDP, RUP and DE levels in diet

Treatments	RDP (%)	RUP (%)	DE (MJ/KgBK)	TDN (%)	Tannin (%)
T1	$59,49 \pm 1,54^{c}$	$40,51 \pm 1,54^{a}$	$3,33 \pm 0,14^{bc}$	$75,57 \pm 3,27^{cd}$	0,00
T2	$59,75 \pm 1,51^{c}$	$40,25 \pm 1,51^{a}$	$3,25 \pm 0,19^{bc}$	$73,61 \pm 44,21^{bc}$	0,15
T3	$58,38 \pm 2,86^{c}$	$41,62 \pm 2,86^{a}$	$3,35 \pm 0,05^{bc}$	$76,03 \pm 2,95^{cd}$	0,20
T4	$55,44 \pm 0,98^{bc}$	$44,56 \pm 0,98^{ab}$	$3,37 \pm 0,13^{bc}$	$76,49 \pm 1,74^{cd}$	0,34
T5	$49,52 \pm 1,05^{ab}$	$50,48 \pm 1,05^{ab}$	$3,14 \pm 0,08^{b}$	$71,15 \pm 1,74^{ab}$	0,30
T6	$53,49 \pm 3,14^{abc}$	$46,51 \pm 3,14^{ab}$	$3,35 \pm 0,19^{bc}$	$76,00 \pm 4,39^{cd}$	0,40
T7	$55,20 \pm 1,27^{bc}$	$44,80 \pm 1,27^{ab}$	$3,31 \pm 0,06^{bc}$	$75,07 \pm 1,29^{cd}$	0,68

T8	46,56 ± 1,75[a]	53,44 ± 1,75[c]	2,83 ± 0,07[a]	65,25 ± 1,69[a]	0,45
T9	52,89 ± 0,55[abc]	47,11 ± 0,55[abc]	3,30 ± 0,10[bc]	74,90 ± 2,22[cd]	0,60
T10	54,91 ± 2,82[bc]	45,09 ± 2,82[ab]	3,46 ± 0,08[c]	78,51 ± 1,75[d]	1,02

The research results indicated that Kaliandra flour was added in treatments P2, P5, and P8, Indigofera flour in treatments P3, P6, and P9, and Gliricidia flour in treatments P4, P7, and P10. The addition of three different legume flours, combined with varying tannin concentrations (5%, 10%, and 15%), led to a reduction in the RDP value. This reduction can be attributed to the inverse relationship between tannin concentration and RDP value. Specifically, lower tannin concentrations resulted in higher RDP values, indicating that the proteins present in *Caliandra*, *Indigofera*, and *Gliricidia* were not maximally protected by tannin. Consequently, rumen microbes degraded the proteins, leading to an increase in N-NH3 levels.This study aims to assess and evaluate feed protein based on rumen degradable protein (RDP) and undegradable protein. RDP, constituting part of the crude protein in the feed, is degraded by rumen microbes into peptides and ammonia, with a significant portion converting into microbial crude protein. The availability of RDP in the rumen serves to increase ammonia nitrogen for the synthesis of microbial protein (Iswanto et al., 2022). Suryani et al. (2014) proposed that the high composition of microbial protein in the rumen is directly proportional to the protein composition in the feed. Since crude protein is a crucial component for microbial protein synthesis, it signifies the availability of nitrogen elements for rumen microbes. However, an excess of N-NH3 production does not confer positive benefits for microbial protein production; instead, it may reduce protein efficiency values as a substantial amount of nitrogen is excreted through urine and feces (Afzalani, 2021).

In this experiment, RUP varied among treatments but constantly ranged from 40,25% to 53,44%. These results concur with a previous study, which stated that the RUP digestibility varied considerably from 25% to 60% in ruminant. According to research conducted by Pangestuti et al. (2023), the consistently highest average UDP value was observed in P4 at 47.66%. The addition of legume flour with varying levels of tannin concentration resulted in a higher UDP value compared to P1, which served as the control and contained complete feed without tannin protection treatment. In this study, the highest UDP results were obtained from the P8 treatment, involving the addition of 15% calliandra flour, with an average value of 53.44%. This outcome is attributed to the fact that calliandra contains condensed tannin, as indicated by Setyawati's research (2014), reporting a condensed tannin content of 7.43 g/100 g and a protein content of 23.24 g/100 g in calliandra leaf flour.

The presence of condensed tannin in calliandra does not interfere with rumen microbial activity; instead, it inhibits nutrient degradation by rumen microbes. The addition of tannin protection demonstrated an ability to increase the UDP value, suggesting that tannin can form protein bonds and create insoluble complex compounds that reduce rumen microbial degradation. As the tannin content in the feed increases, the protein content in the feed becomes protected from rumen microbes, subsequently passing to the feed by-pass where the protein is reabsorbed to enhance feed efficiency.

The data shows that the addition of legume leaves to the feed tends to increase DE values. The data indicates that feed treatment tends to result in increased DE content values. This suggests that substituting concentrate with legumes in complete feed has no significant effect on increasing DE content. However, in the legume substitution treatment at T5, T8, and T10, there are significant differences ($P<0.01$) in DE values. This suggests that adding legumes at these concentrations leads to decreased DE values, thereby impacting a reduction in DE content. This is distinct from the treatment at P5, P8, and P10, where there are very significant differences ($P<0.01$) in DE values. DE values in the treatments involving the addition of calliandra flour at T2, T5, and T8 (5%, 10%, and 15% concentrations), *Indigofera* flour to treatments T3, T6, and T9 (5%, 10%, and 15% concentrations), and *Gliricidia* flour to treatment T4, T7, and T10 (5%, 10%, and 15% concentrations) tend to decrease alongside decreasing TDN values.

4 Conclusion

The conclusion of this study reveals that the addition of 15% calliandra meal reduced RDP by 46.56% and increased RUP up to 53.44%. Furthermore, the inclusion of 15% gliricidia meal led to an increase in DE value by 3.46 MJ/KgDM

The results from this research provide additional information on how rumen degradable protein and rumen undegradable protein can impact nutrient digestion in ruminants consuming leguminous meals.

References

1. Abdullah, L. 2014. Prospektif Agronomi Dan Ekofisiologi *Indigofera Zollingeriana* Sebagai Tanaman Penghasil Hijauan Pakan Berkualitas Tinggi. *Jurnal Pastura*. 3(2) : 79 – 83.
2. Abqoriyah, R. Utomo, dan B. Suwignyo. 2015. Produktivitas Tanaman *Kaliandra* (Calliandra Calothyrsus) Sebagai Hijauan Pakan Pada Umur Pemotongan Yang Berbeda. Buletin Peternakan. 39 (2): 103-108.
3. Addawiyah, R. N, B. Ayuningsih, A. Budiman dan I. Hernaman. 2021. Produksi Gas pada Ransum Domba Berbasis Rumput Gajah cv Mott dan Leguminosa Pohon. *Jurnal Sumber Daya Hewan*. 2(2) : 30 - 34
4. Adinugraha, B. S, dan T. N. Wijayaningrum. 2014. *Rancangan Acak Lengkap dan Rancangan Acak Kelompok pada Bibit Ikan*. Seminar Nasional Pendidikan, Sains dan Teknologi: 47-56.
5. Akbarillah T. D, Kususiyah, dan Hidayat. 2010. Pengaruh penggunaan daun *Indigofera* sp. segar sebagai suplemen pakan terhadap produksi dan warna yolkitik. *Jurnal Sains Peternakan Indonesia*. 5:27-33.
6. Alfiansyah, H. A dan Hartutik. 2021. Tren Produksi Gas, Produksi Gas Total Dan Degradasi Secara *In vitro* Dengan Penambahan Aditif Dengan Level Berbeda Pada Silase Tebon Jagung (Zea mays L.). *Jurnal Nutrisi Ternak Tropis*. 4(2) : 77-87.
7. Anggorodi, R. 1994. *Ilmu Makanan Ternak Umum*. PT. Gramedia Pustaka Utama: Jakarta
8. Angkasa, S. 2017. *Ramuan Pakan Ternak*. Penebar Swadaya : Jakarta.
9. AOAC. 2005. Official Methods of Analyis. 14. Thed. *Association of Analytical Chemist. Washington*
10. Barokah, Y, A. Ali, dan E. Erwan. 2017. Nutrisi Silase Pelepah Kelapa Sawit Yang Ditambah Biomassa *Indigofera* (*Indigofera* Zollingeriana) The Nutrient Content Of Oil Palm Frond Silage Added With *Indigofera* Zollingeriana. *Jurnal Ilmu-Ilmu Peternakan*. 20(2) : 59 – 68.
11. Cahyani, R. D, L. K. Nuswantara dan A. Subrata. 2012. Pengaruh Proteksi Protein Tepung Kedelai Dengan Tanin Daun Bakau Terhadap Konsentrasi Amonia, Undegraded Protein Dan Protein Total Secara *In vitro*. *Animal Agricultural Journal*. 1(1) : 159 – 166.
12. Chuzaemi, S. 2012. *Fisiologi Nutrisi Ruminansia*. UB Press : Malang.

13. Chuzaemi, S., Mashudi, M., Ndaru, P. H., Huda, A. N., & Siswoyo, E. A. (2022, April). Effect of Condensed Tanin and Myristic Acid in Corn Straw-Based Complete Feeds on NH_3 Concentration and Microbial Protein Synthesis. In 6th International Seminar of Animal Nutrition and Feed Science (ISANFS 2021) (pp. 29-34). Atlantis Press.
14. Chuzaemi, S., & Qur'any, I. N. (2022, October). Ruminal profile of completed feed as influenced by myristic and tanins addition. In E3S Web of Conferences (Vol. 335, p. 00028).
15. Hartutik. 2012. *Metode Analisis Mutu Pakan.* UB Press : Malang.
16. Hassen, A., N.F.G. Rethman, W. A. Van Niekerk, & T. J. Tjelele. 2007. Influence of season/year and species on chemical composition and *in vitro* digestibilityof five *Indigofera* accession. *Journal Animal Feed Science and Technology*. 136: 312–322.
17. Hassen, A., N.F.G. Rethman, Z. Apostolides, & W. A. van Niekerk. 2008. Forage production and potential nutritive value of 24 shrubby *Indigofera* accessions under field conditions in South Africa. *J Tropical Grasslands*. 42: 96–103.
18. Heryanto, K. Maaruf, S.S. Malalantang, M. R. Waani. 2016. Pengaruh pemberian rumput raja (Pennisetum purpupoides) dan tebon jagung terhadap performans sapi peranakan onggole (PO) betina. Jurnal Zootek 36(1): 123-130.
19. Huda, A. N., Chuzaemi, S., Mashudi, M., Ndaru, P. H., & Khoirunisa, K. (2022, April). Nutrient Content and Total VFA Concentration Evaluation by Addition of Condensed Tanin and Myristic Acid in Complete Feed Through *In vitro* Method. In 6th International Seminar of Animal Nutrition and Feed Science (ISANFS 2021) (pp. 159-163). Atlantis Press.
20. Indriani, N. P., H. K. Mustafa dan R. Z. Islami. 2022. Introduksi Tanaman Penghasil Jagung (*Zea mays*) dan Hijauan Pakan dengan Berbagai Varietas dan Panen di Desa Cileles di Kabupaten Sumedang. *Media Kontak Tani Ternak*. 4(2): 50-55
21. Jayanegara, A, M. Ridla, E. B. Laconi dan Nahrowi. 2019. *Komponen Antinutrisi Pada Pakan*. IPB Press : Bogor.
22. Jayanegara. A., dan Sofyan. A. 2008. Penentuan aktivitas biologis tanin beberapa hijauan secara *in vitro* menggunakan'Hohenheim gas test'dengan polietilen glikol sebagai determinan. Media Peternakan. 31(1).
23. Liu, H, V. Vaddella and D. Zhou. 2011. Effects of chesnut tanins and cococnut oil on growth performance, methane emission, ruminal fermentation, and microbial populations in sheep. *Journal Dairy Sci*. 9(4) : 6069-6077.
24. Makkar, H. P. S. 2003. Effect and fate of tanins in ruminant animals, adaptation to tanins, and strategis to overcome detrimental effect of feeding tanin – rich feeds. *Small Ruminant Research*. 49: 241–256.
25. Manu, A, E. Marhaeniyanto, Dan E. Fitasari. 2021. Pemanfaatan Daun Tanaman Gamal, Lamtoro Dan Kaliandra Pada Pakan Konsentrat Terhadap Kecernaan Bahan Kering, Bahan Organik, dan Protein Kasar. Fakultas Pertanian. Universitas Tribhuwana Tunggadewi Malang.
26. Mathius, I. M., Martawidjaja, M., Wilson, A., dan Manurung, T. (1996). Studi strategi kebutuhan energi-protein untuk domba lokal: I. fase pertumbuhan. *Jurnal Ilmu Ternak Dan Veteriner*. 2(2), 84–91.
27. Mayasari, D, E. D. Purbajanti, dan Sutarno. Kualitas Hijauan Gamal (Gliricidia Sepium) Yang Diberi Pupuk Organik Cair (Poc) Dengan Dosis Berbeda Forage Quality Of Gamal (Gliricidia Sepium) Advised That Organic Fertilizer Liquid (Poc) With Different Dose. *Jurnal Animal Agriculture*. 1(2) : 293 – 301.

28. Meisari, V. S, H. Sudarwati. 2022. *Pengaruh Penggantian Konsentrat Dengan Tepung Daun Leguminosa Sebagai Sumber Tanin Pada Pakan Lengkap Terhadap Konsentrasi Amonia Dan Sintesis Protein Mikroba Secara In vitro*. Fakultas Peternakan. Universitas Brawijaya.
29. Menke, K. H and H. Steingass. 1988. Estimation of The Energetic Feed Value Obtained from Chemical Analysis and *In vitro* Gas Production Using Rumen Fluid. *Animal Research and Development*. 28: 7-55
30. Mufidah, M S. Chuzaemi, and Mashudi. 2021. The Effect of Addition Myristic acid and Calliandra calothyrsus Leaves Meal Substitution on Feed Consumtion, Feed Conversion and Body Weight Gain of Sheep. *International Journal of Multidisciplinary and Current*. 4(4) : 129-135.
31. Ndaru, P. H, S. Chuzaemi dan M. Mufidah. 2022. In-vitro nutrient degradability of complete feed containing myristic acid and tanins addition. *In E3S Web of Conferences* (Vol. 335, p. 00047). EDP Sciences.
32. Ndaru, P. H, Kusmartono, K, dan Chuzaemi, S. 2014. Pengaruh suplementasi berbagai level daun ketela pohon (Manihot utilissima. Pohl) terhadap produktifitas domba ekor gemuk yang diberi pakan basal jerami jagung (Zea mays). *Jurnal Ilmu-Ilmu Peternakan (Indonesian Journal of Animal Science)*. 24(1) : 9-25.
33. Nugroho, A. P, E. A. Rimbawanto, B. Hartoyo, dan M. Ifani. 2021. Kecernaan Bahan Kering dan Bahan Organik Leguminosa Pohon Sebagai Sumber Protein Pakan Ruminansia Secara *In vitro*. *JITRO (Jurnal Ilmu dan Teknologi Peternakan Tropis)*. 8(2):162-167
34. Parakksi, A. 1999. *Ilmu Nutrisi dan Makanan Ternak Ruminan*. Universitas Indonesia : Jakarta
35. Sadan, M. F., Rahmat, H., & Tanuwiria, U. H. (2021). Pengaruh proteksi protein bungkil kedelai dengan cairan batang pisang terhadap konsentrasi amonia dan undegraded dietary protein (UDP) pada rumen domba (*in vitro*). *Jurnal Nutrisi Ternak Tropis dan Ilmu Pakan*. 3(1) : 10-19.
36. Soeharsono dan B. Sudaryanto. 2006. Tebon Jagung Sebagai Sumber Hijauan Pakan Ternak Strategis di Lahan Kering Kabupaten Gunung Kidul. Prosiding. Lokakarya Nasional Jejaring Pengembangan Sistem Integrasi Jagung Sapi. Puslitbang Peternakan, Bogor. 136 – 141.
37. Steel, R. G. D. and J. H. Torrie. 1983. Prinsip dan Prosedur Statistika Terjemahan Sumantri B. Jakarta: Gramedia.
38. Supriadi, Lutfi, I., dan Oktriani, I. F. (2016). Potensi ketersediaan hijauan pakan limbah tanaman jagung manis di Provinsi Kepulauan Riau. Prosiding Seminar Nasional Membangun Pertanian Modern dan Inovatif Berkelanjutan dalam Rangka Mendukung MEA, 710-716.
39. Sutardi, T. 1981. *Sapi Perah dan Pemberian Makanannya*. Departemen Ilmu Makanan Ternak. Fakultas Peternakan Institut Pertanian Bogor : Bogor.
40. Sutaryono, A. Y. 2021. *Pengelolaan Hijauan Pakan Ternak Dalam Sistem Peternakan Tradisional.* Cv Budi Utama : Sleman.
41. Tilley, J. M. A and R. A. Terry. 1963. a Two Stage Technique for The *In vitro* Digestion of Forage Crops. *Grass and Forage Science*. 18(2): 104-111
42. Tillman, A. D., H. Hartadi, S. Reksohadiprodjo, S. Prawirokusumo, dan S. Lebdosoekojo. 2005. Ilmu Makanan Ternak Dasar. Gadjah Mada University Press: Yogyakarta.

43. Wina, E., Tangendjaja, B and Dumaria. 2010. Effect of (Calliandra calothyrsus) on *in vitro* digestibility of soybean meal and tofu wastes. Proceeding Seminar Nasional Ruminansia 2010. Hal. 11-13.

The effect of isoamylase application in cassava root meal on broiler growth performance, feed retention time, and metabolite profile

Poppy Satya Puspita[1], *Widya* Hermana[2], *Muhammad* Pramujo[1], and *Nahrowi* Nahrowi[2*]

[1]Faculty of Animal Science, Universitas Brawijaya, Jl. Veteran, Kota Malang 65145, Indonesia
[2]Department of Nutrition and Feed Technology, Faculty of Animal Science, IPB University, Jl. Agatis, Kampus IPB Darmaga Bogor 16680, Indonesia

Abstract. The objectives of this study were to evaluate isoamylase application on the cassava root meal (CRM) and identify its effect on broiler growth performance, feed retention time, and metabolite profile. The experiment used 270 sexed broilers strain Lohman with an initial body weight of 45.88 ± 0.41 g were allotted into 3 treatments and 6 replications with 15 birds in each pen for 35 days. Dietary treatments were T0: Basal Ration + 0% CRM, T1: Basal Ration + 50% CRM, T2: Basal Ration + 50% CRM + 0.05% isoamylase. The results showed that the inclusion 50% of CRM both with and or without isoamylase was increased ($P<0.01$) feed intake and feed conversion ratio (FCR) and gave the same effect ($P>0.05$) on body weight, body weight gain, carcass weight, and carcass percentage during the overall period compared to fed T0. The treatment had no significant difference in coefisien total starch digestibility, metabolizable energy, and blood glucose levels. In conclusion, the inclusion of 50% cassava root meal with and or without isoamylase to subtitute corn in ration does not affect the broiler growth performance however FCR was higher than the control.

1 Introduction

Feed costs in poultry production constitute more than 70% of production costs [1], where energy feed sources occupy the largest portion of 70% - 75% in rations [2]. Corn is predominantly used as animal feed raw material, especially poultry feed in Indonesia. However, the availability of maize, especially feed maize, becomes scarce especially in every rainy season and causes an increasing in the price of maize in the market. The increasing in feed raw material costs has accelerated the demand to find alternative feed ingredients that can replace this material with lower production costs. In order to meet the needs of carbohydrates, cassava is the third food crop in Indonesia after rice and maize. Cassava (*Manihot* spp) is the highest carbohydrate source among staple crops and can potentially replace maize as an energy source in poultry feed. Starch is a major component of carbohydrates which is a source of energy. The main composition of cassava starch generally

* Corresponding author: nahrowi2504@yahoo.com

consists of amylose (AMS) and amylopectin (AMP). Cassava starch contains 17% amylose and 83% amylopectin, compared with maize starch which has 28% amylose and 72% amylopectin [3,4]. [5] Reported that high levels of resistant starch in cassava probably consisted of 82.85% amylopectin with 5.79% branch associations and 17.25% amylose with 0.48% branch associations. This causes cassava starch difficult to digest and utilize by poultry. One way to increase the amylose content in cassava root meal (CRM) is adding isoamylase. Isoamylase, also known as a debranched enzyme, hydrolyzes the branch chain α-1.6 glucosidic in glucose, amylopectin, and beta-limit dextrin. An increasing hydrolyzed of branch chains in amylopectin is expected to increase the amylose content. Thus it can increase nutrient availability and broiler growth performance.

Amylase was able to increase starch digestibility as well as reduce the presence of glucose as a potential substrate for non-beneficial bacteria at the end of the digestive tract [6]. [7] Found that feed containing 50% peeled cassava meal suplemented by cocktail enzymes (Maxigrain) produced the same production as a ration containing 100% maize. [8] Reported that the addition of carbohydrase and phytase significantly increased the body weight and metabolizable energy in poultry fed cassava pellets and mash. Supplementation of α-1.6 isoamilase in broiler feed increased nutrient digestibility, metabolizable energy, and broiler productivity [9]. This study was designed to evaluate isoamylase application on the CRM and identify its effect on broiler growth performance, feed retention time, and blood glucose level.

2 Material and Methods

2.1 Material

The main material in this study used cassava root meal from UD. Setia Flour Product (Bogor, Indonesia) and Isoamylase (EC 3.2.1.68) was from Bacillus licheniformis, purchased from Creative Enzymes (USA), Creative Biomart, Inc. This enzyme has optimal activity at pH around 5.6 - 7.0 and at temperatures between 50 - 55 ° C. Enzyme activity ≥ 10,000 Ug-1. The diet was formulated according to the recommendations of Leeson and Summers (2005) for a type of high-density diet for open cages (Table 1). The ration was given in the form of crumble for the starter period (0-21 days) and pellets for the finisher period (22 - 35 days). Feed and drink were given ad libitum. One-day-old sexed Lohmann (270 chicks), obtained from PT Japfa Comfeed Indonesia Hatchery, vaccinated with New castle Disease, Infectious Bursal Disease, and Avian Influenza, and raised for 35 days. The cages (1 m x 1 m x 0.8 m) were randomly alloted of 3 treatments (6 replications with 15 chicks each). Dietary treatments in this study were T0: Basal ration + 0% CRM (control); T1: Basal ration + 50% CRM; T2: Basal ration + 50% CRM + 0.05% Isomylase (500 Ug-1).

2.2 Methods

2.2.1 Growth Performance

Feed intake (FI), body weight (BW), body weight gain (BWG), feed conversion ratio (FCR) were measured weekly. Weight and percentage of digestive tracts and carcass, and intestinal length were observed at 35 days of age.

2.2.2 Feed Retention Time

Observation of retention time at 30th and 90th minutes was carried out at 22 days of age and 60th minutes at 23 days of age. Feed retention time was measured by placing broiler chickens on individual cages and calculating the amount of feed left in the crop from the amount of feed intake for 2 hours before the chicken was fasted for 24 hours, then slaughtered at 30th,60th, and 90th minutes.

$$Feed\ retention\ time\ (\%) = \frac{feed\ in\ the\ crop\ (g)}{feed\ intake\ (g)}\ x\ 100\% \quad (1)$$

Table 1. Feed rations and nutrient compositions of broilers

Ingredients	Starter (0-21 days)		Finisher (22-35 days)	
	T0	T1/T2	T0	T1/T2
Maize	55.00	27.50	55.00	27.50
CRM	0.00	27.50	0.00	27.50
Soy Bean Meal	28.00	28.00	27.35	27.35
Meat Bone Meal	2.00	2.000	2.820	2.820
Corn Gluten Meal	9.590	9.660	6.420	7.420
Crude palm oil	1.650	1.700	4.320	4.630
CaCO3	0.240	0.070	0.250	0.050
DCP	2.460	2.490	2.900	1.850
NaCl	0.060	0.090	0.040	0.040
L-Methionine	0.220	0.190	0.200	0.180
L-Lysine	0.270	0.300	0.200	0.170
Premix[a]	0.500	0.500	0.500	0.500
Isoamylase	-	-/0.05	-	-/0.05
Total (%)	100	100	100	100
Nutrient Composition				
Dry Matter (%)	87.99	88.07	85.72	88.23
Crude Protein (%)	23.04	23.00	21.50	21.50
Crude Fiber (%)	2.41	3.12	2.29	3.08
Extract Ether (%)	4.29	3.59	6.70	6.55
Ash (%)	5.61	5.98	5.31	5.52
Metabolizable Energy (kcalkg^{-1})	3005.88	3000	3150	3150
Ca (%)	0.96	0.96	0.87	0.87
P available (%)	0.48	0.48	0.44	0.44
P total (%)	0.78	0.75	0.72	0.70
Lysine (%)	1.44	1.44	1.29	1.29
Methionine (%)	0.56	0.56	0.51	0.51
Methionine + Cystine (%)	0.97	0.90	0.83	0.82
Salt (%)	0.23	0.22	0.17	0.17

[a]Per kg diet: Copper 16 mg, Iodine 1,25 mg, Iron 40 mg, Manganese 120 mg, Selenium 0,3 mg, Zinc 100 mg. Vitamin A 12000 IU, Vitamin D3 5000 IU, Vitamin E 75 IU, Vitamin K3 3 mg, Riboflavin (B2) 8 mg, Nicotin Acid 60 mg, Panthotenic Acid 15 mg, Pyridoxin (B6) 4 mg, Biotin 0,15 mg, Folic acid 2 mg, Vitamin B12 0,016 mg, Choline 1600 ppm.

2.2.3 Starch digestibility coefficient in small intestine

At the age of 33 days, broiler chickens were placed in individual cages and calculated the amount of feed intake for 2 hours after the chickens were fasted for 24 hours, then slaughtered. Digesta in the small intestine was accommodated in plastic and dried in an oven 60 °C until the weight was constant. The dried samples were analyzed for its starch content. The results of the analysis of starch content in the digesta were going to be compared with the total content of starch consumed by chicken.

$$starch\ digestibility\ coefficien = 1 - \frac{starch\ content\ in\ digesta\ (g)}{starch\ content\ in\ feed\ (g)} \quad (2)$$

2.2.4 Feed metabolizable energy

Metabolizable energy measurements based on [10] were carried out by placing two groups of broilers on individual cages. The first group was fasted for 48 hours, then excreta was collected for 24 hours. The second group was fasted for 24 hours, then fed for 2 hours and calculated the amount of feed intake. Excreta was collected fot 24 hours after completion of feeding. Excreta was dried and analyzed for gross energy content (Calorimetry Bomb).

$$AME\ (kcalkg^{-1}) = \frac{(GEf - FI) - (GEEp - E)}{FI} x\ 1000 \quad (3)$$

$$TME\ (kcalkg^{-1}) = \frac{(GEf - FI) - [(GEEp - E) - (GEEe\ x\ Ee)]}{FI} x\ 1000 \quad (4)$$

*GEf: Groos energy feed (kkalkg^{-1}), GEEp: gross energy excreta (kkalkg^{-1}), GEEe: gross energy endogenous excreta (kkalkg^{-1}), FI: feed intake (g), E: excreta (g), Ee: endogenous excreta (g).

2.2.5 Blood Glucose Levels

Blood sampling was taken at 34 day of age via branchial vein. Blood sampling was taken as ± 2 ml using 3 ml disposable syringe. The blood was put into a vacuum tube that has been filled with Ethylenediaminetetraacetic acid (EDTA) and homogenized. Serum blood was taken for glucose analysis (Enzymatic Colorimetric AKL Method 20101803460).

2.2.6 Statistical Analysis

The dietary experimental design used Completely Randomized Design (CRD) 3 treatments 6 replications. Data were analyzed using analysis of variance (ANOVA) and significantly different data continued with Duncan's multiple range test [11] using SPSS software (SPSS® version 16.0).

3 Result and Discussion

3.1 Broiler Growth Performance and Feed Retention Time

The results on the broiler growth performance showed that the inclusion of CRM up to 50% both with or without isoamylase addition were highly significant ($P < 0.01$) increasing FI, BW, and BWG at starter period. There were no differences on FI, BW, and BWG but FCR highly significant ($P < 0.01$) increased at finisher period. Feed intake and FCR highly significant ($P < 0.01$) increased during the overall period compared to control. The control treatment had feed retention which tended to be higher in the range of 30-90 minutes. In contrast, lower feed retention occurred in treatments T1 and T2.

Table 2. Effect of dietary treatments on the growth performance of broilers

Parameters	Treatments			P Value
	T0	T1	T2	
BW DOC (g bird^{-1})	46.29	45.87	45.76	
Starter (0-21 days)				
FI (g bird^{-1})**	957.54 ± 46.22b	1215.08 ± 35.36a	1219.20 ± 58.83a	0.000
BW(g bird^{-1})**	731.89 ± 50.69b	884.84 ± 41.13a	871.11 ± 19.35a	0.000
BWG (g bird^{-1})**	685.60 ± 51.29b	838.97 ± 41.11a	825.35 ± 19.56a	0.000
FCR (FIBWG^{-1})	1.39 ± 0.06	1.45 ± 0.06	1.47 ± 0.06	0.135
Finisher (22-35 days)				
FI (g bird^{-1})	1769.79 ± 87.59	1897.80 ± 93.89	1845.42 ± 114.84	0.115
BW(g bird^{-1})	1756.38 ± 136.85	1819.82 ± 80.61	1805.66 ± 125.30	0.623
BWG (g bird^{-1})	1024.49 ± 98.38	934.97 ± 47.47	934.55 ± 118.56	0.190
FCR (FIBWG^{-1})**	1.73 ± 0.06b	2.03 ± 0.06a	1.99 ± 0.10a	0.002
Cumulative (0-35 days)				
FI (g bird^{-1})**	2725.11±119.71b	3112.89 ± 121.45a	3064.63 ± 120.70a	0.000
BW(g bird^{-1})	1756.38 ± 136.85	1819.82 ± 80.61	1805.66 ± 125.30	0.623
BWG (g bird^{-1})	1710.09 ± 137.01	1773.95 ± 79.99	1759.91 ± 125.58	0.619
FCR (FIBWG^{-1})**	1.59 ± 0.05b	1.75 ± 0.05a	1.74 ± 0.05a	0.004

** The numbers on the same line and followed by different letters indicate a highly significant difference at the test level of 1%.

FI: Feed Intake; BW: Body Weight; BWG: Body Weight Gain; FCR: Feed Conversion Ratio

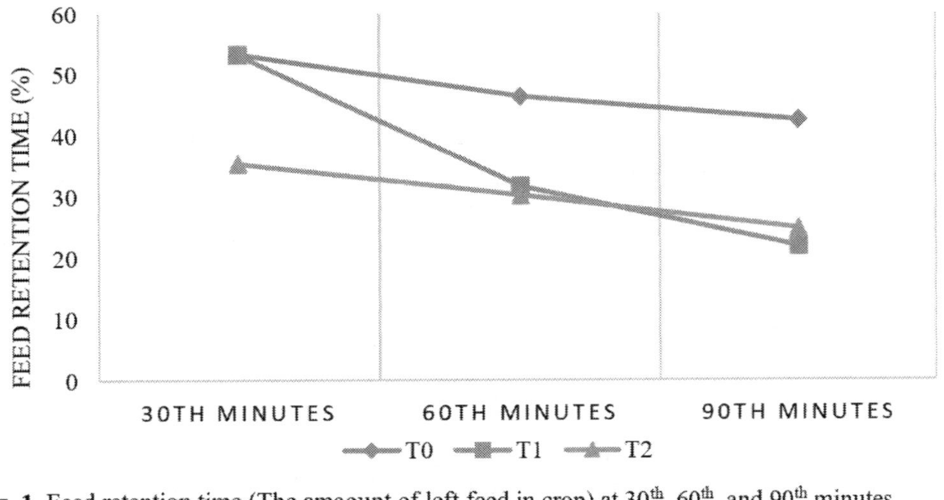

Fig. 1. Feed retention time (The amaount of left feed in crop) at 30th, 60th, and 90th minutes

The highly significant of FCR was due to BW and BWG in the treatment showed the same results with the control. [9] revealed that 300 and 600 Ukg α-1.6 isoamylase were able to increase BW, decrease FI, and achieve the same effect from controls but FCR was much higher than control. The higher of FI in the treatments compared to control was due to faster feed rate in the digestive tract. The higher inclusion of CRM will increase the content of amylopectin in the ration. Amylopectin which has branch chains tends to be less soluble in water. Components of fiber which were insoluble in water cause a faster feed rate [12,13]. Amylopectin in cassava had a relatively long chain length and caused cassava starch to be difficult to digest by poultry [5]. In addition, amylose has a molecular weight of about 100 kDa, amylopectin has a higher molecular weight of 104 - 106 kDa [14]. Increasing of feed rate in this study were supported by a lower feed retention in the treatments (Figure 1).

Higher feed retention rates allow feed to be in contact with digestive tract enzymes more optimally in order to increase feed digestibility. Crop filling was highly dependent on feed volume in the proventriculus and gizzard, whose capacity in chickens was estimated to reach 5-10 g of feed [15]. When these organs were fully filled, feed storage occurred [16]. After a while, when the gizzard was emptied, muscle contraction caused the passage of the digesta moved to a further part of the intestinal tract [17]. Therefore, the function of the anterior digestive tract was closely related, and affected the peristalsis further part of the digestive system. Furthermore, ad libitum feeding caused reducing physical use of crop by chickens [18] because the chance of feed in crop was only about 30 minutes according to digesta rate [19].

Higher feed retention in the control was caused by a slower feed rate compared to the other treatments which can be seen from the highly significant ($P < 0.01$) increasing on gizzard weight (Table 2). Higher gizzard weight indicated a higher process of mechanical digestion. A less developed gizzard would function as a transit organ from the mechanical digestive organs, with the implication of accelerating retention time [20]. Conversely, lower feed retention occured at T1 and T2. This was due to the difference in the AMS / AMP ratio which affected on the starch structure. Starch was arranged in a very complex and large structure, where amorphous and crystalline layers alternately form stiff and semi-crystalline granules, with sizes varying from 1 to 50 μm [21]. The semi-crystalline layer consisted of alternating layers of α-glucan crystals which extended from successive branches of amylopectin and amorf amylopectin branches [22]. The increasing amylopectin content would increase the formation of semi-crystalline layers so that it had the potential to increase the feed rate. [23] stated that the degree of crystallinity was inversely proportional to the stach digestion rate, and [24] proved this by concluding that slow starch digestibility was associated with a regular structure of crystalline and amorphous layers alternating in starch granules. [24] revealed that the digestion of starch granules started from the surface pores and interior channels, which allow amylase to enter the interior and digest granules gradually from the inside.

3.2 Carcass and Visceral Organs

The inclusion of CRM up to 50% subtituted maize was highly significant ($P < 0.01$) reducing both the weight and percentage of gizzard. Addition of isoamylase decreased ($P < 0.05$) the percentage of liver. The treatments did not affect (P> 0.05) on weight and percentage of small intestines, abdominal fat, and carcass as well as lenght of small intestines.

The treatments was highly significant reducing both weight and percentage of gizzard. This was due to the decreasing corn which had a larger particle size than CRM in rations. Thus it reduced mechanical digestive activity in the gizzard and made gizzard less develop. The use of coarse particles or grains had been aimed to stimulate gizzard growth, improve mechanical function, increase digesta peristalsis, and increase gizzard volume, in order to increase retention time [25, 26]. Gizzard would shrink feed particle size to 100-200 μm before entering the small intestine [27, 28]. The higher particle size of the feed in the gizzard, the longer it would take to reach the size of the feed that can enter the small intestine.

The liver was the main organ for synthesis and metabolism, and served for glucose absorption of around 25% to 35% [29]. After consuming carbohydrates, 33% of glucose was absorbed by the liver [30]. Glucose derived from starch degradation was transported to the blood to maintain sufficient concentration to supply energy to other tissues [31, 32]. The results of the observations also showed that the treatment had no effect on the weight and percentage of the proventriculus. This is not in accordance with [33, 34]. The effect of the treatment in this study did not affect the weight, percentage, and length of the small intestine (duodenum, jejenum, ileum). This result was in line with the research [35, 36].

Table 3. Effect of dietary treatments on weight and percentage of carcass, visceral organ, and intestinal length of broilers aged 35 days

Parameters	Treatments			P Value
	T0	T1	T2	
Proventriculus (g)	10.66 ± 2.25	10.50 ± 2.94	7.83 ± 1.32	0.084
Proventriculus percentage (%)	0.61 ± 0.15	0.58 ± 0.18	0.43 ± 0.06	0.097
Gizzard (g)**	23.33 ± 3.50a	18.00 ± 1.67b	14.50 ± 0.06c	0.000
Gizzard percentage (%)**	1.33 ± 0.15a	0.99 ± 0.13b	0.80 ± 0.10c	0.000
Liver (g)*	41.33 ± 4.22ab	44.00 ± 6.63a	36.33 ± 2.87b	0.044
Liver percentage (%)*	2.37 ± 0.15a	2.41 ± 6.63a	2.00 ± 0.11b	0.015
Duodenum (g)	16.66 ± 4.22	18.33 ± 4.58	15.83 ± 1.60	0.511
Duodenum percentage (%)	0.95 ± 0.20	1.00 ± 0.22	0.87 ± 0.10	0.521
Jejenum (g)	26.66 ± 8.01	30.16 ± 6.11	27.77 ± 4.22	0.622
Jejenum percentage (%)	1.52 ± 0.42	1.65± 0.31	1.53 ± 0.23	0.754
Ileum (g)	20.83 ± 4.53	24.16 ± 6.24	20.30 ± 3.26	0.354
Ileum percentage (%)	1.19 ± 0.22	1.32 ± 0.32	1.12 ± 0.19	0.413
Duodenum (cm)	38.00 ± 1.41	39.00 ± 4.42	36.50 ± 4.76	0.539
Jejenum (cm)	81.91 ± 12.69	91.50 ± 9.97	81.91 ± 12.15	0.289
Ileum (cm)	80.16 ± 12.96	89.25 ± 10.74	81.08 ± 13.41	0.401
Abdominal fat (g)	26.50 ± 7.81	28.00 ± 5.29	33.83 ± 5.38	0.136
Abdominal fat percentage (%)	1.51 ± 0.40	1.54 ± 0.31	1.87 ± 0.30	0.175
Carcass weight (g)	1249.3 ± 111.22	1288.16 ± 91.19	1278.00 ± 62.40	0.747
Carcass percentage (%)	71.75 ± 2.26	70.70 ± 2.80	70.67 ± 1.30	0.636

** The numbers on the same line and followed by different letters indicate a highly significant difference at the test level of 1%.

* The numbers on the same line and followed by different letters indicate a significant difference at the test level of 5%.

The results did not affect the weight and percentage both of abdominal fat and carcass. Higher abdominal fat in the treatment group was a factor that contributes to the increasing in carcass weight due to higher body weight. The result of this study was in accordance with [37] found that the use of cassava in the form of pellets and chips were able to replace corn in broiler feed by up to 50%. This was supported by the fact that cassava products did not have an adverse effect on carcass weight, abdominal fat, and carcass composition. The opposite results are shown in the [33] although the result did not record a significant effect of feed treatment on carcass percentage, data showed a tendency to decrease with increasing use of cassava leaf concentrates and cassava tubers.

3.3 Broiler Chicken Metabolite Profile

The inclusion of 50% CRM in rations both with and without isoamylase did not give a significant difference to the starch digestibility coefficient in the small intestine. The results also showed no significant differences in AME, TME, and blood glucose levels in all treatments.

Table 4. Effect of dietary treatments on starch digestibility coeficient in small intestine, metabolizable energy, and blood glucose level on broiler

Parameters	Treatments			P Value
	T0	T1	T2	
Starch digestibility coefficient	0.98 ± 0.010	0.98 ± 0.001	0.97 ± 0.004	0.162
AME (kcal kg^{-1} feed)	3142.89 ± 220.84	3185.42 ± 168.67	3167.95 ± 172.83	0.950
TME (kcal kg^{-1} feed)	3317.73 ± 272.95	3292.95 ± 158.63	3313.50 ± 167.63	0.984
Blood glucose ($mgdL^{-1}$)	169.65 ± 9.78	155.21 ± 16.43	150.54 ± 22.30	0.161

AME: Apparent Metabolizable Energy; TME: True Metabolizable Energy.

The inclusion of 50% CRM in rations both with or without isoamilase did not give a significant difference to the coefficient starch digestibility in small intestine. The results of the study on starch digestibility coefficient in small intestine were 0.97 - 0.98. [38] found that the total starch digestibility coefficient had increased by 0.96 at the age of 3 days of broiler chicken. The same authors also found starch digestibility increased (P <0.01) linearly as grew older in fast-growing broilers but not in slow-growing layer chickens. [39] found that 50% of poultry showed ileal starch digestion above 0.94, when fed with pellet diet containing 38.5% of whole wheat. Better results were obtained from [40] revealed ileal starch digestion was 0.98 in a diet containing 44% of whole wheat with other pellet ingredients. The whole process of starch digestion and absorbtion were carried out during short retention times in the duodenum and jejunum, estimated by [41] for about 1 hour.

Starch digestibility was able to affect AME in feed [42, 43]. The results showed no significant differences in AME and TME in all treatments. This was due to there was no significant difference in starch digestibility coefficient. [37] reported that metabolizable energy (ME) consumption decreased (P <0.001) in the treatment fed cassava pellet and chips without enzyme supplementation, but increased (P <0.001) in all enzyme-supplemented feeds. Supplementation of 200 and 400 Ukg-1 α-1.6 Isoamylase in feed did not have a significant effect on ME compared to control [44]. The results of this study, TME on T1 and T2 tended to be lower compared to control, indicating that to hydrolyze or digest CRM requires higher energy.

The results of blood glucose levels ranged from 150.54 to 169.65 mgdL-1. [45] stated that glucose levels in broilers were higher than mammals, ranging from 180 to 250 mgdL-1.

Chicken in a state of hypoglycemia if the blood glucose level was 137 mgdL-1 and hyperglycemia in blood glucose levels 363 mgdL-1 [46]. The results of this study did not show significant differences in all treatments but the blood glucose levels were still in normal standards. The diet containing CRM with or without isoamylase addition tended to have lower blood glucose levels compared to the control ration (155.21 mgdL-1 and 150.54 mgdL-1). The same result was shown by [44] reported that there was no significant difference in broiler serum glucose concentration between the dietary treatments supplemented with 1500 and 3000 Ukg-1 α-Amylase (A. oryzae), 480 and 960 Ukg-1 α-1.4 Amylase (B. subtilis) and 200 and 400 Ukg-1-1.6 Isoamylase (B. subtilis). Previous research had shown that glucose absorption was not a factor that limits starch utilization [47].

4. Conclusion

It can be concluded that the application of 0.05% isoamylase in CRM was able to reduce the amylose / amylopectin ratio which was close to the amylose / amylopectin ratio in corn and increase the reducing sugar content. The inclusion of 50% CRM with or without isoamylase to subtitute corn in the ration did not affect growth performance, intestinal starch digestibility coefficient, AME, TME, and blood glucose levels of broiler chickens but had a higher FCR compared to control.

This work was financially supported by Indonesia Endowment Fund for Education (LPDP).

References

1. McNab, J. *Advance in poultry nutrition in the world*. Proceedings of the seminar and international poultry show, WPSA, Bnagladesh Branch. p52 (1999)
2. Van der Klis JD, Kwakernaak C, Jansman A, Blok M. *Energy in poultry diets: Adjusted AME or Net Energy*. Proceedings of Australian Poultry Science Symposium. 21: 44-49 (2010)
3. Gomes E, Souza SR, Grandi RP, Silva RD. *Production of thermostable glucoamylase by newly isolated Aspergillus flavus A1.1 and thermomyces Lanuginosus A13.37*. *Braz J Microbiol* 36: 75-82 (2005)
4. Promthong S, Kanto U, TIrawattanawanich C, Tongyai S, Isariyodom S, Markvichitr K. *Comparison of nutrient compositions and carbohydrate fractions of corn, cassava chip and cassava pellet ingredients: animals*. In: Proceedings of 43rd Kasetsart University Annual Conference; p. 211-20 (2005)
5. Kiatponglarp W, Tongta S. *Structural and physical properties of debranched tapioca starch*. Suranaree J Sci Tech 2007;14(2): 195-204 (2007)
6. Anguita M, Gasa J, Martı́n-Orué SM, Perez JF. *Study of the effect of technological processes on starch hydrolysis, non-starch polysaccharides solubilization and physicochemical properties of different ingredients using a two-step in vitro system*. Animal Feed Science and Technology 129, 99 – 115 (2006)
7. Midau A, Augustine C, Yakubu B, Yahaya SM, Kibon A, Udoyong AO. *Performance of broiler chicken fed enzyme supplemented cassava peel meal based diets*. Int J Sust Agric 3(1): 1-4 (2011)
8. Bhuiyan M, Romero LF, Iji P. *Maximising the energy value of cassava products in diets for broiler chickens*. In: Proceedings of the 23rd Annual Australian Poultry Science Symposium; p. 27-25 (2012)
9. Huang Z. *Effects and mechanism of isoamylase additive on growth performance in broilers*. Siliao Gongye. 35:1–5 (2014)

10. Farrell DJ. *Rapid Determination of Metabolizale Energy of Feed Using Cockrerels*. J Poult. Sci. 19: 303-308 (1978)
11. Steel RGD, Torrie JH. *Prinsip dan Prosedur Statistika*. Terjemahan: B. Sumantri. Jakarta (ID): Gramedia Pr. (1995)
12. Cao B, Kumao T, Karasawa Y. *Effects of dietary cellulose levels on growth, nitrogen utilization and retention time of diets on intestine in chicks fed equal amounts of nutrients*. Proceeding of the 6th Asian Pacific Poultry Congress, Nagoya Japan pp. 402-403 (1998)
13. Langhout DJ. *The role of intestinal flora as affected by non-starch polysaccharides in broiler chicks*. PhD Thesis. Wageningen University, Wageningen Netherlands pp 162 (1998)
14. Buléon A, Colonna P, Planchot V, Ball S. *Starch granules: Structure and biosynthesis*. Int. J. Biol. Macromol. 23:85–112 (1998)
15. Svihus B. *Function of the digestive system*. J. Appl. Poultry Res., 23: 306–314 (2014)
16. Jackson S, Duke GE. *Intestine fullness influences feeding behaviour and crop filling in the domestic turkey*. Physiol. Behav., 58: 1027–1034 (1995)
17. Langenfeld M.S. *Systema digestorium, s. apparatus digestorius. In: Chicken Anatomy (in Polish)*. Scientific Publishing Company PWN, pp. 91–117 (1992)
18. Svihus B, Sacranie A, Denstadli V, Choct M. *Nutrient utilization and functionality of the anterior digestive tract caused by intermittent feeding and inclusion of whole wheat in diets for broiler chickens*. Poult. Sci. 89: 2617–2625 (2010)
19. Svihus B, Lund V, Borjgen B, Bedford M, Bakken M. *Effect of intermittent feeding, structural components and phytase on performance and behaviour of broiler chickens*. Brit. Poultry Sci., 54: 222–230 (2013)
20. Ravindran V. *Feed enzyme: The science, practice, and metabolic realities*. J. Appl. Poult. Res. 22 : 628–636 (2013)
21. Svihus B. *Starch digestion capacity of poultry*. Poult. Sci. 93: 2394-2399 (2014)
22. Perez S, Bertoft E. *The molecular structures of starch components and their contribution to the architecture of starch granules: A comprehensive review*. Starch 62:389–420 (2010)
23. Björck I, Liljeberg H, Östman E. *Low glycaemic-index foods*. Br. J. Nutr. 83(Suppl. 1):S149–S155 (2000)
24. Zhang G, Ao Z, Hamaker BR. *Slow digestion property of native cereal starches*. Biomacromolecules 7:3252–3258 (2006)
25. Amerah AM, Ravindran V, Lentle RG, Thomas DG. *Feed particle size: Implications on the digestion and performance of poultry*. World's Poult. Sci. J. 63:439–455 (2007)
26. Svihus B. *The gizzard function, influence of diet structure and effects on nutrient availability*. World's poult. Sci. J., 67: 207-223 (2011)
27. Amerah AM, Ravindran V, Lentle RG, Thomas DG. *Influenced of feed particle size on the performance, energy utilization, digestive tract development, and digesta parameters of broiler starter fed wheat-and-corn-based diets*. Poult. Sci. 87: 2320-2328 (2008)
28. Xu Y, Stark CR, Ferket PR, Williams CM, Pacheco WJ, Brake J. *Effect of dietary coarsely ground corn on broiler live performance, gastrointestinal tract development, apparent ileal digestibility of energy and nitrogen, and digesta particle size distribution and retention time*. Poult. Sci. 94: 53-60 (2015)
29. Sharabi KC, Tavares D, Rines AK, Puigserver P. *Molecular pathophysiology of hepatic glucose production*. Mol. Aspects Med. 46: 21–33 (2015)
30. Moore MC, Coate KC, Winnick JJ, An Z, Cherrington AD. *Regulation of hepatic glucose uptake and storagein vivo*. Adv. Nutr. 3: 286–294 (2012)

31. Denardin CC, Walter M, Silva LPD, Souto GD, Fagundes CA. *Effect of amylose content of rice varieties on glycemic metabolism and biological responses in rats*. Food Chem. 105:1474–1479 (2007)
32. Denardin CC, Boufleur N, Reckziegel P, Silva LPD, Walter M. *Amylose content in rice (Oryza sativa) affects performance, glycemic and lipidic metabolism in rats. Ciˆencia Rural*. 42:381–387 (2012)
33. Eruvbetine D, Tajudeen ID, Adeosun AT dan Olojede AA. Cassava (M. esculenta) leaf and tuber concentrate in diets for broiler chickens. *Biores. Tech.*, 86: 277-281 (2003)
34. Adeyemi OA, Eruvbetine D, Ongutona T, Dipeolu M, Agunbiade JA. *Feeding broiler chicken with diets containing whole cassava root meal fermented with rumen filtrate*. Arch. Zootech. 57 (218): 247-258 (2008)
35. Nwokoro and Ekhosuehi, *Effect of replacement of maize with cassava peel in cockerel diets on performance and carcass characteristics*. Tropical Animal Health and Production, 37: 495-501 (2005)
36. Khempaka S, Molee W, Guillaume M. *Dried cassava pulp as an alternative feedstuff for broilers: Effect on growth performance, carcass traits, digestive organs, and nutrient digestibility*. J. Appl. Poult. Res. 18: 487-493 (2009)
37. Bhuiyan MM, Iji PA. *Energy Value of Cassava Products in Broiler Chicken Diets with or without Enzyme Supplementation*. Asian-Australas. J. Anim. Sci. 28:1317–1326 (2015)
38. Zelenka J, Ceresnakova Z. *Effect of age on digestibility of starch in chickens with different growth rate*. Czech J. Anim. Sci. 50:411–415 (2005)
39. Svihus B, Hetland H. *Ileal starch digestibility in growing broiler chickens fed on a wheat-based diet is improved by mash feeding, dilution with cellulose or whole wheat inclusion*. Br. Poult. Sci. 42:633–637 (2001)
40. Hetland, H., B. Svihus and M. Choctt. *Role of insoluble fiber on gizzard activity in layers*. J. Apply. Poultry Res. 14: 38—46 (2005)
41. Rougiere N, Carré B. *Comparison of gastrointestinal transit times between chickens from D+ and D− genetic lines selected for divergent digestion efficiency*. Animal 4:1861–1872 (2010)
42. Gracia MI, Arani´bar MJ, L´azaro R, Medel P, Mateos GG. *α-Amylase Supplementation of Broiler Diets Based on Corn*. Poult. Sci. 82:436–442 (2003)
43. Kaczmarek SA, Rogiewicz A, Mogielnicka M, Rutkowski A, Jones RO, Slominski BA. *The effect of protease, amylase, and nonstarch polysaccharide-degrading enzyme supplementation on nutrient utilization and growth performance of broiler chickens fed corn-soybean meal-based diets*. Poult. Sci. 93:1745–1753 (2014)
44. Yuan J, Xingyu W, Dafei Y, Maofei W, Xiaonan Y, Zhao L, Yuming G. *Effect of different amylases on the utilization of cornstarch in broiler chickens*. Poult. Sci. 96:1139–1148 (2017)
45. Hazelwood RL. *Pancreatic hormones, insulin, glicagon molar ratios and somatostatin as determinants of avian carbohydrate metabolism*. J Exp Zool. 232:647-652 (2000)
46. Goodwin, Denise, John B, McMurray BL, William IR, Danny LM. *Blood glucose value and definitions of hypoglicemia and hyperglycemia*. Georg Poult Lab (Georgia). The University of Georgia (1994)
47. Suvarna S, Christensen VL, Ort DT, Croom Jr WJ. *High levels of dietary carbohydrate increase glucose transport in poult intestine*. Comp. Biochem. Physiol. A Mol. Integr. Physiol. 141:257–263 (2005)

Effects of egg storage duration prior to incubation on performances of Arab Chickens during the first week post-hatch

Faizal Andri[1*], *Filoza* Marwi[1], *Eka* Nurwahyuni[1], *Dyah Lestari* Yulianti[1], *Heni Setyo* Prayogi[1], and *Edhy* Sudjarwo[1]

[1]Faculty of Animal Science, Universitas Brawijaya, Malang 65145, Indonesia

Abstract. This study investigated the effects of egg storage duration prior to incubation on performances of Arab Chickens during the first week post-hatch. A total of 100 day-old-chicks were distributed among 20 experimental units, each representing a different egg storage duration before incubation (1, 4, 7, and 10 days) with five replicates. There were five birds of each replicate totaling 25 birds per treatment. The variables examined in this study included the growth performance of Arab Chickens during the first week post-hatch, specifically the initial body weight, final body weight, body weight gain, feed intake, water intake, feed conversion ratio, and water conversion ratio. The findings of the study revealed that egg storage duration prior to incubation did not significantly affect ($P>0.05$) feed intake, water intake, and water conversion ratio. Storing eggs for up to 4 days did not result in any adverse effects ($P>0.05$) on initial body weight, final body weight, body weight gain, and feed conversion ratio. However, eggs stored for 7 days or longer exhibited a significant reduction ($P<0.05$) in initial body weight, final body weight, and body weight gain. Additionally, a notable increased ($P<0.05$) in feed conversion ratio observed for eggs stored for 10 days. In conclusion, the hatching eggs of Arab Chickens can be safely stored for up to 4 days without any detrimental effect on their performance during the first week post-hatch.

1 Introduction

Arab Chickens are among the laying-type chickens widely farmed in Indonesia. This breed results from a crossbreeding between Braekel Chickens (an exotic breed from Belgium) and Indonesian local chickens, which was initiated around 1995 [1, 2]. This chicken exhibits the advantage of adapting well to the tropical climate and has a modest brooding trait [3, 4]. Arab Chickens had egg characteristics similar to local chickens but with a higher egg production capability [5-7].

Arab Chicken farming along with other local chickens play a crucial role in meeting the growing national demand for egg. However, the contribution of local chicken eggs to the total national egg production is still low [8]. This condition probably driven by suboptimal

* Corresponding author: f.andri@ub.ac.id

performance in both chick quality and post-hatch management [6, 9]. Addressing these factors becomes essential for enhancing the overall productivity and efficiency of the Arab Chicken farming.

The hatchery aspect is essential in addressing the demand for quality chicks [10, 11]. One of the key aspects in the hatchery management is the determination of optimal storage durations for hatching eggs. This issue is pivotal to ensure robust and healthy flocks in the commercial laying hens [12]. Nevertheless, information on the impact of egg storage duration on the performances of Arab Chickens is currently still limited. Therefore, this study investigated the effects of egg storage duration prior to incubation on performances of Arab Chickens during the first week post-hatch.

2 Materials and Methods

2.1 Experimental design

This experiment was conducted at the Laboratory of Poultry Production, Faculty of Animal Science, Universitas Brawijaya. This study employed a completely randomized design comprising four treatments, each with five replicates. The treatments involved varying durations of egg storage prior to incubation, specifically 1, 4, 7, and 10 days.

2.2 Egg storage and incubation

A total of 200 eggs (50 eggs per storage duration) were used in this experiment. Eggs were stored in egg trays with the large end positioned upward. The temperature and relative humidity during storage were 24-26 °C and 60-70%, respectively. At the end of each storage period, eggs were subsequently transferred to a forced-draft incubator. The temperature and relative humidity for incubation were set at 37.7-37.9 °C and 50-60%, respectively, throughout the whole period. Regular hourly turning at a 90° angle occurred during the setter phase. At the 18th day of embryonic development, candling was performed to remove clear eggs, while those with viable embryos were transitioned from turning trays to hatching baskets and subsequently moved to the hatcher.

2.3 Rearing phase

Upon reaching 21.5 days of incubation, five healthy day-old chicks from each replicate, totaling 25 chicks per treatment, were selected based on dry fluff and absence of abnormalities. These chicks were reared for one week and fed a commercial diet with crude protein, ether extract, crude fibre, and ash contents were 20.00, 5.00, 5.00, and 8.00%. Feed and drinking water were provided ad libitum throughout the experimental period. The house temperature was maintained at 30-35°C using a gas brooder. The growth performance of chicks during the first week post-hatch was assessed, focusing on variables such as initial body weight, final body weight, body weight gain, feed intake, water intake, feed conversion ratio, and water conversion ratio.

2.4 Data analysis

Statistical analysis involved one-way analysis of variance, with significance declared at $P<0.05$. Post hoc Duncan tests were conducted to differentiate means in variables exhibiting significant effects. The entire data analysis process was performed using IBM SPSS version 25.

3 Results and discussion

Effects of egg storage duration prior to incubation on initial body weight, final body weight, body weight gain, feed intake, water intake, feed conversion ratio, and water conversion ratio of Arab Chickens during the first week post-hatch is presented on Table 1. The findings of the study revealed that egg storage duration prior to incubation did not significantly affect (P>0.05) feed intake, water intake, and water conversion ratio. Conversely, the study highlights significant differences (P<0.05) in initial body weight, final body weight, body weight gain, and feed conversion ratio among the treatments. Notably, storing eggs for up to 4 days did not yield adverse effects (P>0.05) on these parameters, implying that this duration is within an acceptable range for maintaining optimal growth during the first week post-hatch. However, the findings take a noteworthy turn for eggs stored for 7 days or longer. In these cases, a significant reduction (P<0.05) in initial body weight, final body weight, and body weight gain was observed. This suggests that prolonged egg storage durations, beyond the 4-day threshold, negatively impact the overall growth trajectory of Arab Chickens during the early stages of development. Furthermore, a notable increase (P<0.05) in feed conversion ratio was identified for eggs stored for 10 days. This heightened feed conversion ratio implies an increased amount of feed required to achieve a unit of body weight gain, suggesting a less efficient conversion of feed into body weight.

Table 1. Effects of egg storage duration prior to incubation on performances of Arab Chickens during the first week post-hatch[1].

Variables[2]	**Egg storage duration**				**P value**
	1 day	**4 days**	**7 days**	**10 days**	
IBW (g)	30.68 ± 0.33^{c}	29.88 ± 0.88^{c}	28.28 ± 0.46^{b}	26.00 ± 1.02^{a}	0.001
FBW (g)	70.56 ± 1.42^{c}	68.36 ± 1.63^{c}	64.04 ± 1.71^{b}	58.43 ± 1.89^{a}	0.001
BWG (g)	39.88 ± 1.69^{c}	38.48 ± 1.33^{c}	35.76 ± 1.49^{b}	32.43 ± 0.96^{a}	0.001
FI (g)	63.91 ± 5.95	64.39 ± 6.03	63.35 ± 6.46	63.83 ± 4.54	0.994
WI (ml)	118.97 ± 13.36	113.18 ± 16.25	111.57 ± 10.85	109.72 ± 11.33	0.711
FCR (g/g)	1.60 ± 0.09^{a}	1.68 ± 0.19^{a}	1.77 ± 0.20^{ab}	1.97 ± 0.19^{b}	0.021
WCR (ml/g)	2.98 ± 0.34	2.95 ± 0.50	3.12 ± 0.31	3.38 ± 0.27	0.279

[1]Data were presented as mean ± standard deviation from five replicates with each replicates consisted of five Arab Chickens

[2]IBW: initial body weight, FBW: final body weight, BWG: body weight gain, FI: feed intake, WI: water intake, FCR: feed conversion ratio, WCR: water conversion ratio

[a-c]Uncommon superscript indicates a significant difference (P<0.05)

The reduction in growth performance of Arab chickens during the critical post-hatch phase with prolonged egg storage probably due to the complex interplay of factors encompassing nutrient degradation and oxidative stress. The egg protein and lipids may degrade over an extended storage period. This degradation can result in a less nutritionally robust environment for the developing embryo, impacting the chick's ability to access vital nutrients during its early growth phase [13, 14]. Extended storage also deplete the energy reserves within the egg, which may compromise the embryo ability to access sufficient energy stores, leading to reduced vigor and growth potential post-hatch [15]. Moreover, longer storage periods also expose the eggs to an increased risk of oxidative stress. Oxidative stress can negatively impact cellular functions and overall health [16]. The developing embryo may face challenges in combating oxidative stress [17], potentially leading to suboptimal growth performance.

4 Conclusion

In conclusion, this study underlines the significance of carefully managing egg storage durations for Arab Chickens, with a particular emphasis on minimizing adverse effects on growth-related parameters. The hatching eggs of Arab chickens can be safely stored for up to 4 days without any detrimental effect on their performance during the first week post-hatch.

The authors acknowledged financial support from Faculty of Animal Science Universitas Brawijaya under research grant number 1873.6/UN10.F05/PN/2023.

References

1. B. Indarsih, M.H. Tamzil, Asian J. Anim. Sci. **16**, 62-67 (2022)
2. T. Sartika, F. Saputra, H. Takahashi, HAYATI J. Biosci. **30**, 122-130 (2023)
3. H. Nurcahya, D.N. Liarachma, S. Darwati, J. Trop. Biodivers. **3**, 91-97 (2023)
4. Iswati, G. Ciptadi, E3S Web Conf. **335**, 00030 (2022)
5. F. Andri, A. Sukoco, T. Hilman, E. Widodo, Livest. Res. Rural Dev. **28**, 221 (2016)
6. Iswati, M.H. Natsir, G. Ciptadi, T. Susilawati, J. World's Poult. Res. **11**, 73-82 (2021)
7. Syafwan, A. Budiansyah, U. Haroen, K. Simanungkalit, L.A.B. Sembiring, I.L. Aritonang, J. Adv. Vet. Anim. Res. **10**, 30-41 (2023)
8. DGLAH, Livestock and Animal Health Statistics 2022 (Directorate General of Livestock and Animal Health, Jakarta, 2022)
9. Syafwan, Noferdiman, S. Zubaida, T.M. Pasaribu, Adrizal, Trop. Anim. Sci. J. **44**, 462-477 (2021)
10. G. Underwood, D. Andrews, T. Phung, L.E. Edwards, Anim. Prod. Sci. **61**, 867-875 (2021)
11. K. Tona, K. Voemesse, O. N'nanlé, O.E. Oke, Y.A.E. Kouame, A. Bilalissi, H. Meteyake, O.M. Oso, Front. Physiol. **13**, 895854 (2022)
12. A.H. Maman, I. Yildirim, Eur. Poult. Sci. **86**, 353 (2022)
13. H. Nasri, H. van den Brand, T. Najjar, M. Bouzouaia, Poult. Sci. **99**, 4607-4615. (2020)
14. Y. Meng, N. Qiu, V. Guyonnet, Y. Mine, Food Chem. **393**, 133403 (2022)
15. P.E.N. Givisiez, A.L.B.M. Filho, M.R.B. Santos, H.B. Oliveira, P.R. Ferket, C.J.B. Oliveira, R.D. Malheiros, Poult. Sci. **99**, 6774-6782 (2020)
16. P.F. Surai, V.I. Fisinin, F. Karadas, Anim. Nutr. **2**, 1-11 (2016)
17. J. Yang, X. Ding, S. Bai, J. Wang, Q. Zeng, H. Peng, Y. Xuan, Z. Su, K. Zhang, Animals, **10**, 1409 (2020)

Roles of tree and shrub legume leaves as protein sources for cattle raised by small farmers in Bangkalan Regency

Mashudi Mashudi[1], *Poespitasari* Hazanah Ndaru[1], *Wike* Andre Septian[1], *Kusmartono* Kusmartono[1*], *and Aprilia* Dwi Kartika[1]

[1]Faculty of Animal Science, Universitas Brawijaya, Malang 65145, Indonesia

Abstract. Inclusion of tree or shrub legume leaves in the ration of beef cattle improved quality and the main indicators are the increase in crude protein content and digestibility. The present study was done from March to September 2023 involving 30 farmers in Bangkalan regency who raised madura cattle and the information gathered were type and number of feeds offered daily. Tree and shrub legume leaves such as *Gliricidia*, *Moringa*, *Indigofera*, and jackfruit are the most common protein sources used to supplement field grass or straws. Samples of forage were taken and the following rations were formulated to simulate cattle feed: A (grass only); B (60% grass + 40% gliricidia); C (60% grass + 40% moringa); D (60% grass + 40% Indigofera); E (60% grass + 40% jackfruit) leaves. Parameters measured using an in vitro gas production technique showed that supplementation with tree and shrub legume leaves on field grass basal diet improved dry matter digestibility (DMD) and organic matter (OMD) values and this was closely related to improvements of rumen NH_3 concentration which also contributed to efficiency of microbial protein synthesis (ESPM) value improvement. It can be concluded that tree and shrub legume leaves have shown important roles in improving the quality of field grass based diet for Madura cattle.

1 Introduction

Poor nutrition due to imbalance of energy and protein content of ration can be a major cause of low beef cattle productivity in Bangkalan regency, especially in the seasonally dry areas where there is a severe shortage of feed during the dry season [1]. Concentrate or supplemental feeds are sometimes used by farmers to improve overall animal efficiency, and [2] reported that by supplementing 0.5 kg rice bran in comparison to 2.5 kg concentrate containing 40% rice bran, 19% tofu waste, 31% maize bran and 10% soybean hulls increased dry matter intake (56.1 vs 89.3 g/kg0.75/d) in Madura cows fed on field grass based-diet. However, the high price of those concentrate as energy and protein supplements (e.g., coconut cake meal, palm kernel cake, brand pollard, rice brand) limits their widespread use

* Corresponding author: kusmartono_ansi@ub.ac.id

[3]. One of possible ways to improve protein content of the ration is by giving tree or shrub legume leaves as it has been reported that feeding leguminous fodder that is high in protein can improve protein content leading to improvements of rumen fermentation parameters and digestibility of low quality fibrous feeds, and hence improved animal production [4].Tree or shrub legumes may help to increase beef cattle productivity in Bangkalan regency by providing a high protein supplement. When fed to supplement low quality roughages such as field grass or rice straw, farmers would have another reason to grow and use leguminous trees in their agricultural systems. This paper aimed to explain roles of tree/shrub legume foliage as protein supplements on gas production, rumen fermentation parameters and microbial protein synthesis in the ration that contained low quality roughage as a basal diet.

2 Materials and Methods

2.1 Experimental design

This study was conducted in Bangkalan regency from March to September 2023 as part of in vivo study involving 30 farmers who were raising Madura bulls for fattening purposes. Chemical analysis of feeds identified was done in the Animal Nutrition laboratory, Faculty of Animal Science, Brawijaya University, Malang. Materials used for formulating ration were local feeds namely field grass, *Gliricidia*, *Moringa*, *Indigofera*, and jackfruit leaves collected from beef cattle farmers in Bangkalan regency. The rations were then formulated in such a way to mimic the beef cattle ration under small farmer's condition and the use of tree/shrub legume leaves were used to improve the protein contents of the ration. Rumen liquids were taken from the slaughterhouse in the morning at three different times as replication and it was done due the absence of rumen fistulated animals.

2.2 Methods

Feed samples included field grass, gliricidia, moringa, indigofera and jackfruit leaves were taken weekly, and sub-samples were brought to the laboratory for proximate analysis [5]. Based on the chemical analysis data, the ration was then formulated as the following treatments: A = field grass (control); B = field grass + gliricidia leaves; C = field grass + moringa leaves; D = field grass + indigofera leaves; E = field grass + jackfruit leaves. The ratio between field grass and the tree/shrub leaves was 60:40 on a dry matter basis. Parameters measured were in vitro gas production (IVGP) following the procedure of [6], dry matter digestibility (DMD) and organic matter digestibility (OMD). Observation of gas production was carried out at 2, 4, 6, 8, 12, 16, 24, 36, 48, and 72 hour and the calculation of gas production was executed using the Neway (Excel) program [7] using the following equation of [8]: p = b (1-e-ct), where p = the volume of gas produced at time t, b = fraction of potential degraded or gas production potential, c = rate of gas production. Rumen ammonia concentration was measured according to [9], whilst efficiency of microbial protein synthesis (EMPS) was measured as guided by [10].

2.3 Data analysis

Data obtained were subjected to statistical analyses using a Randomised Block Design (RBD) according to [11] whereas times of rumen liquid taking were used as a block. Statistical analyses were performed on all data by following the general linear procedure in PROC GLM from SAS. The data obtained were analysed for variance at the 5% significance level. If the data shows a significant difference as an effect of the treatment, then continue with Duncan's new multiple range test (DMRT).

3 Results and discussion

3.1 Chemical composition of forages used in this study

Data on chemical composition of field grass and tree and shrub legume leaves collected from cattle farmers in Bangkalan regency is presented in Table 1.

Table 1. Type of feeds commonly used by farmers in Bangkalan regency

No	Feeds	Nutrient contents					
		DM (%)	Ash	CP	CF	EE	NFE
			(% DM)				
1	Field grass	25.3	13.5	9.2	31.9	0.9	59.9
2	Rice bran	89.2	16.9	8.4	28.9	4.0	41.8
3	Glyricidia	22.5	8.1	20.4	7.7	12.1	58.7
4	Moringa leaves	24.2	12.8	27,4	6.8	10.4	42.5
5	Indigofera	26.3	7.9	18.4	15.6	2.1	56.0
6	Jackfruit leaves	35.4	7.6	15.4	32.6	2.6	42.4

CF – crude fibre; CP – crude protein; DM – dry matter; EE – ether extract; NFE – nitrogen free extract

During the execution of the study, it was found that field grass being a majority of the feed used as a sole diet by the farmers and as seen in Table 1 crude protein (CP) content of field grass was 9.2%, which was a little bit higher than CP content of rice bran (8.4%). A small number of farmers occasionally provide those two feeds for their cattle as rice bran is considered an expensive stuff in Bangkalan regency (IDR 4,000-4,200). Crude protein contents of gliricidia, moringa, indigofera and jackfruit leaves were in the range of values reported elsewhere by [1, 4, 12].

3.2 IVGP, DMD, OMD, rumen NH_3 and EMPS

Data on IVGP, DMD, OMD, rumen NH_3 concentration and EMPS is presented in Table 2. Statistical analysis showed that feed treatments significantly affected IVGP, DMD, OMD, rumen NH3 concentration EMPS ($P<0.05$). The highest IVGP values were recorded in treatment C (132.6 ml/500 mg DM) followed by treatments E, D, B and A (121.8, 119.4, 112.9 and 99.6 ml/500 mg DM, respectively). Similar trends were also observed in DMD, and OMD values, in which the DMD and MD values increased by the presence of tree or shrub legume leaves in the ration. This evidence clearly shows that increasing CP content of the ration due to the supplementation of tree or shrub legume leaves has allowed rumen microbes to grow and do the digestion process better.

Table 2. IVGP, NH3 and EMPS

Parameters observed	Treatments					
	A	B	C	D	E	SEM
IVGP (ml/500 mg DM)	99.6^a	112.9^b	132.6^d	119.4^b	121.8^c	4.34

b (ml/500 mg DM)	12.3^{a}	21.4^{b}	23.4^{c}	20.5^{b}	21.6^{b}	1.23
c (ml/hour)	0.001^{a}	0.0017^{b}	0.0045^{d}	0.0021^{c}	0.0019^{c}	0.0002
DMD (%)	56.2^{a}	63.7^{b}	66.6^{c}	64.2^{b}	62.1^{b}	3.12
OMD (%)	61.4^{a}	69.2^{b}	74.1^{d}	71.6^{c}	68.3^{b}	2.45
NH_3 (mg N-NH_3/l)	79.8^{a}	102.6^{b}	113.8^{d}	104.5^{bc}	106.8^{c}	4.12
ESPM (g N/kg OMDR)	19.4^{a}	24.6^{b}	32.4^{d}	28.5^{c}	30.2^{c}	2.32

DMD – dry matter digestibility; ESPM – efficiency of microbial protein synthesis; IVGP – in vitro gas production OMD – organic matter digestibility; Different superscripts in the same row showed significant difference (P<0.05).

Ammonia concentration required by microbes in the rumen were at the normal level for all treatments ranging from 79.8 mg N-NH3/l (in rice bran supplementation) to 113.8 mg N-NH3/l (in moringa leaves supplementation). ESPM values also went on the same trend as NH3 concentration value in which by supplementing moringa leaves (treatment C) on field grass produced the highest ESPM value (32.4 g N/kg DOMR) followed by treatments E, D, B and A (30.2, 28.5, 24.6 and 19.4 mg N/kg DOMR respectively). [13] stated that the protein in Moringa has good rumen bypass properties and in the study of [14] showed that by feeding dairy cows 2 kg or 3 kg DM of Moringa leaves increased milk yield (4.9 and 5.1 kg/day) compared to cows fed only Brachiaria brizantha hay; this improvement was related to an increase in feed intake and in the digestibility of nutrients (dry matter, organic matter, crude protein, and dietary fibre) in Moringa leaves.

4 Conclusion

In summary, tree or shrub legume leaves have potential roles in improving quality of ration based on field grass. The adequate concentration of NH3 in the rumen and improvement on ESPM values when tree or shrub legume leaves are supplemented has become an important finding of this study.

The authors wish to thank the Ministry of Education, culture, research and technology for funding this study through "Matching Fund scheme"

References

1. Kusmartono, Mashudi, P. H. Ndaru. In Proceedings of the 3rd International Conference on Environmentally Sustainable Animal Industry ICESAI, 13 Oktober, Malang, Indonesia (2023)
2. Kusmartono, Mashudi., P. H. Ndaru, W. A. Septian., A. D. Kartika, In Proceedings of 5th International Conference on Agriculture and Bioindustry ICAGRI, 11-12 October 2023, Aceh, Indonesia (2023) (to be published).
3. E. H. Cabezas-Garcia, D. Lowe, F. Lively, Animals. **11** (2021)
4. R. A. Leng. (Food and Agriculture Organization of the United Nations, Rome, 1997)
5. AOAC. (Oxford University Press, England 2019)
6. K. H. Menke, L. Raab, A. Salewski, H. Steingass, D. Fritz, W. Schneider, J. Agric. Sci. **92** (1979)
7. X. Chen. (Unit. Rowett Research Institute, Backburn, Aberdeen 1994)
8. E. R. Ørskov, I. McDonald, J. Agric. Sci. **92** (1970)
9. E. J. Conway. (Crosby Lookwood, London, 1962)

10. M. Blümmel, R. Mgomezulu, X. B. Chen, H. P. S. Makkar, K. Becker, E. R. Ørskov, J. Agric. Sci. **133**, (1999)
11. H. Sudarwati, M. H. Natsir, V. A. Nurgiartiningsih, (UB Press, Indonesia, 2021)
12. E. Marhaeniyanto, S. Susanti, B. Siswanto, A. T. Murti. Ternak Trop. **20**, (2019)
13. A. A. Amad, J. Zentek, Front. in Anim. Sci. **4**, (2023)
14. N. R. Sanchez, E. Sporndlyn, I. Ledin, Livest. Sci. **101**, (2006)

Grafting worker larvae to obtain uniform *Apis mellifera* queen bees in support of the evaluation and identification of forage plant species

Sri Minarti[1*], *Lilik Eka* Radiati[1], *Dewi* Masyitoh[2], *Eka* Nurwahyuni[1], and *Muhammad Luthfi Hakim* Asy'ari[1]

[1]Faculty of Animal Science, Universitas Brawijaya, Malang 65145, Indonesia
[2]Faculty of Agriculture and Animal Husbandry, Islamic University of Malang, Malang 65144, Indonesia

Abstract. The productivity of bee colonies is greatly influenced by the performance of the queen in production. This research aims to obtain uniform *Apis mellifera* queen bees, both in terms of age and morphometrics, which will be developed into colonies to evaluate quality and identify bee forage plant species in grazing areas. The study was conducted in two different locations: Sukorame Village, Tumpang Subdistrict, Malang Regency, and Pajaran Village, Poncokusumo Subdistrict, Malang Regency. Pollen sample analysis was carried out at the Genetech Saraswanti Indo Surabaya Laboratory. The research method involved field observations and direct experiments, focusing on the research locations and grafting techniques. Data collected included both primary and secondary data, and data analysis was performed using unpaired t-tests to compare data from the two different research locations. The research results indicate that the difference in location has no significant effect ($P>0.05$) on the *Apis mellifera* queen bee production using the grafting method in terms of the measured morphological variables. The average abdomen length in Sukorame and Pajaran locations was 9.08±1.65 mm and 9.11±1.37 mm, respectively. The average abdomen width in Sukorame and Pajaran locations was 4.99±0.41 mm and 4.96±0.36 mm, respectively. The average abdomen height in Sukorame and Pajaran locations was 4.18±0.51 mm and 4.35±0.57 mm, respectively. The difference in location with the grafting method also had no significant effect ($P>0.05$) on the proboscis length, with average values of 1.30±0.39 mm for Sukorame and 1.45±0.43 mm for Pajaran locations. The use of different locations in the production of *Apis mellifera* queen bees using the grafting method provides consistent performance in terms of abdomen length, abdomen width, abdomen height, and proboscis length.

* Corresponding author: minartiherman@ub.ac.id

1 Introduction

The diversity of vegetation in Indonesia, supported by its tropical climate, has the potential to provide flowers throughout the year. This can be utilized for honeybee cultivation because these plants contain pollen and nectar that can be utilized as food for bees to sustain their lives. Similar to other animal species, bees also require energy from food throughout their lives. Honeybee cultivation can be considered an alternative business to enhance income, especially for communities around forested areas [1].

Honey bees are insects that offer numerous benefits, with their products being highly beneficial to health. The products derived from bee cultivation include honey, pollen, beeswax, propolis, royal jelly, and queen bees. The *Apis mellifera* bee is one of the most cultivated honey bee species in Indonesia due to its high honey production, ranging from 35 to 50 kg per colony per year, and its docile nature.

The queen bee plays a crucial role in the success of honey bee cultivation. The queen bee can produce larvae that develop into new queen bees, worker bees, and drones. If the queen bee becomes less productive, replacement becomes necessary. The technique used in the queen bee replacement process involves grafting or transplanting worker bee larvae because both are fertile, allowing for the quick and abundant production of new queens.

Given this background, research is needed in different grazing locations with varying plant support capacities as sources of quality nutrition. This is essential for obtaining uniform and optimal morphometric performance in *Apis mellifera* queen bees. The objective of this research is to determine the effects of forage quality from plants in different locations on the production of *Apis mellifera* queen bees using the grafting method, specifically assessing the morphological quality of the bees, including abdomen length, abdomen width, abdomen height, and proboscis length.

2 Materials and Methods

2.1 Research time and location

This research was conducted for 45 days, starting from May to June 2023, at the Honeybee Farm 'PT. Kembang Joyo Sriwijaya,' located in two different places: Sukorame Village, Tumpang Subdistrict, with an elevation of 536 meters above sea level, and Pajaran Village, Poncokusumo Subdistrict, Malang Regency, also at an elevation of 536 meters above sea level. The distance between location 1 and location 2 is 3.3 km. Pollen samples were analyzed at the Saraswanti Indo Genetech Laboratory in Surabaya. The environmental conditions at location 1 are much more enclosed compared to location 2, which will affect the sunlight entering the grazing area, thereby impacting temperature and humidity conditions at the research sites.

2.2 Research materials

The materials used in this research were 30 one-day-old *Apis mellifera* bee larvae obtained from the honeybee farm PT. Kembang Joyo. The larvae were collected from comb cells containing honeybee eggs by drawing a square line. This method was employed to identify the larvae of the desired age for grafting.

2.3 Research method

The research method employed involved direct field observation and experimentation. This approach included observing the research locations and techniques used in grafting, as well as collecting data by gathering all necessary information for the study. The collected data comprised both primary and secondary data. Primary data were obtained from temperature, humidity, and vegetation in two different locations. Pollen samples were collected for proximate analysis, and morphological observations of grafted queen bees were conducted. A comparison was then made between the locations, which are in Sukorame Village at an elevation of 536 meters above sea level and Pajaran Village at an elevation of 535 meters above sea level, with a distance of 3.3 km. Temperature, humidity, and pollen sample checks were performed every three days during monitoring. Secondary data were obtained from rainfall data provided by BMKG Malang and vegetation data obtained from BPS Malang.

2.4 Research procedure

2.4.1 Research preparation

- Conducting a survey of location and plant vegetation.
- Preparing equipment: smoker, thermometer, bar frames, queen cells, grafting tools.
- Selecting colonies and comb nests as the source of larvae.
- Marking comb nests containing eggs using thread.
- Preparing builder colonies (Cell builders).

2.4.2 Research implementation

- Pollen Sampling Stage

1. Installing a pollen trap at the entrance and exit of bee colonies every 3 days during the research.
2. Placing a collecting tray at the bottom of the entrance for pollen storage. Storing collected pollen in a refrigerated cabinet.

- Larva Grafting Stage

1. Preparing tools and materials
2. Selecting combs with larvae aged less than 24 days that have been marked.
3. Retrieving Queen cells installed on bar frames and left undisturbed for 1 day in the cell builder.
4. Applying royal jelly to the queen cell using a brush.
5. Transferring larvae from worker bee cells to queen cells using a grafting tool.
6. Inserting the grafted queen cell into the cell builder

- Larva Inspection Stage

1. Observing grafted larvae the next day.
2. Subsequent observations conducted every 3 days.
3. Performing queen pupa caging once they are covered, on the 13th day

- Future Queen Bee Rearing Stage
  1. Preparing a queen cage based on the success of grafting
  2. Gradually transferring the queen cell to the queen cage.
  3. Placing the queen cage in the hatching colony until the new queen hatches
  4. Harvesting the new queen

- Morphology Measurement of Queen Bees
  1. Placing the newly hatched queen in a container sprayed with alcohol.
  2. Separating each part of the queen's body after it has fainted or died.
  3. Placing body parts on millimeter paper, followed by photography.
  4. Observing the photographed body parts using Photoshop.
  5. Measuring the morphology according to the predetermined variables.

2.5 Measured variables

The research method employed involved direct field observation and experimentation. This approach included observing the research locations and techniques used in grafting, as well as collecting data by gathering all necessary information for the study. The collected data comprised both primary and secondary data. Primary data were obtained from temperature, humidity, and vegetation in two different locations. Pollen samples were collected for proximate analysis, and morphological observations of grafted queen bees were conducted. A comparison was then made between the locations, which are in Sukorame Village at an elevation of 536 meters above sea level and Pajaran Village at an elevation of 535 meters above sea level, with a distance of 3.3 km. Temperature, humidity, and pollen sample checks were performed every three days during monitoring. Secondary data were obtained from rainfall data provided by BMKG Malang and vegetation data obtained from BPS Malang.

2.5.1 Abdomen length

The separated abdomen is then measured in length by placing it on a millimeter block paper. Subsequently, photographs are taken from the ventral side, and the images are imported into the Photoshop application for the abdomen length measurement stage, involving the conversion from pixel units to millimeter units [2]. The measurement of abdomen length starts from point A to point B, as illustrated in Figure 1.

Source: Personal Documentation (2023)

Fig. 1. Abdomen Length Measurement

2.5.2 Abdomen width

The separated abdomen is then measured in width by placing it on a millimeter block paper. Subsequently, photographs are taken from the dorsal side, and the images are imported into the Photoshop application for the abdomen width measurement stage, involving the conversion from pixel units to millimeter units [2]. The measurement of abdomen width starts from point A to point B, as illustrated in Figure 2.

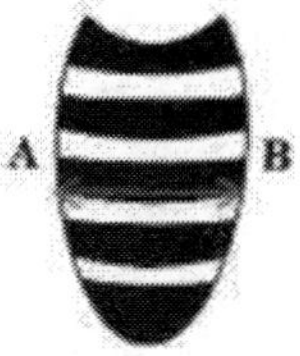

Source: Personal Documentation (2023)

Fig. 2. Abdomen Width Measurement

2.5.3 Abdomen height

The separated abdomen is then measured for its height by placing it on a millimeter block paper. Subsequently, photographs are taken from the ventral side, and the images are imported into the Photoshop application for the abdomen height measurement stage, involving the conversion from pixel units to millimeter units [2]. The measurement of abdomen height starts from point A to point B, as illustrated in Figure 3.

Source: Personal Documentation (2023)

Fig. 3. Abdomen Height Measurement

2.5.4 Proboscis length

The separated head is then measured for its proboscis length by placing it on a millimeter block paper. Subsequently, photographs are taken of the rostral part, and the images are imported into the Photoshop application for the proboscis length measurement stage, involving the conversion from pixel units to millimeter units [2]. The measurement of proboscis length starts from point A to point B, as illustrated in Figure 4.

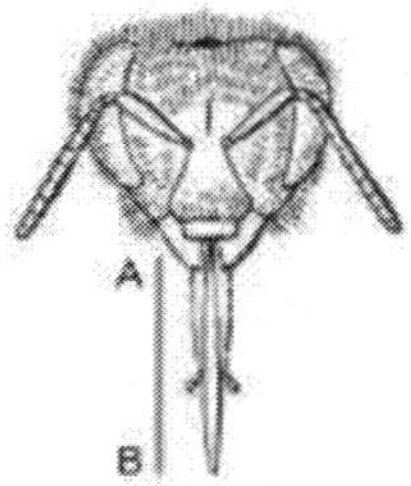

Source: Personal Documentation (2023)

Fig. 4. Proboscis Length Measurement

2.6 Data analysis

The data obtained from the research will be processed using statistical analysis, specifically an unpaired t-test, to compare data from two different research locations: the grazing locations in Sukorame Village, Tumpang Subdistrict, and Pajaran Village, Poncokusumo Subdistrict, Malang Regency.

3 Results and discussion

The variety of vegetation is a crucial factor in the world of beekeeping. This is because the overall species richness of plants has a strongly positive correlation with the abundance of bees in general. Diverse plant species have the potential to provide greater diversity and continuity in food sources such as nectar and pollen throughout the bee activity season [3]. The presence of an adequate supply of nectar and pollen is essential to avoid impacting the productivity of honeybees in producing various products. Insufficient food can weaken the bee colony, ultimately affecting the morphometric quality of the bees themselves. The measurement results from the observation of the impact of different locations on the production of *Apis mellifera* queen bees using the grafting method can be seen in Table 1.

Table 1. The average morphological measurements of *Apis mellifera* queen bees in the research locations.

Parameter	Average ± SD (mm)	
	Location 1	**Location 2**
Abdomen Length	9.10 ± 1.70	9.17 ± 1.35
Abdomen Width	5.00 ± 0.43	4.86 ± 0.38
Abdomen Height	4.20 ± 0.52	4.36 ± 0.56
Proboscis Length	1.29 ± 0.40	1.47 ± 0.44

3.1 The length of the abdomen of the queen honeybee (*Apis mellifera*)

The results of observing the influence of different locations with the grafting method on the abdomen length of *Apis mellifera* queen bees can be seen in Table 1. Subsequently, after being analyzed using an unpaired t-test, it was found that the different locations and grafting methods did not show a significant difference ($P>0.05$) in the abdomen length of queen bees. The average results of the abdomen length of *Apis mellifera* queen bees are shown in Table 1. The research findings indicate an average abdomen length of 9.10 ± 1.70 mm at location 1 and 9.17 ± 1.35 mm at location 2. These results reveal that the highest average abdomen length of queen bees is found at location 2 compared to location 1. This is due to the prolonged rainy season, during which a bee colony may face difficulties in finding food sources as pollen and nectar become damaged. Consequently, the population may decrease, affecting the process of providing royal jelly to potential queen bees [4]. According to Kuntadi [5], larvae will develop into queen bees if they receive more royal jelly consumption, primarily from the secretion of the mandibular gland. Conversely, larvae receiving a small amount of royal jelly, with a higher content primarily from the hypopharyngeal gland, will develop into worker bees.

The length of a queen bee's body can be influenced by its species. The bee's body serves as a crucial space for vital organs to develop, and one of the ideal factors determining egg storage and spermatheca capacity can be seen in the body's length. This aligns with Gabka [6], who states that a queen bee's egg-laying ability depends on the capacity of the egg sac (ovarium) and the volume of the sperm sac (spermatheca) in its reproductive organs. The larger the volume of the egg sac, the more egg tubes (ovariols) it contains, thus affecting the quantity of eggs produced. This is further supported by Floris's statement that queen bees have a larger body size than other bee castes due to the developed ovaries functioning for egg-laying. One of the criteria for assessing the quality of a queen bee is the length of the abdomen. This is because one of the main tasks of a queen bee is to produce eggs. If a queen bee has a long, large, and proportional abdomen, it will be capable of accommodating a large quantity of spermatozoa.

3.2 The abdomen width of the queen bee *Apis mellifera*

The results of observing the influence of different locations using the grafting method on the abdomen width of *Apis mellifera* queen bees are presented in Table 1. Subsequently, after being analyzed using an unpaired t-test, it was found that the different locations and grafting methods did not show a significant difference ($P>0.05$) in the abdomen width of queen bees. The average results of the abdomen width of *Apis mellifera* queen bees are shown in Table 1. The research findings indicate an average abdomen width of 5.00 ± 0.43 mm at location 1 and 4.86 ± 0.38 mm at location 2. These results reveal that the highest average abdomen width of queen bees is found at location 1 compared to location 2. This is due to the similar environmental conditions and vegetation variety in the two research locations, which correlates with the average morphological body size. Windarsih, et al [7], stated that environmental conditions such as temperature, humidity, wind speed, altitude, and light intensity can influence the morphometric characteristics of insects. Another factor that can affect it is the number of nurse bees, as a low number of nurse bees may make it impossible to provide royal jelly. Widowati [8] stated that royal jelly is the food for honeybee larvae and queen bees throughout their lives. This is what makes queen bees true egg producers with a larger body size than other bee castes.

Other fundamental factors that can influence the acceptance of grafted larvae in a colony include the selection of a strong colony, the availability of food reserves, the age of the larvae used, the technique of grafting, and the presence of a queen bee in the colony. Minarti and Akhiroh [9] added that the cup used in the grafting process is one of the factors where the diameter size directly correlates with the development of the potential queen. Therefore, if the queen is formed with the same diameter, it will result in the same abdomen width. Abdomen width can be used to assess productivity in egg-laying. The larger the abdomen of the queen bee, the more productive it is in laying eggs, as a large abdomen allows a queen bee to accommodate a larger spermatheca from the male bees.

3.3 The abdomen height of the queen bee *Apis mellifera*

The observations on the influence of different locations using the grafting method on the abdomen height of *Apis mellifera* queen bees can be seen in Table 1. Furthermore, after being analyzed using an unpaired t-test, it was found that the different locations and grafting methods did not show a significant difference ($P>0.05$) in the abdomen height of queen bees. The average results of the abdomen height of *Apis mellifera* queen bees are shown in Table 1. The research findings indicate an average abdomen height of 4.20 ± 0.52 mm at location 1 and 4.36 ± 0.56 mm at location 2. These results reveal that the highest average abdomen height of queen bees is found at location 2 compared to location 1. This condition can be influenced by the surrounding environmental conditions at the honeybee grazing location. This is reinforced by Budiwijono's [4] opinion that different locations' elevations in the grazing process can be influenced by microclimate conditions, namely temperature, humidity, wind speed, rainfall, and light intensity. In humid and wet environmental conditions, honeybee colonies may face difficulties in obtaining food in the form of nectar and pollen for their living and colony development needs. Seeing this condition, beekeepers usually provide artificial feed consisting of sugar and water to maintain the survival of a bee colony. However, in general, this condition can lead to a decrease in the number of bee colonies and no harvesting process, making them susceptible to pests and diseases.

The larvae used in this research are 1-day-old larvae because in the process of providing royal jelly, 1-day-old larvae have a longer time frame compared to larvae aged 2 or 3 days. As a result, the proportion of royal jelly obtained by young larvae will be higher than that of older larvae. This is supported by Pavel, stating that differences in body size can be caused by the food provided by nurse bees to the larvae. This statement is reinforced by previous research indicating that changes in feeding patterns will affect larval development into new individuals. Factors influencing abdomen height include the age of the grafted larvae, the provision of royal jelly, and the presence of the old queen in the colony. This aligns with Kuntadi's [5] assertion that grafting techniques using young larvae will produce potential queen bees with higher quality.

3.4 The length of the proboscis of the queen bee *Apis mellifera*

The results of observing the influence of different locations with grafting methods on the proboscis length of *Apis mellifera* queen bees are presented in Table 1. Furthermore, after being analyzed using an unpaired t-test, it shows that the different locations with different grafting methods are not significant ($P>0.05$) regarding the proboscis length of queen bees. The average proboscis length of *Apis mellifera* queen bees can be seen in Table 1. The research results obtained an average proboscis length at location 1 of 1.29 ± 0.40 mm, while at location 2 it was 1.47 ± 0.44 mm. These results indicate that the highest average proboscis length of queen bees is found at location 2 compared to location 1. The differences in overall variability may be caused by differences in the quality of pollen throughout the time, especially related to the protein content that influences the development of the hypopharyngeal gland responsible for royal jelly production by nurse bees. If this requirement is met, it will result in high-quality queens in larger quantities.

In general, a longer proboscis is more suitable for flowers with deep nectar positions, while a shorter proboscis is used for flowers with shallow nectar positions. This indicates that the proboscis size within a bee colony is a population strategy to expand opportunities in obtaining sufficient nectar from various flowers with diverse nectar positions [10]. This differs from the conditions of queen bees because the queen's duties, unlike worker bees, do not involve various tasks during her lifetime due to morpho-functional differences among bee castes, including proboscis length. The proboscis of queen bees is shorter than that of worker bees, making it less suitable for exploring the inside of flowers to collect nectar. While some morphometric measurements of queen bees may be superior to those of worker bees, the proboscis length of worker bees is longer, as they are responsible for collecting food. Based on previous research, it is mentioned that a queen bee has been observed engaging in flower probing activities. This likely occurs during orientation or mating flights.

4 Conclusion

The use of different locations in the production of *Apis mellifera* queen bees using the grafting method provides consistent performance in terms of abdomen length, abdomen width, abdomen height, and proboscis length.

References

1. B. Rahmad, N. Damir, Mulawarman. Types of honey bees and feed source plants. J. Forestry Research Faloak, 5(1), pp.47-61 (2021)
2. Ustadi, N. Ikhsan, M. Junus, Suyadi. Improves morphometric characteristics of queen Apis cerana javana. Biodiversitas 23 (1): 166-170 (2022)
3. AP. Wu, J. Liu, FF. He, YH. Wang, XJ. Zhang, XD. Duan, ZY. Qian. Negative relationship between diversity and productivity (2018)
4. T. Budiwijono. Identification of Bee Colony Productivity Apis Mellifera. J. Gamma. 7(2): 111-123 (2012)
5. Kuntadi. Effect of Larval Age on the Quality of Queen Produced in Queen Bees Apis cerana. J. Indonesian Entomology. 10(1): 1-6 (2013)
6. J. Gabka, M. Ochnio, Z. Kaminski, B. Madras majewska. Effect of Age of Eggs Used on The Number of Received Queen Cells. J. Apicultural Science. 55(1): 47-51 (2011)
7. Windarsih, G, DW. Utami, S. Yuriyah. Morphological characteristics of Zingiberaceae. Biodiversitas 22 (12): 5507-5529 (2021)
8. R. Widowati. Pollen Substitute Natural Pollen Substitute. EJ. Widya Health and Environment ISSN 2338-7793, Volume 1 (1) (2013)
9. Minarti, S dan P. Akhiroh. Study of The Age of Grafted Larvae on Some Morphological Characteristics of Queen Honey Bees (Apis mellifera). J. Animal Sciences 31(1): 66-73 (2021)
10. Novita, R. Saepudin, Sutriyono. Morphometric Analysis of Apis cerana Worker Honey Bees. J. livestock Science Vol. 8 No 1 (2013)

Performance of carotenoid contents in colostrum of dairy cows during the initial to five days of lactation

Puguh Surjowardojo[1], *Hanum* Muarifah[1], *Zia Ul Rahman* Fithron[2*], and *Rifa'i* Rifa'i[3]

[1]Department of Animal Production, Faculty of Animal Science, Universitas Brawijaya Malang, Indonesia
[2]Department of Social Economic, Faculty of Animal Science, Universitas Brawijaya Malang, Indonesia
[3]Department of Animal Production, Faculty of Animal Science, Universitas Kahuripan Kediri, Indonesia

Abstract. Colostrum is the main food for newborn calves, due to contain immunoglobulin, vitamin, zinc, magnesium, riboflavin, folic acid and carotene. This research was conducted to performance of carotene contents in the colostrum of lactating dairy cows. This research was held on testing laboratory of food quality and food safety Faculty of Agriculture Technology Universitas Brawijaya. The material of this research had using total of 10 pregnant dairy cows. After giving birth to the calf/parturition, colostrum samples were taken from all cows on different milking days (milking fifth days). The method in this research was used case study an data obtained with exploratory descriptive. Based on the research results, the highest carotene level was 4.12 µg/g on the first day of milking

1 Introduction

Colostrum is the first milk produced by dairy cows immediately after calving. It is crucial for the health and development of newborn calves due to its high nutrient content, including essential bioactive compounds like carotenoids. Carotenoids play a vital role in enhancing the immune system and overall health of both cows and their offspring. They are known for their antioxidant properties and their potential to boost calf immunity, which is especially important during the initial days of life when the calf's immune system is not yet fully developed. Furthermore, carotenoids contribute to the distinctive yellow color of milk [1]. Carotene is a precursor of vitamin A. Carotenoids can be categorized into two main groups, namely carotene and xanthophyll. Carotenes consist of nonpolar hydrocarbons and include compounds such as β-carotene, α-carotene, and lycopene. On the other hand, xanthophyll is characterized by having a hydroxyl or keto end group, making it a more polar compound. This group includes lutein, zeaxanthin, canthaxanthin, and β-cryptoxanthin [2]. β-carotene has the potential as an antioxidant and also has the potential to help improve the immune system for newborn calves [3]. Apart from that, β-carotene has an important role in vision, growth and reproduction [4].

Cows that are fed on pasture and feed rich in carotenoids (grass silage) three weeks before giving birth, produce colostrum which is rich in milk fat and high in lutein, β-carotene, vitamin A and vitamin E [5]. Colostrum has much higher levels of beta-carotene than milk,

* Corresponding author: zfithron@ub.ac.id

and during the first week postpartum, these levels decrease rapidly [6]. The importance of maintaining feed intake for the mother before and after giving birth. According to [7], supplementation of beta-carotene in cow feed can increase the concentration of beta-carotene in colostrum, change the color of colostrum, and increase serum beta-carotene in newborn calves in pregnant cows that have received sufficient intake vitamin A. The aim of this research is to determine the performance of carotene contents in cow colostrum during the first to five days of lactation.

2 Materials and Methods

The research method was used a case study and data was collecting field data at KPSP Setia Kawan Nongkojajar, Pasuruan, and the data obtained with exploratory descriptive. The sampel was used Peranakan Friesien Holstein (PFH) dairy cows with purposive sampling. The criteria of postpartum cows producing colostrum on days 1-5 with a lactation period of 1 - 6. The feed ingredients used for steaming up in the form of forage and concentrate were given more than normal feeding. The material used for the research using total of 10 pregnant dairy cows. After giving birth to the calf/parturition, colostrum samples were taken from all cows on different milking days (milking fifth days). After the mother gives birth, the colostrum is milked before being fed to the calf at ± 30 - 45 minutes after parturition. Colostrum was milked from day 1st to day 5th. The colostrum was put into a 100 ml sterile glass bottle, then put into a cooler bag filled with ice gel to be taken to the laboratory. Bottles containing colostrum will be stored at -20 °C until analyzed for carotene. Testing for colostrum carotene content used a UV-VIS Spectrophotometer (Shimadzu 1601) in the food quality and safety laboratory, Faculty of Agricultural Technology, Brawijaya University.

3 Results and discussion

The carotene contents in the colostrum who had steaming up treatment had higher carotene content, this was due to the fact that the amount of forage and concentrate given was higher than that given to normal cows feed in the dry period before calving. According to [5], it was explained that feeding grass silage (high carotenoid) vs corn silage (low carotenoid) to PFH dairy cows during the dry period, the result was in a higher concentrations of carotenoids in plasma, it was affected by the characteristic of forage was gave during the dry period. So that, indicated that plasma concentrations of fat-soluble micronutrients involved in the nutritional. Additional, steaming feed up should be added to dry period cows to increase carotenoid levels in colostrum.

Table 1. Caroten content in different treatmens

	Category											
	Steaming Up (SU)						Without Steaming Up/Control (CO)					
Milking day	1	2	3	4	5	Avg	1	2	3	4	5	Avg
Total of Caroten Content (µg/g)	**4.12**	**0.83**	**1.26**	**0.23**	**0.16**	**1.32**	**0.10**	**0.03**	**0.02**	**0.04**	**0.02**	**0.042**

The result of table 1 showed that the average carotene content that were given steaming up was 1.32 µg/g, while the average carotene content that were not given steaming up was

0.042 µg/g, and the highest carotene content on day 1 of milking was 4.12 µg/g and the lowest carotene content was 0.16 µg/g in cows that were steamed up. Meanwhile, the dairy cow that were not given steaming up, the highest and lowest carotene levels were 0.10 µg/g and 0.02 µg/g. These results show that the carotene content on day 1 had increased and the day 2 – 5 had decreased, due to the steaming feed up in the form the forage and concentrate were given more than regular feeding. That's In line with [8] during the first few days of lactation, the levels of carotenoids decrease dramatically and then stabilize around day five as udder gland secretion changes to regular milk. Calderon et al. [5] reported that the concentrations of the carotenoids content in colostrum were initially highest and decreased sharply during the first week of breastfeeding. According to [4] dairy cows have a higher milk fat concentration in the first day of milking colostrum due to micronutrients that can dissolve in milk product fat, namely β-carotene. Colostrum contains higher levels of beta-carotene compared to milk, and these levels decrease rapidly during the first week after postpartum. This is associated with a rapid decrease in peripartum plasma concentrations, particularly evident one week before birth and reaching their lowest point at birth for retinol or one week after birth for carotene, as described by [6] This decline is likely due to not only a minor reduction in dietary intake but also an increased uptake by the mammary glands. When cows are provided with dietary supplements during this period, it can help limit the decrease in plasma carotene and retinol levels, as observed in a study by [9]. According to Brandano, Rassu, and Lanzu [10], various factors affect the quality and quantity of colostrum, including individual characteristics, breed, the feed consumed during the dry period, the duration of the dry period in confinement, the timing of colostrum collection after birth, and the body condition score (BCS).

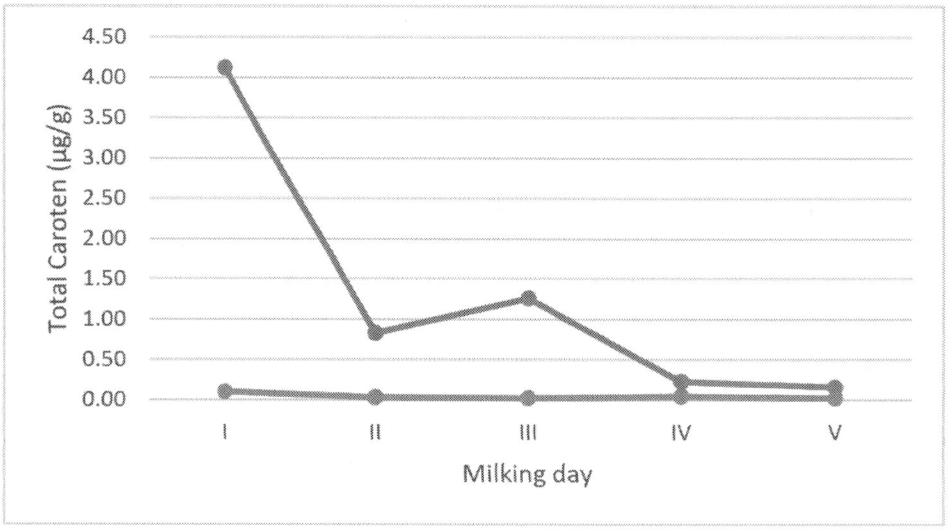

Fig 1. Carotene content on different milking days

4 Conclusion

The conclusion of this research:

1. the average carotene content that were given steaming up was 1.32 µg/g, while the average carotene content that were not given steaming up was 0.042 µg/g, that means the steaming feed up in the form the forage and concentrate were given more than normal feeding.

2. the highest carotene content of colostrum on day 1 of milking was 4.12 µg/g and the lowest carotene content was 0.16 µg/g in cows that were steamed up, due to carotene has the potential as an antioxidant and also has the potential to help improve the immune system for newborn calves.
3. The dairy cow that was not given steaming up had the highest and lowest carotene levels were 0.10 µg/g and 0.02 µg/g, because of steaming up is giving feed with high nutritional content aims to supply cows to have sufficient energy during and after giving parturition.

Funding for this study was provided by professors grand of the faculty of Animal Science Universitas Brawijaya with the number contract of No. 2044.10/UN.F05/PN/2023.

References

1. C. Agabriel, A. Cornu, C. Journal, C. Sibra, P. Grolier, B. Martin. Tanker milk variability according to farm feeding practices: Vitamins A and E, carotenoids, color, and terpenoids. J. Dairy Sci. **90**, 4884–4896 (2007)
2. D. M. Deming and J. W. Erdman Jr., "Mammalian carotenoid absorption and metabolism," Pure Appl. Chem., **71**(12),2213-2223 (1999)
3. F. Kondorik, M. Martosupono, A.B. Susanto. Peranan beta-carotene dalam sistem imun untuk mencegah kanker. Jurnal Biologi & Pembelajarannya. **4** (1), 1-8 (2017)
4. P. Noziere, P. Graulet, A. Lucas, B. Martin, P. Grolier, M. Doreau. Carotenoids for ruminants: From forages to dairy products. Animal Feed Science and Technology. **131**, 418–450 (2006)
5. F. Caldero´n, B. Chauveau-Duriot, B. Martin, B. Graulet, M. Doreau, and P. Nozie`re.Variations in Carotenoids, Vitamins A and E, and Color in Cow's Plasma and Milk During Late Pregnancy and the First Three Months of Lactation. J. Dairy Sci. **90**, 2335–2346. (2007)
6. Johnston, L.A., Chew, B.P. Peripartum changes of plasma and milk Vitamin A and beta-carotene among dairy cows with or without mastitis. J. Dairy Sci. **67**, 1832–1840 (1984)
7. C.M Prom., M.A Engstrom and J.K Drackley. Effects of prepartum supplementation of β-carotene on colostrum and calves. J. Dairy Sci. **105**, 8839–8849 (2022).
8. D.B. Parrish, G.H. Wise, JS Hughes. Properties of the colostrum of the dairy cow. II. Effect of prepartal rations upon nitrogenous constituents. J Dairy Sci **31**,889–895 (1948)
9. Michal, J. J., L. R. Heirman, T. S. Wong, and B. P. Chew. Modulatory effects of dietary β-carotene on blood and mammary leukocyte function in periparturient dairy cows. J. Dairy Sci. **77**:1408–1421 (1994)
10. P. Brandano, S. P. G. Rassu, dan A. Lanzu. Feeding Dairy Lamb. (G. Pulina & R. Bencini, Eds.) (Dairy Sheep Nutrition). CABI Publishing. Walingford. (2004)

AUTHOR INDEX

AUTHOR INDEX